Drug Delivery Technology

Also of interest

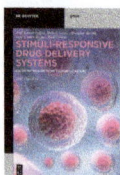

Stimuli-Responsive Drug Delivery Systems.
From Introduction to Application
Kumar Bajpai A, Saini, Mishra, Kumar Bajpai J, Tiwari, 2023
ISBN 978-1-5015-1883-6, e-ISBN (PDF) 978-1-5015-1884-3

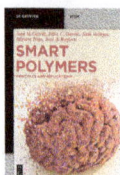

Smart Polymers.
Principles and Applications
García JM, García FC, Reglero Ruiz, Vallejos, Trigo-López, 2022
ISBN 978-1-5015-2240-6, e-ISBN (PDF) 978-1-5015-2246-8

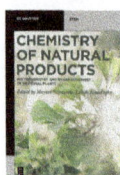

Chemistry of Natural Products.
Phytochemistry and Pharmacognosy of Medicinal Plants
Napagoda, Jayasinghe (Eds.), 2022
ISBN 978-3-11-059589-5, e-ISBN (PDF) 978-3-11-059594-9

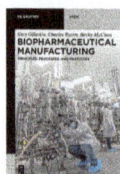

Biopharmaceutical Manufacturing.
Principles, Processes, and Practices
Gilleskie, Rutter, McCuen, 2021
ISBN 978-3-11-061687-3, e-ISBN (PDF) 978-3-11-061688-0

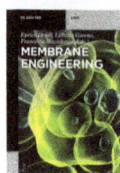

Membrane Engineering
Drioli, Giorno, Macedonio (Eds.), 2018
ISBN 978-3-11-028140-8, e-ISBN (PDF) 978-3-11-028139-2

Contents

List of contributing authors

Chapter 1
Namita D. Desai
C. U. Shah College of Pharmacy
S.N.D.T Women's University
Santacruz (W)
Mumbai 400049
Maharashtra
India

Niserga D. Sawant
C. U. Shah College of Pharmacy
S.N.D.T Women's University
Santacruz (W)
Mumbai 400049
Maharashtra
India

Pratima A. Tatke
C. U. Shah College of Pharmacy
S.N.D.T Women's University
Santacruz (W)
Mumbai 400049
Maharashtra
India
pratima.tatke@cushahpharmacy.sndt.ac.in,
drpratimatatke@gmail.com

Chapter 2
Surendra Agrawal
Department of Pharmaceutical Chemistry
Datta Meghe College of Pharmacy
Datta Meghe Institute of Medical Sciences
(Deemed to be University)
Sawangi (Meghe)
Wardha 442004
Maharashtra
India
sagrawal80@gmail.com

Pravina Gurjar
Sharadchandra Pawar College of Pharmacy
Otur 412409
Pune
Maharashtra
India

Ayushi Agarwal
SPPSPTM
SVKM's NMIMS
Mumbai 400056
Maharashtra
India

Chapter 3
Amarjitsing Rajput
Department of Pharmaceutics
Poona College of Pharmacy
Bharti Vidyapeeth Deemed University
Erandwane, Pune 411038
Maharashtra
India

And

Department of Biosciences and
Bioengineering
Indian Institute of Technology Bombay
Powai 400076
Maharashtra
India

Satish Mandlik
Department of Pharmaceutics
Poona College of Pharmacy
Bharti Vidyapeeth Deemed University
Erandwane, Pune 411038
Maharashtra
India

Shipa Dawre
Department of Pharmaceutics
School of Pharmacy and Technology
Management
SVKM's Narsee Monjee Institute of
Management and Studies (NMIMS)
Mukesh Patel Technology Park
Shirpur 425405
Maharashtra
India

https://doi.org/10.1515/9783110746808-203

Deepa Mandlik
Department of Pharmaceutics
Poona College of Pharmacy
Bharti Vidyapeeth Deemed University
Erandwane, Pune 411038
Maharashtra
India

Prashant Pingale
Department of Pharmaceutics
GES's Sir Dr. M. S. Gosavi College of
Pharmaceutical Education and Research
Nashik 42200
Maharashtra
India

Shital Butani
Department of Pharmaceutics and
Pharmaceutical Technology
Institute of Pharmacy
Nirma University
S.G. Highway
Ahmedabad 382 481
Gujarat
India
Email: shital_26@yahoo.com

Chapter 4
Shashikant B. Bagade
Department of Pharmaceutical Chemistry,
SVKM's NMIMS School of Pharmacy and
Technology Management 425405, Shirpur,
Dist. Dhule, Maharashtra, India
Phone No: +91 9637474753
shashikant.bagade@nmims.edu

Shivanee Vyas
SVKM's NMIMS School of Pharmacy and
Technology Management
425405 Shirpur
Dist. Dhule
Maharashtra
India

Amit B. Page
SVKM's NMIMS School of Pharmacy and
Technology Management
425405 Shirpur
Dist. Dhule
Maharashtra
India

Kiran D. Patil
SVKM's Institute of Pharmacy
Dist. Dhule
Maharashtra 424001
India

Chapter 5
Madhur Kulkarni
SCES's Indira College of Pharmacy
New Mumbai Pune Highway
Tathwade 411033
Pune, Maharashtra
India
madhur.kulkarni@indiraicp.edu.in

Roopal Bhat
SCES's Indira College of Pharmacy
New Mumbai Pune Highway
Tathwade 411033
Pune, Maharashtra
India

Suvarna Ingale
SCES's Indira College of Pharmacy
New Mumbai Pune Highway
Tathwade 411033
Pune, Maharashtra
India

Abhijit Date
Department of Pharmaceutics
The Daniel K. Inouye College of Pharmacy
University of Hawaii at Hilo
200 W. Kawili Street
Hilo, HI 96720
USA

Chapter 6
Rupali A. Patil
GES's Sir Dr. M. S. Gosavi College of
Pharmaceutical Education and Research
Nashik 425005
Maharashtra
India
ruupalipatil@gmail.com

Krutika H. Pardeshi
GES's Sir Dr. M. S. Gosavi College of
Pharmaceutical Education and Research
Nashik 425005
Maharashtra
India

Harshal P. Chavan
GES's Sir Dr. M. S. Gosavi College of
Pharmaceutical Education and Research
Nashik 425005
Maharashtra
India

Sunil V. Amrutkar
GES's Sir Dr. M. S. Gosavi College of
Pharmaceutical Education and Research
Nashik 425005
Maharashtra
India

Chapter 7
Vaibhav Bhamare
Department of Pharmaceutics
K. K. Wagh College of Pharmacy
Nashik 422006
Maharashtra
India
vaibhav.bhamre@gmail.com

Rakesh Amrutkar
K. K. Wagh College of Pharmacy
Nashik 422006
Maharashtra
India

Vinod Patil
SPH College of Pharmacy
Malegaon 423105
Maharashtra
India

Chandrashekhar Upasani
SNJB's SSDJ College of Pharmacy
Chandwad 423101
Maharashtra
India

Chapter 8
Deepa Mandlik
Department of Pharmaceutics
Bharti Vidyapeeth Deemed University
Poona College of Pharmacy
Bharti Vidyapeeth Educational Complex
Erandwane, Pune 411038
Maharashtra
India

Satish Mandlik
Department of Pharmaceutics
Bharti Vidyapeeth Deemed University
Poona College of Pharmacy
Bharti Vidyapeeth Educational Complex
Erandwane, Pune 411038
Maharashtra
India

Amarjitsing Rajput
Department of Pharmaceutics
Bharti Vidyapeeth Deemed University
Poona College of Pharmacy
Bharti Vidyapeeth Educational Complex
Erandwane, Pune 411038
Maharashtra
India
amarjit.rajput@yahoo.com

Chapter 9
Dattatraya M. Shinkar
GES's Sir Dr. M. S. Gosavi College of
Pharmaceutical Education and Research
Nashik 42200
Maharashtra
India
drdmshinkar@gmail.com

Sunil V. Amrutkar
GES's Sir Dr. M. S. Gosavi College of
Pharmaceutical Education and Research
Nashik 42200
Maharashtra
India

Prashant L. Pingale
Department of Pharmaceutics
GES's Sir Dr. M. S. Gosavi College of
Pharmaceutical Education and Research
Nashik 42200
Maharashtra
India

Chapter 10
Deepak Kulkarni
Srinath College of Pharmacy
Bajajnagar, Aurangabad
Maharashtra
India 431136

Sachin Surwase
BIO-IT Foundry Technology Institute
Pusan National University
Busan
Republic of Korea

Shubham Musale
Department of Pharmaceutics
Dr. D.Y. Patil Institute of Pharmaceutical
science and Research
Pimpri-Pune
Maharashtra
India 411018

Prabhanjan Giram
Department of Pharmaceutics
Dr. D.Y. Patil Institute of Pharmaceutical
science and Research
Pimpri-Pune
Maharashtra
India 411018
prabhanjanpharma@gmail.com;
prabhanjan.giram@dypvp.edu.in

Chapter 11
Swarnali Das Paul
Shri Shankaracharya College of
Pharmaceutical Sciences
Bhilai 490020
Chhattisgarh
India
swarnali34@gmail.com

Harish Sharma
Shri RawatpuraSarkar Institute of Pharmacy
Kumhari (C.G.)
India

And
Rungta Institute of Pharmaceutical Sciences
Kohka
Bhilai, (C.G.)
India

Gyanesh Sahu
Rungta Institute of Pharmaceutical Sciences
Kohka
Bhilai (C.G.)
India

And
Rungta College of Pharmaceutical Sciences
and Research
Kohka
Bhilai (C.G.)
India

Chanchal Deep Kaur
Shri RawatpuraSarkar Institute of Pharmacy
Kumhari (C.G.)
India

Sanjoy Kumar Pal
SSIT
Skyline University
Nigeria

Chapter 12
Rupali Patil
GES's Sir Dr. M. S. Gosavi College of
Pharmaceutical Education and Research
Nashik 425005
Maharashtra
India
ruupalipatil@gmail.com

Pratiksha Aher
GES's Sir Dr. M. S. Gosavi College of
Pharmaceutical Education and Research
Nashik 425005
Maharashtra
India

Punam Bagad
GES's Sir Dr. M. S. Gosavi College of
Pharmaceutical Education and Research
Nashik 425005
Maharashtra
India

Shraddha Ekhande
GES's Sir Dr. M. S. Gosavi College of
Pharmaceutical Education and Research
Nashik 425005
Maharashtra
India

Namita D. Desai, Niserga D. Sawant, Pratima A. Tatke
Chapter 1
Biopotentiation using herbals: novel approach for poorly bioavailable drugs

Abstract: Owing to rapid developments in drug design technologies, many drugs have been progressively introduced each year. However, poor pharmacokinetic profiles of these drugs remain an area of concern. Bioavailability is an integral part of the pharmacokinetics paradigm. Many pharmaceutical methods such as nanonization, micronization, and prodrug approach have been applied in the past to enhance bioavailability of drugs. However, the use of herbal bioenhancers (biopotentiaters) has recently evoked interest. Interestingly, the principle of biopotentiation was first recorded in Ayurveda as "Yogvahi," in ancient times. Herbal bioenhancers are phytomolecules that may improve the drugs' or nutrients' biological activity, such as bioavailability and bioefficacy at low doses, at which they themselves have no other pharmacological activity. This chapter aims to consolidate scientific reports on various herbal bioenhancers, together with a brief description of their mode of action and pharmacotherapeutic applications. This chapter also highlights the groundbreaking combination of natural bioenhancers and novel techniques like liposomes, nanoparticles, transferosomes, and ethosomes.

This chapter emphasizes on the promising potential of herbal bioenhancers in improving bioavailability of drugs. The role of these bioenhancers in novel drug delivery approaches is also highlighted. The proposed approach is a combination of traditional knowledge of Ayurvedic system of medicine and the novel techniques in pharmaceutical technology like liposomes, nanoparticles, transferosomes, and ethosomes for effectively delivering drugs.

1.1 Introduction

With the rapid advancements in drug design technologies, many researchers have been developing number of newer molecules that show impressive *in vitro* performance but their low solubility and/or poor permeation characteristics, result in poor *in vivo* bioavailability. Hence, there is an increasing demand to improve bioavailability of

Namita D. Desai, Niserga D. Sawant, C. U. Shah College of Pharmacy, S.N.D.T Women's University, Santacruz (W), Mumbai, 400049, Maharashtra, India
Pratima A. Tatke, C. U. Shah College of Pharmacy, S.N.D.T Women's University, Santacruz (W), Mumbai, 400049, Maharashtra, India, e-mail: pratima.tatke@cushahpharmacy.sndt.ac.in, drpratimatatke@gmail.com

https://doi.org/10.1515/9783110746808-001

drugs to meet the growing requirements of the medical field and pharmaceutical industry. Further, the development of newer drug delivery approaches and novel drug technologies has significantly impacted the cost of drug therapy. Hence, increasing the drug bioavailability is one of the feasible means to achieve affordable treatment with reduction in drug dosage and hence the cost of therapy [1]. Broadly, the approaches in overcoming bioavailability concerns include pharmaceutical, pharmacokinetic and the biological approaches. The pharmacokinetic approach deals with the development of new chemical entities or prodrugs and is very expensive and time-consuming, requiring repetition of clinical trials and long time for regulatory approval. The biological approach involves changing the route of administration from oral to parenteral. Hence, the pharmaceutical approach in modulating the drug solubility and/or permeability holds promise in improving bioavailability [2]. Many pharmaceutical methods such as nanonization, micronization, eutectic mixtures, solid solutions, and solid dispersions have been used extensively but the use of bioenhancers or biopotentiers, especially those derived from natural origin, has recently evoked interest due to ease of availability, simplicity, and scalability of the approach.

Bioenhancement or biopotentiation promotes decrease in the usual daily dose of the drugs leading to reduced side and toxic effects, enhances therapeutic efficacy, reduces development of resistance, reduces the raw materials required in manufacturing and ultimately provides economic advantage by reducing the cost of therapy. Herbal bioenhancers or biopotentiators are agents of herbal origin deprived of their own pharmacological activity at the prescribed dose and capable of increasing bioavailable fraction and therapeutic effectiveness of coadministered drugs. Ayurveda conceptualizes the role of biopotentiers in the theory of "Yogavahi." The concepts and methods such as "Yogavahi," "Anupana," Bhaishajya Kala," "Bhavana" (trituration), "Rasayana," "Yoga" (formulations) and "Kalpanas" (various dosage forms), the concept of "Purana Aushadhies" (old drugs) and the concept of action-augmenting drugs are being used since time immemorial in the Ayurvedic system of medicine [3, 4]. In addition, "Samshodhana" (bio-purification) can also be considered. A very common example of "Yogavahi" in Ayurveda includes Pippali (*Piper longum*) and Maricha (*Piper nigrum*) that contain an important active constituent named "piperine" (1-piperonyl piperidine), which is responsible for the bioenhancing effect. Other common examples of Yogavahi include Ghrita, Swarna (gold) preparation, Guggulu preparation, and Bhasmas. In the year 1979, piperine of the black pepper series was proven globally as the first bioavailability enhancer by scientists at the Indian Institute of Integrative Medicine (formerly known as Regional Research Laboratory), located in Jammu [5]. The practice of using the Ayurvedic preparation, "Trikatu," which in Sanskrit means "three acrids" is well known. This Ayurvedic preparation contains ginger, long pepper and black pepper for potentiating bioavailability of drugs, vitamins and nutrients. [6] The renewed interest in biopotentiation is mainly because the poor bioavailability of existing and newly developed drugs that require administration for longer periods of time to produce desired therapeutic effects, results in greater side effects and expensive treatment regimen. Biopotentiation is therapeutically beneficial, because it directly correlates with the plasma concentration and effectiveness of the drugs and in turn, can lead to reduction in the cost of treatment as well as reduce the toxicity of drugs [7]. Examples of drugs that have shown increased bioavailability in the presence of herbal bioenhancers are metformin, phenobarbitone, rifampicin, tetracyclines, pyrazinamide, sulfadiazine, secnidazole, ethambutol, phenytoin, vasicine, carbamazepine, nimesulide, indomethacin [8–10], labetalol, coenzyme Q10, β-carotene, dapsone, ciprofloxacin, curcumin and amino acids [11].

Synthetic absorption enhancers such as surfactants, fatty acids, cyclodextrins, and chelating agents though reported to enhance the permeability of drugs across biomembranes have their own shortcomings. They are associated with toxicity to the biological membranes on repeated use, are required in higher concentrations, and are often inefficient in inhibiting enzymes that metabolize drugs or transporters involved in drug efflux that play a role in reducing drug bioavailability, thus hindering their use as efficient bioavailability enhancers [12, 13].

The novel delivery approaches available in literature for improving the solubility and/or permeability of drugs include polymeric and lipid systems such as self-emulsifying drug delivery systems, nanoparticles, microparticles, liposomes, phytosomes, and micelles. However, the challenges faced in a majority of these approaches include inefficient drug loading and use of synthetic surfactants and co-surfactants in relatively high concentrations and inclusion of exogenous compounds such as charge inducers to increase drug entrapment that can cause irritation to biomembranes [14].

The greater safety potential of herbal bioenhancers can be proposed as favorable solution to the toxicity concerns of synthetic absorption enhancers. Moreover, ease of availability and formulation, improved uptake by biomembranes, synergizing the activity of drugs and reducing the use of synthetic surfactants in novel drug delivery systems are the benefits associated with the use of herbal bioenhancers and highlight their promising potential in drug delivery [15, 16].

1.2 Mechanism of action of herbal bioenhancers

Herbal bioenhancers generally enhance the absorption of orally administered drugs by exerting action on the gastrointestinal tract (GIT) and subsequently improving the bioavailability by exerting on the drug absorption, metabolism process and also act on drug targets [17]. The elaborate mechanism by which herbal bioenhancers act is still not well understood. However, some of the suggested mechanisms (Figure 1.1) are as follows:

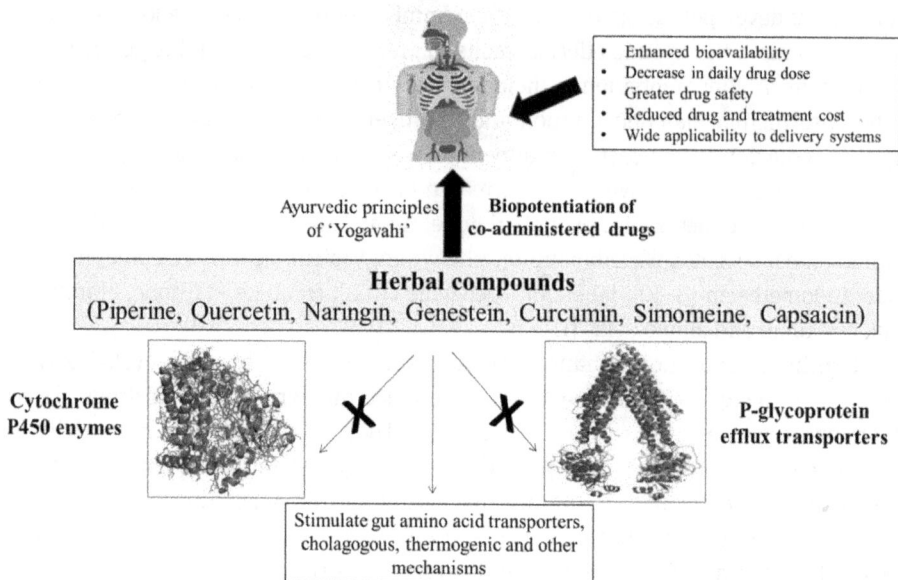

Figure 1.1: Mechanisms of action of herbal bioenhancers.

1.2.1 Modifications in drug metabolizing enzymes

Bioenhancers inhibit important phase I enzymes that metabolize drugs, namely cytochrome P-450 enzymes and its isoenzymes and uridine 5′-diphospho (UDP)–glucuronyl transferase and also inhibit glucuronic acid production in intestines. Thus, the untransformed drugs can enter into the systemic circulation in larger amounts from the GIT. CYP1 A1, CYP1 B1, CYP1 B2, CYP2 E1 and CYP3 A4 are examples of metabolizing enzymes reported to be inhibited by piperine. Piperine, when administered orally in rats showed strong inhibition of aryl hydrocarbon hydroxylase and UDP–glucuronyl transferase activities in the hepatic tissue [18]. Piperine has also been shown to lower endogenous UDP–glucuronic acid levels, altering glucuronidation rates and limiting transferase activity [19]. Hence, the drugs subjected to metabolism by these enzymes will show bioenhancement due to potentiation of activity by piperine [20, 21].

1.2.2 Modifications in transporter proteins

P-glycoproteins (P-gp), transporter proteins are found on endothelial cells of the blood–brain barrier, adrenal glands, apical surfaces of epithelial cells and on many neoplastic cells. Inhibition of this efflux mechanism can be proposed as an attractive strategy to increase the intra cellular concentrations of drugs and hence, their therapeutic activity. Interference with protein binding sites, adenosine triphosphate hydrolysis or changes in cell membrane lipid integrity could be the mechanisms. Bioenhancers also inhibit the cell pumps such as breast cancer resistance proteins responsible for drug elimination from cells and enhance absorption by stimulating gut amino acid transporters, thus improving bioavailability of coadministered drugs [22, 23].

1.2.3 Thermogenic/bioenergetic effects

Bioenhancers can also act through thermogenic/ bioenergetic mechanisms that are supposed to be triggered by stimulation of thermoreceptors and/or act directly on adrenoreceptors. Stimulation of dopaminergic and serotinergic systems and purinergic receptors of P2-type can also facilitate the release of catecholamines. Further, β-3 adrenoceptors stimulation augments thermogenesis, reduces the white adipose tissues without affecting the intake of food, increases the amounts of insulin receptors and reduces insulin and glucose levels in the blood, thus promoting anti-obesity and antidiabetic effects, contributing to the mechanism of thermogenesis. Increased activity of thyroid peroxidase leads to increased levels of triiodothyronine and thyroxine in plasma, thus increasing thermogenesis due to the concurrent increase in tissue oxygen uptake [23]. All these effects along with contributing to weight loss and increased cellular energy levels help utilization of nutrients and drugs by promoting gastrointestinal absorption.

1.2.4 Other possible mechanisms

The other mechanisms include cholagogous effects, increasing the responsiveness of target receptors to drugs molecules, increased vasodilating effect on GIT to increase drug absorption and modulating dynamics of cell membrane to increase drug transport across the membranes [20, 21].

Herbal bioenhancers can act by one or more of these mechanisms and different bioenhancers may show similar or varying mechanisms of action [23].

1.3 Herbal bioenhancers in drug delivery

1.3.1 Piperine

Piperine, a plant alkaloidal compound present in *Piper longum* (long pepper) as well as in *Piper nigrum* (black pepper), is among the earliest and the most extensively studied bioenhancer. The estimated percent of piperine present in black pepper is 5–9% and falls under GRAS category of the Food and Drug Administration for its utility in seasoning, flavoring or as spice. It is reported to display number of beneficial properties like antidiarrheal [23], antipyretic [24], fertility enhancement [25], anti-inflammatory [26, 27], antifungal [28], antimetastatic [29], antioxidant [30, 31], antithyroid [32], antitumor [33, 34], antiplatelet [35], antimutagenic [36, 37], analgesic [38], antidepressant [39], hepatoprotective [40], antiasthmatic [41], and antihypertensive activities [42]. The reported dosage of piperine as bioenhancer is about 15 mg/day, but not exceeding 20 mg/day when administered in divided doses, although it may vary in the presence of drug compounds. However, earlier study suggested that the suitable dose of piperine is about 10% by weight of the active drug compound [23]. Piperine showed LD_{50} of 330 and 514 mg/kg, respectively, when studied in mice and rats, and oral dose of 100 mg/kg was reported to be nontoxic during subacute toxicity studies [43]. There are two possible mechanisms by which piperine enhances the bioavailability of drugs, which also include nonspecific mechanisms that primarily stimulate quick absorption of drugs and nutrient compounds. The first mechanism may be by increasing blood supply, emulsifying the contents and increasing the activity of γ-glutamyl transpeptidases; the enzymes that are involved in active and passive transport process of nutrients, decreasing hydrochloric acid secretion. The second mechanism may be by inhibition of enzymes contributing to the drug metabolism process in the GIT, thus preventing inactivation and elimination of drugs [23]. Piperine was reported to enhance oral exposure of drugs probably by inhibiting cellular efflux mediated by P-gp during intestinal absorption. The bioenhancement of fexofenadine, when coadministered with piperine, either by peroral route (10 mg/kg) or intravenous route (5 mg/kg) was demonstrated. Pharmacokinetic studies revealed that the bioavailability of fexofenadine almost doubled on oral administration. Piperine, on the other hand, had no effect on the pharmacokinetics of intravenously administered fexofenadine, implying that piperine boosted gastrointestinal absorption rather than decreasing hepatic extraction [44]. Piperine's effects on resveratrol plasma levels when coadministered orally to C57BL mice were studied. After coadministration of piperine, the area under the curve (AUC) and C_{max} of resveratrol increased by 229% and 1,544%, respectively [45]. Table 1.1 represents the examples of drugs that showed bioenhancement in the presence of piperine.

1.3.2 Gingerols

Zingiber officinale (ginger), belonging to the family Zingiberaceae contains gingerols as the potentially active constituent isolated from the rhizomes. They can be transformed to shogaols, gingeroney and paradol [54, 55] and the presence of volatile oils (1–3%) in the ginger is responsible for its characteristic odor [56]. It has been reported that gingerols increase the gastrointestinal motility in experimental

Table 1.1: Studies on bioenhancement of drugs after coadministration of piperine.

Drug	Activity	Model	Inference	References
Metformin	Antidiabetic	Rats	Combination of half the standard subtherapeutic oral dose of metformin; 125 mg/kg with piperine showed 17% lowering of blood glucose, as compared to 12.5% with 250 mg/kg metformin	[46]
Nimesulide	Anti-inflammatory	Swiss albino mice and male Wistar rats	The median effective oral dose of combination of nimesulide and piperine was significantly lowered (1.5 mg/kg) when compared with nimesulide (11.2 mg/kg) in the writhing test	[47]
Nevirapine	Non-nucleoside reverse transcriptase inhibitor	Human volunteers	$AUC_{0-\infty}$ value of oral nevirapine was increased by 170% in the presence of piperine	[48]
Carbamazepine	Antiepileptic	Human volunteers	$AUC_{0-12\ h}$ of oral carbamazepine alone and in the presence of piperine was found to be 81.37 ± 4.90 µg/mL h and 91.70 ± 5.02 µg/mL h, respectively	[49]
Nateglinide	Antidiabetic	Male Wistar rats	Concentration of oral nateglinide in plasma at the end of 180 min was increased by 40% in rats treated with combination of nateglinide, 50 mg/kg and piperine, 10 mg/kg	[50]
Rifampicin	Antitubercular	Rats	Piperine enhanced the oral bioavailability of coadministered rifampicin	[51]
Propranolol	Antiasthmatic	Human volunteers	$AUC_{0-\infty}$ value of oral propranolol was increased by 203.2% in the presence of piperine	[52]
Saquinavir mesylate	HIV protease inhibitor	Caco-2 human cell lines and male Sprague Dawley rats	Piperine increased Saquinavir's oral bioavailability by nearly 10-fold by inhibiting the P-gp-mediated drug efflux	[53]

animals and possess antipyretic, analgesic, antibacterial, and sedative properties [57]. The bioenhancing activity of Z. officinale has been estimated to be between 10 and 30 mg/kg of body weight. The effects of orally Z. officinale with piperine on drug bioavailability were studied and the results are reported in Table 1.2 [58].

Table 1.2: Studies on bioenhancement of drugs after coadministration of Z. officinale and Z. officinale–piperine combination.

Category	Drug	Percent bioavailability enhancement	
		Z. officinale	Z. officinale in combination with piperine
Fluoroquinolones	Norfloxacin	49	67
	Ciprofloxacin	68	70
Macrolides	Erythromycin	68	105
	Azithromycin	78	85
	Roxithromycin	72	93
Cephalosporins	Cefadroxil	68	65
	Cefalexin	75	85
Penicillin	Cloxacillin	76	90
	Amoxycillin	80	90
Aminoglycosides	Ciprofloxacin	68	70
Fluoroquinolones	Kanamycin	65	92
	Pefloxacin	53	69
Antiviral	Zidovudine	105	126
Cardiovascular	Amlodipine	68	95
	Lisinopril	76	105
	Propranolol	76	104
Anti-inflammatory/antiarthritic	Piroxicam	86	134
	Diclofenac	90	140
	Nimesulide	144	165
Antihistamines	Theophylline	76	80
	Bromhexine	46	75
	Salbutamol	78	92
Immunosuppressant	Tacrolimus	75	117

1.3.3 Glycyrrhizin

Glycyrrhizin, a glycosidic compound isolated from stolons and roots of *Glycyrrhiza glabra* (liquorice), is used as expectorant for treating sore throat, asthma, bronchitis, allergies, reducing inflammation and also in gastritis, peptic ulcers and rheumatism. It is used in treating liver diseases by helping the liver detoxify drugs. It builds up immunity, stimulates adrenal glands and possesses laxative and diuretic properties. Sucrose is 50 times less sweeter than glycyrrhizin. The action of anticancer agent, paclitaxel, on cellular division was significantly enhanced in the presence of glycyrrhizin. Results of *in vitro* study performed in MCF-7 cancer cell line indicated that in the presence of glycyrrhizin, the cancerous cell growth inhibition improved 5-fold. It also shows antibacterial and antifungal properties and is known to enhance activity of antibiotics such as tetracycline, ampicillin, rifampicin and nalidixic acids acting against gram-negative bacteria like *E. coli* and gram-positive bacteria like *Bacillus subtilis* and *Mycobacterium smegmatis* [59].

1.3.4 Niaziridin

It is a type of nitrile glycoside obtained from the plant parts such as pods, bark, and leaves of *Moringa oleifera* (drumsticks). *Moringa oleifera* displays properties such as anticancer [60], diuretic, [61], antimicrobial [62], antifertility [63], spasmolytic and hypotensive [64], antioxidant [65], antifungal [66], anti-inflammatory [67], antiulcer [68], antiteratogenic [69], antiarthritic [70], hypolipidemic [71] and hepatoprotective [72]. It has been reported that niaziridin improves activity of ampicillin, rifampicin, nalidixic acids and tetracycline 1.2- to 19-fold against gram-positive organisms, and antifungal activity of clotrimazole (10 µg/mL) against *Candida albicans* 5- to 6-fold, respectively [73].

1.3.5 Quercetin

Quercetin, a flavonoid, is found in grains, leaves, vegetables, and fruits. Quercetin exhibits radical scavenging, anti-inflammatory, antioxidant, anticancer, antiatherosclerotic and antiviral properties. It is found in abundance in apples, berries, broccoli and onions; also citrus fruits are a rich source of this phytoconstituent [74]. Quercetin is a P-gp modulator and inhibits CYP3A4 enzyme [75–77]. Rabbits pretreated with 15 mg/kg of quercetin, 30 min before verapamil administration, showed considerably altered pharmacokinetic behavior of verapamil (10 mg/kg). C_{max} and AUC of verapamil increased approximately 2-fold after pretreatment with quercetin as compared to verapamil administered alone, perorally. Absolute bioavailability and relative bioavailability of verapamil in the rabbits pre-treated with quercetin were significantly higher ($P < 0.05$) as compared to the control group [78]. The absolute bioavailability of oral diltiazem, when studied in rabbits pre-treated with quercetin was significantly higher, showing 9.10–12.81% increase ($p < 0.05$ at dose of 2 mg/kg, $p < 0.01$ at doses of 10 and 20 mg/kg) than 4.64% shown by the control group [74]. Another study was performed for 7 days on 12 healthy volunteers, each administered 500 mg quercetin or placebo, thrice daily. Later, single dose of fexofenadine (60 mg) was administered orally on day 7 and the results indicated an increase in the AUC of fexofenadine by 55% in the presence of quercetin [79]. One study suggested that quercetin increased the AUC and C_{max} of ranolazine due to inhibition of P-gp and CYP 3A4 [80]. The various drugs coadministered with quercetin that showed enhanced oral bioavailability are listed in Table 1.3.

Table 1.3: Studies on bioenhancement of drugs after coadministration of quercetin.

Drug	Category	Study model	Mechanism of action	References
Epigallocatechin-3-gallate	Phenolic antioxidant	*In vivo* (rats)	Inhibition of CYP 3A4 and P-gp	[74]
Tamoxifen	Selective estrogen receptor modulator	*In vivo* (rats)	Inhibition of CYP 3A4 and P-gp	[77]
Clopidogrel	Platelet aggregation inhibitor	*In vivo* (rats, dogs)	Inhibition of P-gp	[81]
Etoposide	Podophyllotoxin derivative	*In vivo* (rats)	Inhibition of CYP 3A4 and P-gp	[82]
Doxorubicin	Daunorubicin precursor	*In vivo* (human MCF-7 ADRr cells)	Inhibition of CYP 3A4 and P-gp	[83]
Pioglitazone	Thiazolidinedione	*In vivo* (rats)	Inhibition of CYP 3A	[84]

1.3.6 Genistein

Genistein is an isoflavone and a well-known phytoestrogen found in a variety of food sources like *Glycine max* (soybean) and *Pueraria lobata* (kudzu) and is studied widely for a number of health benefits such as anti-inflammatory and anticancer activities [85, 86]. The effects of oral genistein on pharmacokinetics of orally and intravenously administered paclitaxel in rats were demonstrated. 10 mg/kg genistein significantly increased the AUC of oral paclitaxel by 54.7% and was attributed to the significantly ($p < 0.05$) reduced total plasma clearance (Cl/F) of paclitaxel, which was lowered by 35.2% [87]. It was also reported that the genistein co-treatment of HT-29 human colon cancer cells increased the cytosolic epigallocatechin-3-gallate

(EGCG) by 2- to 5-fold, compared to EGCG alone. Intragastric coadministration of EGCG, 75 mg/kg and genistein, 200 mg/kg to CF-1 mice showed higher $AUC_{0-\infty}$ of EGCG (183.9 ± 20.2 mg/mL versus 125.8 ± 26.4 mg/mL, respectively, at 3 min). Co-treatment with genistein also increased the half-life, C_{max} and $AUC_{0-6\ h}$ of EGCG in the small intestine by 1.4-, 2.0-, and 4.7-folds, respectively, as compared with mice treated with EGCG alone [88].

1.3.7 Curcumin

Curcumin, the primary curcuminoid of *Curcuma longa* (turmeric), which is a popular Indian spice, possesses several biological and pharmacological properties [89]. It has been reported that potential interactions between conventional drugs and curcuminoids are due to modulation of CYP 450 and phase II enzymes, along with effects on organic anion transporting polypeptides and inhibitory effect on P-gp efflux transporters. The changes in the pharmacokinetic profiles of oral celiprolol and midazolam due to effects on CYP3A expression and P-gp induced by curcumin were evaluated in male Sprague Dawley rats. Celiprolol and midazolam showed significant increase in bioavailability in the presence of curcumin (60 mg/kg) in pretreated rats, but not in rats where it was coadministered only once. The expression of intestinal P-gp proteins and CYP 3A protein content was reduced by 49% and 42%, respectively, after pretreatment for 4 days with curcumin, when evaluated by Western blot analyses. It was understood that celiprolol is a substrate for P-gp and midazolam shows extensive metabolism by CYP 3A enzymes. Hence, the pharmacokinetics of these two drugs have been modified by curcumin due to the downregulation of intestinal P-gp and CYP 3A in the small intestine [90]. In another study, after co-administration of curcumin and tamoxifen orally to rats, the AUC and C_{max} values of tamoxifen were considerably increased by 64% and 71%, respectively [91, 92].

1.3.8 Capsaicin

Capsaicin, the phytoconstituent of *Capsicum annum* (chili peppers), causes irritation in humans and other mammals, producing burning in tissues, which come in contact with it. The effect on theophylline absorption and bioavailability in rabbits after oral administration (20 mg/kg), including and excluding the ground capsicum fruit suspension, was investigated. Comparing the pharmacokinetic parameters indicated that the concomitant administration of capsicum fruit suspension increases AUC and C_{max} from 86.06 ± 9.78 to 138.32 ± 17.27 mg /L h and 6.65 ± 0.76 to 8.78 ± 0.98 mg/L, respectively [93]. Fexofenadine showed significant increase in intestinal transport and apparent permeability (P_{app}) by 2.8- and 2.6-folds, respectively, in the ileum of capsaicin treated rats, as compared to the control group. The *in vivo* studies revealed that maximum plasma concentration (C_{max}) and area under the concentration–time curve (AUC) were increased considerably by 2.3- and 2.4-folds, respectively, in rats pre-treated with capsaicin, when compared with the control group. Hence, capsaicin pretreatment was found to enhance the intestinal absorption and oral bioavailability of fexofenadine in rats, likely due to inhibitory effect on P-gp mediated cellular efflux [94].

1.3.9 Lysergol

Ipomoea muricata and *Ipomoea violacea* contain the phytomolecule Lysergol, as the active constituent. Lysergol (9,10-didehydro-6-methylergoline-8-*O*-methanol) is also found in *Rivea corymbosa* and fungi, e.g., *Claviceps, Rhizopus, and Penicillium* [6, 95]. Lysergol has been utilized since long as hypotensive, analgesic, psychotropic and immunostimulant; it helps maintain regular blood flow due to vasoactive properties and promotes gastrointestinal drug absorption [96]. Lysergol as a bioenhancer has been shown to increase bioavailability of certain antibiotics, and the bioavailability enhancing potential is accredited to the inhibition of metabolic enzymes as well as P-gp efflux transporters. Lysergol isolated from *I. muricata* seeds at a concentration of 2–10 µg/mL has been shown to enhance bioavailability of rifampicin and tetracycline 2.96- to 8.53-fold [6]. The bioenhancing potential of lysergol on curcumin was studied. *In vitro* phase I and II metabolic stability evaluation of curcumin following preincubation with lysergol was performed using rat liver microsomes. Additionally, the effect on major efflux transporters using breast cancer resistance protein (BCRP) and human P-gp membrane preparations were also studied. The results were compared with the P-gp inhibitory potential of verapamil and BCRP inhibitory potential of pantoprazole. Subsequently, the studies gave strong evidence about the involvement of BCRP inhibition, ruling out the possibility of P-gp inhibition, and a remarkable increase in the *in vitro* half-life of curcumin was seen [97]. Lysergol (20 mg/kg) has also been known to potentiate the activity of quaternary protoberberine alkaloid, berberine 2-fold, when studied in male Sprague Dawley rats [98].

1.3.10 Naringin

Phytoconstituent naringin is an example of a flavonoid isolated from grapefruit. There are reports regarding pharmacokinetic interactions of grapefruit juice coadministered with drugs such as ketoconazole, verapamil, erythromycin, and midazolam [99]. The effect on bioavailability and pharmacokinetics of oral paclitaxel coadministered with naringin to rats was demonstrated. The blood levels of paclitaxel coadministered with naringin increased significantly when compared with the control. After coadministration of naringin, paclitaxel showed 1.35- to 1.69-fold higher relative bioavailability than the control. Additionally, after coadministration with naringin, the absolute bioavailability of paclitaxel increased 2-fold. Naringin increased the bioavailability of paclitaxel by means of inhibition of P-gp efflux transporters and CYP 3A [100]. Similarly, diltiazem showed improvement in bioavailability in the presence of naringin, following oral administration to rats [101].

1.3.11 Senomenine

It is an alkaloidal compound isolated from *Sinomenium acutum*, extensively used for treating arthritic and rheumatic pain in Japan and China. It has been reported that sinomenine could play a significant role in improving the bioavailability of paeoniflorin when studied in rats. The absorption properties and the intestinal kinetic absorption characteristics of paeoniflorin in the *in vitro* everted rat gut sac model during coadministration of sinomenine were investigated. Results of the study demonstrated that 16 and 136 μM sinomenine significantly increased the absorption of 20 μM paeoniflorin by 1.5- and 2.5-folds, respectively [102]. A 12-fold enhancement in bioavailability of paeoniflorin (150 mg/kg) in the presence of sinomenine (90 mg/kg) was observed when studied in rats [103].

1.3.12 Allicin

It is an active bioenhancer phytomolecule found in *Allium sativum* and has been reported to enhance the fungicidal properties of amphotericin B against *Candida albicans*, *Aspergillus fumigatus* and *Saccharomyces cerevisiae*. Amphotericin B, coadministered with allicin, exhibited greater antifungal activity against *Saccharomyces cerevisiae*. In the study, it was suggested that Amphotericin B which is known for its fungicidal activity when coadministered with allicin was also involved in vacuole disruption and it was also understood that expression of allicin-mediated activity of Amphotericin B required cell membrane ergosterol [104].

1.3.13 Other plant extracts studied for bioenhancement activity

Nigella sativa (black cumin) was evaluated for bioenhancing effects for amoxicillin. Extracts of seeds of *N. sativa* were obtained after extractions with hexane and methanol for 6 h. Results of *in vitro* study performed using everted rat intestinal sacs demonstrated an increased permeation of amoxicillin with hexane extract of *N. sativa* seeds. Results of *in vivo* study demonstrated that *N. sativa* extract improved the bioavailability of amoxicillin by increasing the C_{max} and AUC_{0-t} from $4{,}138.251 \pm 156.93$

to 5,995.045 ± 196.28 ng/mL and 8,890.40 ± 143.33 to 13,483.46 ± 152.45 ng/mL h, respectively [105]. The effect of *N. sativa* on pharmacokinetics of cyclosporine in rabbits was investigated. The results of *in vivo* studies showed significant increase in clearance of cyclosporine by about 2-fold, suggesting that in the presence of *N. sativa*, the expression of intestinal P-gp and/or CYP3A4 were activated [106].

The dried ripe fruit of *Carum carvi* is the source of caraway. It has been reported that caraway increases the bioavailability of antibiotics, antifungal, antiviral, and anticancer drugs [107]. The bioenhancing effect of 100 mg *C. carvi* extract on the pharmacokinetics of rifampicin, isoniazid and pyrazinamide; fixed-dose combination, when administered to 20 healthy human volunteers, was studied. Administration of caraway extract displayed an increase in C_{max}, which was as follows, 36.01%, 32.22%, and 33.22%, respectively, and the increase in $AUC_{0-24\,h}$ was 29.06%, 32.16%, and 27.92% for isoniazid, rifampicin and pyrazinamide respectively. It has been suggested that enhanced bioavailability in the presence of caraway extract was due to enhancement of mucosal and serosal permeation of drugs along with the inhibitory influence on P-gp efflux [108].

The effect of *Aloe vera* preparations on bioavailability of vitamin C and vitamin E in healthy human subjects was studied. The bioavailability of vitamin C and vitamin E were determined in normal fasting subjects, with eight subjects studied for vitamin C and ten subjects for vitamin E. The gels and whole leaf extracts of *Aloe vera* improved absorption of vitamin C and vitamin E and prolonged its plasma concentration [109]. Therefore, *Aloe vera* was suggested to supplement the increase in absorption of vitamin E and vitamin C and could be considered as a nutritional bioenhancer in the near future [110]. The permeability study was performed using Franz diffusion cells, and the results of the study indicated that *Aloe vera* gel significantly enhanced the buccal permeability of diadanosine, with enhancement ratio ranging from 5.09 (0.25% w/v of *A. vera* gel) to 11.78 (2% w/v of *A. vera* gel), respectively. Therefore, *A. vera* could be proposed for use as buccal penetration enhancer for diadanosine for treatment of HIV and AIDS [111].

1.4 Role of herbal extracts in biosynthesis of silver and gold nanoparticles

Plants have been used to synthesize gold and silver nanoparticles by reducing ions Ag^+ and Au^{3+} in aqueous form using the broth of *Azadirachta indica* leaves. The approach can be utilized for commercial syntheses of nanoparticles, since scaling up of the operations is feasible [112]. Gold nanotriangles and silver nanoparticles employing *Aloe vera* extract were also synthesized [113].

1.5 Recent advances in herbal drug delivery systems

A well-designed formulation approach for optimization of the pharmacokinetics of the herbal actives is essential to potentiate their activity *in vivo* and for further maximization of their potential in bioavailability enhancement (Figure 1.2).

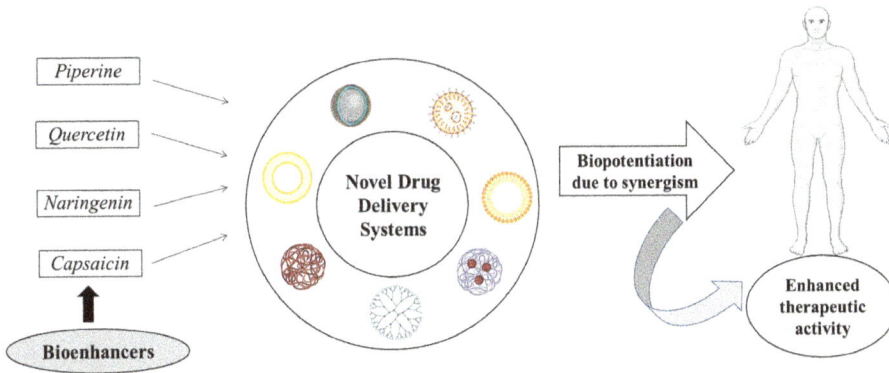

Figure 1.2: Recent advances in herbal drug delivery systems.

These bioenhancers have their own therapeutic activity and also show synergistic activity with other drugs. This activity can be enhanced by incorporating them into novel drug delivery systems like liposomes, nanoparticles, microspheres and transferosomes that can be administered by various routes. Table 1.4 represents the various novel formulations of herbal bioenhancers.

An example in which bioenhancer entrapped in liposomes is discussed for improving the bioavailability of domperidone is mentioned. The study was aimed at synthesizing phytosomes made of piperine and phosphatidylcholine by salting-out method [126]. Phytosomes imply molecular complexes of phytoconstituents and phospholipids of natural origin that show improved bioavailability as compared to liposomes [22]. These engineered phytosomes with piperine showed significant improvement in bioavailability of oral domperidone (79.5%), when compared with pure drug suspension in rats. Pharmacokinetic parameters of domperidone such as maximum plasma concentration and $AUC_{0-24\ h}$ were potentiated by this approach. The improved drug absorption was attributed to inhibitory effects on P-gp transporters. The involvement of piperine as bioenhancer in phytosomes suggests a roadmap for the improvement of oral absorption of many other poorly bioavailable drugs.

Table 1.4: Novel drug delivery systems of herbal bioenhancers.

Drug delivery	Active	In vitro/ in vivo study	Biological activity	Application	Physicochemical property	References
					Percent entrapment efficiency	
Herbal liposomes	Quercetin	Intranasal	Antioxidant and anticancer	Reduced dose, enhanced penetration across the blood–brain barrier	60%	[114]
	Silymarin	Buccal	Hepatoprotective	Improved bioavailability	70%	[115]
	Artemisinin	In vitro	Antiviral	Targetability of essential oils to cells, enhanced permeation into cytoplasm	74%	[116]
	Paclitaxel	In vitro	Anticancer	Liposomal formulation of paclitaxel showed 96% drug release at pH 5 within 15 min	94%	[117]
					Particle size range	
Nanoparticles	Triptolide	Topical	Anti-inflammatory	Increased hydration and enhanced drug penetration through stratum corneum	123 ± 0.9 nm	[118]
	Naringenin	Oral	Hepatoprotective	Protective effects on liver with substantial increase in antioxidant enzyme levels, considerable reduction in liver function index, and lipid peroxidation	457.10 ± 18.49 nm	[119]
Microspheres	Rutin	In vitro	Cerebrovascular and cardiovascular	Targeting the cerebrovascular and cardiovascular systems	165–195 μm	[120]
	Zedoary	Oral	Hepatoprotective	Sustained release and higher bioavailability	100–600 μm	[121]

Transferosomes	Capsaicin	Topical	Analgesic	Increased skin penetration	150.6 nm	[122]
	Colchicine	In vitro	Anti-gout	Increased skin penetration in ex vivo studies	100–200 nm	[123]
Ethosomes	Piperine	Topical	Atopic dermatitis	Ethosomal topical cream showed high skin penetration in in vivo studies	318.1 nm	[124]
	Quercetin	Topical	Antioxidant	Increased skin penetration and bioavailability in in vivo studies	96.99–476.74 nm	[125]

1.6 Conclusions

In developing countries, with the discovery of newer drug molecules and delivery approaches, the increasing treatment cost remains a matter of concern. Hence, an innovative but systematic approach is required to reduce these costs. Biopotentiation using herbal drugs constitutes an innovative and integrative concept, the discovery of which was based on traditional Indian medicine system (mentioned by Charaka, Sushruta, and other apothecaries). Application of these Ayurvedic principles to drugs and delivery systems provides the advantages of integrating ancient and indigenous systems of medicine with contemporary medicine. This novel concept would be useful in decreasing treatment cost, toxicity and other adverse effects, thus ultimately influencing the national economy.

1.7 Future prospects

Biopotentiators present the required criteria to be considered as an ideal drug. Additionally, these can be easily procured, are safe, effective and nonaddictive, having widely based effects on several classes of active compounds. Drug discovery process has been greatly strengthened by Ayurveda using reverse pharmacology approach and proposes innovative means of identifying active molecules and reducing the developmental cost. The traditional wisdom of Ayurveda has immense utility in increasing the bioavailability of drugs and could open new avenues in health care, for making treatment of diseases more economical and available to wider sections to the society. Literature reports the use of herbal compounds in oral delivery for biopotentiation of drugs, but limited research is reported for the other routes such as transdermal, ocular, nasal and vaginal. This area of research can focus more stringently upon identification of herbal active compounds, study of their mechanisms of actions, evaluation of safety and toxicity, regulatory concerns, development of viable combinations with other drugs, and delivery systems by various routes for enhancing bioavailability.

List of abbreviations

Area under plasma concentration–time curve extrapolated to infinity	$AUC_{0-\infty}$
Breast cancer resistance protein	BCRP
Maximum plasma concentration	C_{max}
Epigallocatechin-3-gallate	EGCG
Gastrointestinal tract	GIT
P-glycoprotein	P-gp
Uridine 5′-diphospho-glucuronyl transferase	UDP-glucuronyl transferase

References

[1]   Atal N, Bedi KL, Bioenhancers: Revolutionary concept to market, Journal of Ayurveda and Integrative Medicine, 2010, 1, 96–99.

[2]   Kang MJ, Cho JY, Shim BH, Kim DK, Lee J, Bioavailability enhancing activities of natural compounds from medicinal plants, The Journal of Medicinal Plants Research, 2009, 3(13), 1204–1211.

[3]   Sutrasthan AH Chapter 1: Ayushkameeya आयु कामीयं Adhyaya "Desire for long life.": 352.

[4]   Jaiswal YS, Williams LL, A glimpse of Ayurveda – The forgotten history and principles of Indian traditional medicine, Journal of Traditional and Complementary Medicine, 2016, 7, 50–53.

[5]   Atal CK, A breakthrough in drug bioavailability-a clue from age old wisdom of Ayurveda, IDMA Bulletin, 1979, 10, 483–484.

[6]   Khanuja S, Arya J, Srivastava S, Shasany A, Kumar TS, Darokar M, et al. Antibiotic pharmaceutical composition with lysergol as bio-enhancer and method of treatment [Internet]. 2007 [cited 2021 Oct 15]. Available from: https://patents.google.com/patent/US20070060604A1/en

[7]   Jhanwar B, Gupta S, Biopotentiation using herbs: Novel technique for poor bioavailable drugs, International Journal of Medicine and Pharmaceutical Research, 2014, 6, 443–454.

[8]   Oberoi K, Tatke P Piperine Enhances the Bioavailability of Secnidazole in Rats. MOJBB [Internet]. 2017 [cited 2021 Oct 20];3. Available from: https://medcraveonline.com/MOJBB/piperine-enhances-the-bioavailability-of-secnidazole-in-rats.html

[9]   Khedekar K, Tatke P, 2018, 'Enhancement of bioavailability of metformin using phytoconstituents', M.Pharm Thesis, S.N.D.T. Women's University, Mumbai.

[10]   Vijaylakshmi B, Tatke P, 2014, 'Evaluation of Piperine for bioenhancement of Metformin', M.Pharm Thesis, S.N.D.T. Women's University, Mumbai.

[11]   Khan IA, Mirza ZM, Kumar A, Verma V, Qazi GN, Piperine, a Phytochemical Potentiator of Ciprofloxacin against *Staphylococcus aureus*, Antimicrobial Agents and Chemotherapy, 2006, 50, 810–812.

[12]   Beg S, Swain S, Rizwan M, Irfanuddin M, Malini DS, Bioavailability enhancement strategies: Basics, formulation approaches and regulatory considerations, Current Drug Delivery, 2011, 8, 691–702.

[13]   Lemmer HJR, Hamman JH, Paracellular drug absorption enhancement through tight junction modulation, Expert Opinion on Drug Delivery, 2013, 10, 103–114.

[14]   Javed S, Kohli K, Ali M, Reassessing bioavailability of silymarin, Alternative Medicine Review, 2011, 16(3), 239–249.

[15]   Porter CJH, Pouton CW, Cuine JF, Charman WN, Enhancing intestinal drug solubilisation using lipid-based delivery systems, Advanced drug delivery reviews, 2008, 60, 673–691.

[16]   Javed S, Ahsan W, Kohli K, Pharmacological influences of natural products as bioenhancers of silymarin against carbon tetrachloride-induced hepatotoxicity in rats, Clinical Phytoscience, 2018, 4, 18.

[17]   M AS, Mahurkar N The Mechanism of Actions for Herbal Bioenhancers. RGUHS JPharm Sci [Internet]. 2021 [cited 2021 Oct 27];11. Available from: https://journalgrid.com/view/article/rjps/322.

[18]   Atal C, Dubey R, Singh J, Biochemical basis of enhanced drug bioavailability by piperine: Evidence that piperine is a potent inhibitor of drug metabolism, The Journal of Pharmacology and Experimental Therapeutics, 1985, 232(1), 258–262.

[19]   Singh J, Dubey RK, Atal CK, Piperine-mediated inhibition of glucuronidation activity in isolated epithelial cells of the guinea-pig small intestine: Evidence that piperine lowers the

endogenous UDP-glucuronic acid content, The Journal of Pharmacology and Experimental Therapeutics, 1986, 236, 488–493.

[20] Khajuria A, Zutshi U, Bedi KL, Permeability characteristics of piperine on oral absorption–an active alkaloid from peppers and a bioavailability enhancer, Indian Journal of Experimental Biology, 1998, 36, 46–50.

[21] Bajad S, Bedi KL, Singla AK, Johri RK, Piperine inhibits gastric emptying and gastrointestinal transit in rats and mice, Planta Medica, 2001, 67, 176–179.

[22] Yurdakok-Dikmen B, Turgut Y, Filazi A, Herbal Bioenhancers in Veterinary Phytomedicine, Frontiers in Veterinary Science, 2018, 5, 249.

[23] Majeed M, Badmaev V, Rajendran R Use of piperine as a bioavailability enhancer [Internet]. 1998 [cited 2021 Oct 15]. Available from: https://patents.google.com/patent/US5744161A/en

[24] Parmar VS, Jain SC, Bisht KS, Jain R, Taneja P, Jha A, et al., Phytochemistry of the genus Piper, Phytochem, 1997, 46, 597–673.

[25] Pholpramool C, Piyachaturawat P Enhancement of fertilization by piperine in hamsters. Cell Biology International [Internet]. [cited 2021 Oct 15]; Available from: https://www.academia.edu/20809409/Enhancementof_fertilization_by_piperine_in_hamsters

[26] Mujumdar AM, Dhuley JN, Deshmukh VK, Raman PH, Naik SR, Anti-Inflammatory Activity of Piperine, Japanese Journal of Medical Science & Biology, 1990, 43, 95–100.

[27] Kumar S, Singhal V, Roshan R, Sharma A, Rembhotkar GW, Ghosh B, Piperine inhibits TNF-alpha induced adhesion of neutrophils to endothelial monolayer through suppression of NF-kappaB and IkappaB kinase activation, European Journal of Pharmacology, 2007, 575, 177–186.

[28] Navickiene HMD, Alécio AC, Kato MJ, Bolzani VDS, Mcm Y, Aj C, et al, Antifungal amides from Piper hispidum and Piper tuberculatum, Phytochem, 2000, 55, 621–626.

[29] Pradeep CR, Kuttan G, Effect of piperine on the inhibition of lung metastasis induced B16F-10 melanoma cells in mice, Clinical & Experimental Metastasis, 2002, 19, 703–708.

[30] Mittal R, Gupta RL, In vitro antioxidant activity of piperine, Methods and Findings in Experimental and Clinical Pharmacology, 2000, 22, 271.

[31] Akhilender Naidu K, Thippeswamy NB, Inhibition of human low density lipoprotein oxidation by active principles from spices, Molecular and Cellular Biochemistry, 2002, 229, 19–23.

[32] Panda S, Kar A, Piperine Lowers the Serum Concentrations of Thyroid Hormones, Glucose and Hepatic 5′d Activity in Adult Male Mice. Horm Metab Res.©, New York, Georg Thieme Verlag Stuttgart, 2003, 35, 523–526.

[33] Sunila ES, Kuttan G, Immunomodulatory and antitumor activity of Piper longum Linn and Piperine, Journal of Ethnopharmacology., 2004, 90, 339–346.

[34] Manoharan S, Balakrishnan S, Menon VP, Alias LM, Reena AR, Chemopreventive efficacy of curcumin and piperine during 7, 12-dimethylbenz (a) anthracene-induced hamster buccal pouch carcinogenesis, Singapore Medical Journal, 2009, 50, 139.

[35] Park B-S, Son D-J, Park Y-H, Kim TW, Lee S-E, Antiplatelet effects of acidamides isolated from the fruits of Piper longum L, Phytomed, 2007, 14, 853–855.

[36] El Hamss R, Idaomar M, Alonso-Moraga A, Muñoz Serrano A, Antimutagenic properties of bell and black peppers, Food and Chemical Toxicology, 2003, 41, 41–47.

[37] Selvendiran K, Banu SM, Sakthisekaran D, Oral supplementation of piperine leads to altered phase II enzymes and reduced DNA damage and DNA-protein cross links in Benzo(a)pyrene induced experimental lung carcinogenesis, Molecular and Cellular Biochemistry, 2005, 268, 141–147.

[38] Pooja S, Agrawal RP, Nyati P, Savita V, Phadnis P, Analgesic activity of Piper nigrum extract per se and its interaction with diclofenac sodium and pentazocine in albino mice, The Internet Journal of Pharmacology, 2007, 5(1), 3.

[39] Lee SA, Hong SS, Han XH, Hwang JS, Oh GJ, Lee KS, et al., Piperine from the Fruits of *Piper longum* with Inhibitory Effect on Monoamine Oxidase and Antidepressant-Like Activity, Chemical and Pharmaceutical Bulletin, 2005, 53, 832–835.

[40] Matsuda H, Ninomiya K, Morikawa T, Yasuda D, Yamaguchi I, Yoshikawa M, Protective effects of amide constituents from the fruit of Piper chaba on d-galactosamine/TNF-α-induced cell death in mouse hepatocytes, Bioorganic & Medicinal Chemistry Letters, 2008, 18, 2038–2042.

[41] Kim S-H, Lee Y-C, Piperine inhibits eosinophil infiltration and airway hyper responsiveness by suppressing T cell activity and Th2 cytokine production in the ovalbumin-induced asthma model, The Journal of Pharmacy and Pharmacology, 2009, 61, 353–359.

[42] Taqvi SIH, Shah AJ, Gilani AH, Blood Pressure Lowering and Vasomodulator Effects of Piperine, Journal of Cardiovascular Pharmacology, 2008, 52, 452–458.

[43] Piyachaturawat P, Glinsukon T, Toskulkao C, Acute and subacute toxicity of piperine in mice, rats and hamsters, Toxicology Letters, 1983, 16, 351–359.

[44] Jin M-J, Han H-K, Effect of Piperine, a Major Component of Black Pepper, on the Intestinal Absorption of Fexofenadine and Its Implication on Food–Drug Interaction, Journal of Food Science, 2010, 75, H93–6.

[45] Johnson JJ, Nihal M, Siddiqui IA, Scarlett CO, Bailey HH, Mukhtar H, et al., Enhancing the bioavailability of resveratrol by combining it with piperine, Molecular Nutrition & Food Research, 2011, 55, 1169–1176.

[46] Atal S, Atal S, Vyas S, Phadnis P, Bio-enhancing Effect of Piperine with Metformin on Lowering Blood Glucose Level in Alloxan Induced Diabetic Mice, Pharmacognosy Research, 2016, 8, 56–60.

[47] Gupta SK, Bansal P, Bhardwaj RK, Velpandian T, Comparative anti-nociceptive, anti-inflammatory and toxicity profile of Nimesulide vs Nimesulide And Piperine Combination, Pharmacological Research: The Official Journal of the Italian Pharmacological Society, 2000, 41, 657–662.

[48] Kasibhatta R, Naidu MUR, Influence of Piperine on the Pharmacokinetics of Nevirapine under Fasting Conditions: A Randomised, Crossover, Placebo-Controlled Study, Drugs in R & D, 2007, 8, 383–391.

[49] Pattanaik S, Hota D, Prabhakar S, Kharbanda P, Pandhi P, Pharmacokinetic interaction of single dose of piperine with steady-state carbamazepine in epilepsy patients: Carbamazepine kinetics with piperine in epilepsy, Phytotherapy research: PTR, 2009, 23, 1281–1286.

[50] Sama V, Nadipelli M, Yenumula P, Bommineni M, Mullangi R, Effect of Piperine on Antihyperglycemic Activity and Pharmacokinetic Profile of Nateglinide, Arzneimittelforschung, 2012, 62, 384–388.

[51] Kapil RS, Zutshi U, Bedi KL, Singh G, Johri RK, Dhar SK, et al. Process for preparation of pharmaceutical composition with enhanced activity for treatment of tuberculosis and leprosy [Internet]. 1995 [cited 2021 Oct 16]. Available from: https://patents.google.com/patent/US5439891/en

[52] Bano G, Raina RK, Zutshi U, Bedi KL, Johri RK, Sharma SC, Effect of piperine on bioavailability and pharmacokinetics of propranolol and theophylline in healthy volunteers, European Journal of Clinical Pharmacology, 1991, 41, 615–617.

[53] Sudipta B, Himanshu R, Patel Vandana B, Hitesh P, Effect of herbal bio-enhancers on Saquinavir in human caco-2 cell monolayers and pharmacokinetics in rats, International Journal of Medicine and Pharmaceutical Research, 2012, 2, 27–41.

[54] Jolad SD, Lantz RC, Solyom AM, Chen GJ, Bates RB, Timmermann BN, Fresh organically grown ginger (Zingiber officinale): Composition and effects on LPS-induced PGE2 production, Phytochem, 2004, 65, 1937–1954.

[55] Govindarajan VS, Ginger-chemistry, technology, and quality evaluation: Part 2, Critical Reviews in Food Science and Nutrition, 1982, 17, 189–258.

[56] Evans WC, Trease and Evans' Pharmacognosy, Elsevier Health Sciences, Saunders Ltd., 2009.

[57] O'Hara M, Kiefer D, Farrell K, Kemper K, A review of 12 commonly used medicinal herbs, Archives of Family Medicine, 1998, 7, 523–536.

[58] Qazi G, Bedi K, Johri R, Tikoo M, Tikoo A, Sharma S, et al. Bioavailability enhancing activity of Zingiber officinale Linn and its extracts/fractions thereof [Internet]. 2003 [cited 2021 Oct 16]. Available from: https://patents.google.com/patent/US20030170326A1/en

[59] Khanuja SPS, Kumar S, Arya JS, Shasany AK, Singh M, Awasthi S, et al. Composition comprising pharmaceutical/nutraceutical agent and a bio-enhancer obtained from Glycyrrhiza glabra [Internet]. 2005 [cited 2021 Oct 16]. Available from: https://patents.google.com/patent/US6979471B1/en

[60] Aruna K, Sivaramakrishnan VM, Anticarcinogenic effects of some Indian plant products, Food and Chemical Toxicology, 1992, 30, 953–956.

[61] Cáceres A, Saravia A, Rizzo S, Zabala L, De Leon E, Nave F, Pharmacologie properties of Moringa oleifera. 2: Screening for antispasmodic, anti-inflammatory and diuretic activity, Journal of Ethnopharmacology, 1992, 36, 233–237.

[62] Caceres A, Cabrera O, Morales O, Mollinedo P, Mendia P, Pharmacological properties of Moringa oleifera. 1: Preliminary screening for antimicrobial activity, Journal of Ethnopharmacology, 1991, 33, 213–216.

[63] Shukla S, Mathur R, Prakash AO, Antifertility profile of the aqueous extract of Moringa oleifera roots, Journal of Ethnopharmacology, 1988, 22, 51–62.

[64] Gilani AH, Aftab K, Suria A, Siddiqui S, Salem R, Siddiqui BS, et al., Pharmacological studies on hypotensive and spasmolytic activities of pure compounds from Moringa oleifera, Phytotherapy research: PTR, 1994, 8, 87–91.

[65] Siddhuraju P, Becker K, Antioxidant Properties of Various Solvent Extracts of Total Phenolic Constituents from Three Different Agroclimatic Origins of Drumstick Tree (Moringa oleifera Lam.) Leaves, Journal of Agricultural and Food Chemistry ACS, 2003, 51, 2144–2155.

[66] Nwosu MO, Okafor JI, Preliminary studies of the antifungal activities of some medicinal plants against Basidiobolus and some other pathogenic fungi, Mycoses, 1995, 38, 191–195.

[67] Guevara AP, Vargas C, Uy M, Anti-inflammatory and antitumor activities of seed extracts of malunggay, Moringa oleifera L. (Moringaceae), Philippine Journal of Science (Philippines) [Internet], 1996, [cited 2021 Oct 16]; Available from, https://scholar.google.com/scholar_lookup?title=Antiinflammatory+and+antitumor+activities+of+seed+extracts+of+malunggay%2C+Moringa+oleifera+L.+%28Moringaceae%29&author=Guevara%2C+A.P.&publication_year=1996.

[68] Pal SK, Mukherjee PK, Saha BP, Studies on the antiulcer activity of Moringa oleifera leaf extract on gastric ulcer models in rats, Phytotherapy Research: PTR, 1995, 9, 463–465.

[69] Saravillo KB, Herrera AA Biological activity of Moringa oleifera Lam. (Malunggay) crude seed extract. Philippine Agricultural Scientist (Philippines) [Internet]. 2004 [cited 2021 Oct 16]; Available from: https://scholar.google.com/scholar_lookup?title=Biological+activity+of+Moringa+oleifera+Lam.+%28Malunggay%29+crude+seed+extract&author=Saravillo%2C+K.B.&publication_year=2004

[70] Mahajan SG, Mali RG, Mehta AA, Protective Effect of Ethanolic Extract of Seeds of Moringa oleifera Lam. Against Inflammation Associated with Development of Arthritis in Rats, Journal of Immunotoxicology, Taylor & Francis, 2007, 4, 39–47.

[71] Mehta K, Balaraman R, Amin AH, Bafna PA, Gulati OD, Effect of fruits of Moringa oleifera on the lipid profile of normal and hypercholesterolaemic rabbits, Journal of Ethnopharmacology, 2003, 86, 191–195.

[72] Pari L, Kumar NA, Hepatoprotective Activity of Moringa oleifera on Antitubercular Drug-Induced Liver Damage in Rats, Journal of Medicinal Food, Mary Ann Liebert, Inc., Publishers, 2002, 5, 171–177.

[73] Khanuja SPS, Arya JS, Tiruppadiripuliyur RSK, Saikia D, Kaur H, Singh M, et al. Nitrile glycoside useful as a bioenhancer of drugs and nutrients, process of its isolation from moringa oleifera [Internet]. 2005 [cited 2021 Oct 16]. Available from: https://patents.google.com/patent/US6858588B2/en

[74] Nijveldt RJ, van Nood E, van Hoorn DE, Boelens PG, van Norren K, van Leeuwen PA, Flavonoids: A review of probable mechanisms of action and potential applications, The American Journal of Clinical Nutrition, 2001, 74, 418–425.

[75] Hsiu S-L, Hou Y-C, Wang Y-H, Tsao C-W, Su S-F, Chao P-DL, Quercetin significantly decreased cyclosporin oral bioavailability in pigs and rats, Life Sciences, 2002, 72, 227–235.

[76] Choi J-S LX, Enhanced diltiazem bioavailability after oral administration of diltiazem with quercetin to rabbits, International Journal of Pharmaceutics, 2005, 297, 1–8.

[77] Shin S-C, Choi J-S LX, Enhanced bioavailability of tamoxifen after oral administration of tamoxifen with quercetin in rats, International Journal of Pharmaceutics, 2006, 313, 144–149.

[78] Choi J-S, Han H-K, The effect of quercetin on the pharmacokinetics of verapamil and its major metabolite, norverapamil, in rabbits, The Journal of Pharmacy and Pharmacology, 2010, 56, 1537–1542.

[79] Kim K-A, Park P-W, Park J-Y, Short-term effect of quercetin on the pharmacokinetics of fexofenadine, a substrate of P-glycoprotein, in healthy volunteers, European Journal of Clinical Pharmacology, 2009, 65, 609–614.

[80] Babu PR, Babu KN, Peter PLH, Rajesh K, Babu PJ, Influence of quercetin on the pharmacokinetics of ranolazine in rats and in vitro models, Drug Development and Industrial Pharmacy, Taylor & Francis, 2013, 39, 873–879.

[81] Lee JH, Shin Y-J, Oh J-H, Lee Y-J, Pharmacokinetic interactions of clopidogrel with quercetin, telmisartan, and cyclosporine A in rats and dogs, Archives of pharmacal research, 2012, 35, 1831–1837.

[82] Li X, Choi J-S, Effects of Quercetin on the Pharmacokinetics of Etoposide after Oral or Intravenous Administration of Etoposide in Rats. Anticancer Research, International Institute of Anticancer Research, 2009, 29, 1411–1415.

[83] Scambia G, Ranelletti FO, Panici PB, De Vincenzo R, Bonanno G, Ferrandina G, et al., Quercetin potentiates the effect of adriamycin in a multidrug-resistant MCF-7 human breast-cancer cell line: P-glycoprotein as a possible target, Cancer Chemotherapy and Pharmacology, 1994, 34, 459–464.

[84] Umathe SN, Dixit PV, Kumar V, Bansod KU, Wanjari MM, Quercetin pretreatment increases the bioavailability of pioglitazone in rats: Involvement of CYP3A inhibition, Biochemical Pharmacology, 2008, 75, 1670–1676.

[85] Lambert JD, Hong J, Yang G, Liao J, Yang CS, Inhibition of carcinogenesis by polyphenols: Evidence from laboratory investigations, The American Journal of Clinical Nutrition, 2005, 81, 284S–291S.

[86] Kurzer MS, Xu X, Dietary Phytoestrogens, Annual Review of Nutrition, 1997, 17, 353–381.

[87] Li X, Choi J-S, Effect of genistein on the pharmacokinetics of paclitaxel administered orally or intravenously in rats, International Journal of Pharmaceutics, 2007, 337, 188–193.

[88] Lambert JD, Kwon S-J, Ju J, Bose M, Lee M-J, Hong J, et al., Effect of genistein on the bioavailability and intestinal cancer chemopreventive activity of (-)-epigallocatechin-3-gallate, Carcinogenesis, 2008, 29, 2019–2024.

[89] Cavaleri F, Jia W, The true nature of curcumin's polypharmacology, Journal of Preventive Medicine, 2017, 2 (2:5), 1–11.

[90] Zhang W, Tan TMC, Lim L-Y, Impact of Curcumin-Induced Changes in P-Glycoprotein and CYP3A Expression on the Pharmacokinetics of Peroral Celiprolol and Midazolam in Rats, Drug metabolism and Disposition: The Biological Fate of Chemicals, 2007, 35, 110–115.

[91] Cho YA, Lee W, Choi JS, Effects of curcumin on the pharmacokinetics of tamoxifen and its active metabolite, 4-hydroxytamoxifen, in rats: Possible role of CYP3A4 and P-glycoprotein inhibition by curcumin, Pharmazie, 2012, 67, 124–130.

[92] Pavithra BH, Prakash N, Jayakumar K, Modification of pharmacokinetics of norfloxacin following oral administration of curcumin in rabbits, Journal of Veterinary Science. The Korean Society of Veterinary Science, 2009, 10, 293–297.

[93] Bouraoui A, Toumi A, Ben Mustapha H, Brazier JL, Effects of capsicum fruit on theophylline absorption and bioavailability in rabbits, Drug-nutrient interactions, 1988, 5, 345–350.

[94] Bedada SK, Appani R, Boga PK, Capsaicin pretreatment enhanced the bioavailability of fexofenadine in rats by P-glycoprotein modulation: *In vitro, in* situ and *in vivo* evaluation, Drug Development and Industrial Pharmacy, 2017, 43, 932–938.

[95] Kesarwani K, Gupta R, Bioavailability enhancers of herbal origin: An overview, Asian Pacific Journal of Tropical Biomedicine, 2013, 3, 253–266.

[96] Ajazuddin AA, Qureshi A, Kumari L, Vaishnav P, Sharma M, et al., Role of herbal bioactives as a potential bioavailability enhancer for Active Pharmaceutical Ingredients, Fitoterapia, 2014, 97, 1–14.

[97] Shukla M, Malik MY, Jaiswal S, Sharma A, Tanpula DK, Goyani R, et al., A mechanistic investigation of the bioavailability enhancing potential of lysergol, a novel bioenhancer, using curcumin, RSC Advance the Royal Society of Chemistry, 2016, 6, 58933–58942.

[98] Patil S, Dash RP, Anandjiwala S, Nivsarkar M, Simultaneous quantification of berberine and lysergol by HPLC-UV: Evidence that lysergol enhances the oral bioavailability of berberine in rats, Biomedical Chromatography, 2012, 26, 1170–1175.

[99] Mertens-Talcott SU, Zadezensky I, De Castro WV, Derendorf H, Butterweck V, Grapefruit-Drug Interactions: Can Interactions With Drugs Be Avoided?, The Journal of Clinical Pharmacology, 2006, 46, 1390–1416.

[100] Choi J-S, Shin S-C, Enhanced paclitaxel bioavailability after oral coadministration of paclitaxel prodrug with naringin to rats, International Journal of Pharmaceutics, 2005, 292, 149–156.

[101] Choi J-S, Han H-K, Enhanced oral exposure of diltiazem by the concomitant use of naringin in rats, International Journal of Pharmaceutics, 2005, 305, 122–128.

[102] Chan K, Liu ZQ, Jiang ZH, Zhou H, Wong YF, Xu H-X, et al., The effects of sinomenine on intestinal absorption of paeoniflorin by the everted rat gut sac model, Journal of Ethnopharmacology, 2006, 103, 425–432.

[103] Liu ZQ, Zhou H, Liu L, Jiang ZH, Wong YF, Xie Y, et al., Influence of co-administrated sinomenine on pharmacokinetic fate of paeoniflorin in unrestrained conscious rats, Journal of Ethnopharmacology, 2005, 99, 61–67.

[104] Ogita A, Fujita K, Taniguchi M, Tanaka T, Enhancement of the Fungicidal Activity of Amphotericin B by Allicin, an Allyl-Sulfur Compound from Garlic, against the Yeast *Saccharomyces cerevisiae* as a Model System, Planta Medica, 2006, 72, 1247–1250.

[105] Ali B, Amin S, Ahmad J, Ali A, Ali M, Mir SR, Bioavailability enhancement studies of amoxicillin with Nigella, The indian journal of Medical Research, 2012, 135, 555–559.

[106] Al-Jenoobi FI, Al-Suwayeh SA, Muzaffar I, Alam MA, Al-Kharfy KM, Korashy HM, et al., Effects of Nigella sativa and Lepidium sativum on cyclosporine pharmacokinetics, BioMed Research International, 2013, 2013, 953520.

[107] Qazi G, Bedi K, Johri R, Tikoo M, Tikoo A, Sharma S, et al. Bioavailability enhancing activity of Carum carvi extracts and fractions thereof [Internet]. 2003 [cited 2021 Oct 17]. Available from: https://patents.google.com/patent/US20030228381A1/en

[108] Choudhary N, Khajuria V, Gillani ZH, Tandon VR, Arora E, Effect of Carum carvi, a herbal bioenhancer on pharmacokinetics of antitubercular drugs: A study in healthy human volunteers, Perspectives in Clinical Research, 2014, 5, 80–84.

[109] Kubitzki K editor, Flowering Plants. Monocotyledons: Lilianae (Except Orchidaceae) [Internet], Berlin Heidelberg, Springer-Verlag, 1998, [cited 2021 Oct 17]. Available from:, https://www.springer.com/gp/book/9783540640608.

[110] Vinson JA, Al Kharrat H, Andreoli L, Effect of Aloe vera preparations on the human bioavailability of vitamins C and E, Phytomedicine, 2005, 12, 760–765.

[111] Ojewole E, Mackraj I, Akhundov K, Hamman J, Viljoen A, Olivier E, et al., Investigating the Effect of Aloe vera Gel on the Buccal Permeability of Didanosine, Planta Med. © Georg Thieme Verlag KG Stuttgart · New York, 2012, 78, 354–361.

[112] Shankar SS, Rai A, Ahmad A, Sastry M, Rapid synthesis of Au, Ag, and bimetallic Au core–Ag shell nanoparticles using Neem (Azadirachta indica) leaf broth, Journal of Colloid and Interface Science, 2004, 275, 496–502.

[113] Chandran SP, Chaudhary M, Pasricha R, Ahmad A, Sastry M, Synthesis of gold nanotriangles and silver nanoparticles using Aloe vera plant extract, Biotechnology Progress, 2006, 22, 577–583.

[114] Ajazuddin Null SS, Applications of novel drug delivery system for herbal formulations, Fitoterapia, 2010, 81, 680–689.

[115] El-Samaligy MS, Afifi NN, Mahmoud EA, Evaluation of hybrid liposomes-encapsulated silymarin regarding physical stability and in vivo performance, International Journal of Pharmaceutics, 2006, 319, 121–129.

[116] Sinico C, De Logu A, Lai F, Valenti D, Manconi M, Loy G, et al., Liposomal incorporation of Artemisia arborescens L. essential oil and *in vitro* antiviral activity, European Journal of Pharmaceutics and Biopharmaceutics, 2005, 59, 161–168.

[117] Rane S, Prabhakar B, Influence of liposome composition on paclitaxel entrapment and pH sensitivity of liposomes, International Journal of Medicine and Pharmaceutical Research, 2009, 1(3), 914–917.

[118] Mei Z, Chen H, Weng T, Yang Y, Yang X, Solid lipid nanoparticle and microemulsion for topical delivery of triptolide, European Journal of Pharmaceutics and Biopharmaceutics: Official Journal of Arbeitsgemeinschaft fur Pharmazeutische Verfahrenstechnik e.V, 2003, 56, 189–196.

[119] Yen F-L, Wu T-H, Lin L-T, Cham T-M, Lin -C-C, Naringenin-loaded nanoparticles improve the physicochemical properties and the hepatoprotective effects of naringenin in orally-administered rats with CCl(4)-induced acute liver failure, Pharmaceutical Research, 2009, 26, 893–902.

[120] Xiao L, Zhang Y, Xu J, Jin X Preparation of floating rutin-alginate-chitosan microcapsule. Chinese Traditional and Herbal Drugs [Internet]. 1994 [cited 2021 Oct 17];0. Available from: http://wprim.whocc.org.cn/admin/article/articleDetail?WPRIMID=577724&articleId=577724

[121] You J, Cui F, Han X, Wang Y, Yang L, Yu Y, et al., Study of the preparation of sustained-release microspheres containing zedoary turmeric oil by the emulsion-solvent-diffusion method and evaluation of the self-emulsification and bioavailability of the oil, Colloids and Surfaces. B, Biointerfaces, 2006, 48, 35–41.

[122] Xiao-Ying L, Luo J-B, Yan Z-H, Rong H-S, Huang W-M, [Preparation and in vitro and in vivo evaluations of topically applied capsaicin transferosomes], Yao Xue Xue Bao, 2006, 41, 461–466.

[123] Singh HP, Utreja P, Tiwary AK, Jain S, Elastic Liposomal Formulation for Sustained Delivery of Colchicine: In Vitro Characterization and *In Vivo* Evaluation of Anti-gout Activity, The AAPS journal, 2009, 11, 54–64.

[124] Kumar P, Sharma DK, Ashawat MS, Topical creams of piperine loaded lipid nanocarriers for management of atopic dermatitis: Development, characterization and in vivo investigation using BALB/c mice model, Journal of Liposome Research Taylor & Francis, 2021, 4, 1–9, 10.1080/08982104.2021.1880436.

[125] Ramadon D, Anwar E, Harahap Y, *In vitro* Penetration and Bioavailability of Novel Transdermal Quercetin-loaded Ethosomal Gel, Indian Journal of Pharmaceutical Sciences. OMICS International, 2018, 79, 948–956.

[126] Islam N, Irfan M, Hussain T, Mushtaq M, Khan IU, Yousaf AM, et al., Piperine phytosomes for bioavailability enhancement of domperidone, Journal of Liposome Research, Taylor & Francis, 2021, 4, 1–9, 10.1080/08982104.2021.1918153z.

Surendra Agrawal, Pravina Gurjar, Ayushi Agarwal

Chapter 2
Herbal bioenhancers in microparticulate drug delivery

Abstract: Review of microparticulate drug delivery unveils the boons that the system has to offer, yet only minimal knowledge of the role of herbal bioenhancers is available. Scientists focus intensely on ameliorating the bioavailability, while simultaneously downsizing adverse events to ensure safe drug delivery, directly drifting towards deploying plant-derived bioenhancers. Microparticulate drug delivery employs disintegrants and superdisintegrants for a sophisticated release, but dismally, hindrances such as the stagnant hydrophilic layer around a particle, the size of the drug particles released, and the P-glycoprotein efflux pumps pose complications to not only absorption but bioavailability as well. The bioenhancers that have been utilized in several novel drug deliveries have introduced serious compatibility and performance issues. Microparticulate drug delivery is best suited for use in plant bioenhancers such as quercetin, sinomenine, piperine, naringin, glycyrrhizin, genistein, and nitrile glycoside due to the simplified formulation and superlative stability. Modification of intestinal motility, stimulation of intestinal amino acid transporters, inhibition of cellular efflux pumps, inhibition of liver metabolism, and inhibition of intestinal enzymatic degradation are a handful of the proposed mechanisms. This chapter vividly puts forth the mechanisms and the studies on the use of plant bioenhancers for microparticulate drug delivery, the regulatory concerns, and the prospects in this discipline.

2.1 Introduction to microparticulate drug delivery and the challenges in bioavailability with the same

The world is exerting to beat the therapeutic needs of the population by introducing novel therapeutic agents. The rate and extent to which a therapeutic agent reaches circulation and is distributed to the target site are important in defining its bioavailability

Surendra Agrawal, Department of Pharmaceutical Chemistry, Datta Meghe College of Pharmacy, Datta Meghe Institute of Medical Sciences (Deemed to be University), Sawangi (Meghe), Wardha 442004, Maharashtra, India, e-mail: sagrawal80@gmail.com
Pravina Gurjar, Sharadchandra Pawar College of Pharmacy, Otur 412409, Pune, Maharashtra, India
Ayushi Agarwal, Research Scholar, SPPSPTM, SVKM's NMIMS, Mumbai 400056, Maharashtra, India

https://doi.org/10.1515/9783110746808-002

[1]. Despite their potential therapeutic benefits, these neophytes exhibit restricted bioavailability due to their stunted permeation across the GI epithelia. This truncated permeability could even be attributed to minimal zwitterionic and lipophilicity characteristics at physiological conditions [2], and efflux by P-glycoprotein (P-gp) or poor water solubility [3]. Use of permeability-enhancing dosage forms are observed to be a practical approach to spice up the intestinal absorption of poorly absorbed drugs. Diverse dosage formulations such as emulsions [4], liposomes [5], and particle size reduction techniques such as liquid crystalline phases, microparticulated carriers, complexation, and micronization are employed as effective tools to ameliorate drug absorption [6, 7].

Microparticulate drug delivery, including microencapsulation, demonstrates a meticulous approach with a profound therapeutic impact, which is in demand, worldwide. Modulation of biocompatibility, target specificity, controlled and sustained release patterns, and stability, along with uniform encapsulation, better compliance, and reduced dosage frequency, are some of the superior properties of micro particulates. In microparticles, both water-insoluble and sparingly water-soluble agents are encapsulated intelligently to upgrade their efficacy. The distinct properties of microparticulates that are controlled prudently, include particle shape, structure, particle size, drug loading, porosity, entrapment efficiency, and release profile. Several available microparticle-based formulations are marketed, including buserelin, risperidone, and octreotide acetate, while others are in clinical trials. However, the microparticulate drug delivery poses a plethora of challenges [8, 9].

A key challenge is the penetration through epithelial membrane [10]. Barriers include an aqueous stagnant layer, characteristic of its hydrophilic nature. The membranes enveloping the cells are lipid bilayers, housing proteins such as carrier molecules and receptors, through which lipophilic drugs are easily bioavailable via passive diffusion. However, small water-soluble molecules (about 0.4 nm size), such as ethanol, use the aqueous channels within the proteins [11].

Pioneers are enhancing the absorption of those microparticulate drug deliveries by incorporating absorption enhancers such as surfactants, bile salts, chelating agents, salicylates, fatty acids, and polymers that exaggerate intestinal absorption [12, 13]. Chitosan, predominantly trimethylated chitosan, opens the tight junctions by the redistribution of cytoskeletal F-actin in paracellular route, thereby enhancing drug absorption. Increasing the aqueous solubility of hydrophobic drugs could be considered as one of the mechanisms for the action of bile salts and fatty acids as absorption enhancers. The other mechanism may be by aggravating the fluidity of the apical and basolateral membranes. The disruption of cell–cell contacts by modifying the extracellular calcium concentration is attributed to Ca^{2+} ion chelators such as ethylene glycol-bis (β-aminoethylether)-N, N'-tetraacetic acid and ethylenediaminetetraacetic acid [14].

Paolino et al. investigated the effect of methacrylate copolymer during microencapsulation of curcumin and, subsequently, studied the bioavailability of curcumin.

The microencapsulated curcumin revealed a septuple in bioavailability, as compared to unadorned curcumin, with a slumped Tmax and a quintuple in the peak of plasma concentration [15].

In Ayurveda, the concept of bioenhancers has been employed for over a century using "Trikatu." The safety of Trikatu was proved by acute and subacute toxicity studies administered in Charles Foster rats by Kesarwani et al. Trikatu revealed no significant alterations in normal physiological parameters and was well tolerated by the animals [15, 16]. A plenty of natural compounds have the capacity to amplify the bioavailability when coadministered. Bioenhancers are chemical entities that intensify the bioavailability of pharmaceuticals, but do not yield a synergistic impact with the therapeutic agents, as shown in Figure 2.1 [17].

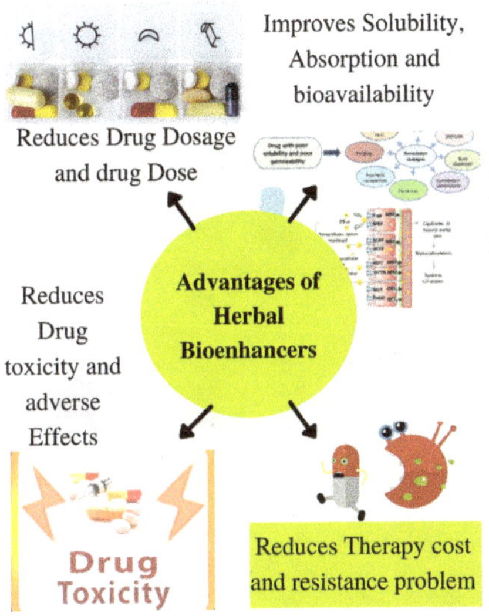

Figure 2.1: Advantages of herbal bioenhancers in drug delivery.

2.2 The proposed mechanism of bioenhancing action with different phytochemicals

2.2.1 Increased gastrointestinal absorption

The mechanism for increased gastrointestinal absorption includes increased free diffusion and reducing the steric and interactive barriers in the gastrointestinal tract as mentioned in Figure 2.2.

A. **By enhancing the solubility:** Bile acid assists in the production of micelles, which are imperative for lipid and lipid-soluble medication absorption. Piperine hikes bile acid secretion while also impeding bile acid metabolism, which helps in increased micelle production. This refines not only the absorption but also solubility [18].

B. **Increased permeability due to epithelial cell modification:** Piperine enhances gamma-glutamyl transpeptidase activity and elevates amino acid uptake by epithelial cells, by interacting with the intestinal epithelial cells [19].

C. **Inhibition of solubilizer attachment-** The entry of chemicals into cells is hindered when they are chemically conjugated to a highly water-soluble substance. This is called solubilizer attachment. The chemicals attached to the glucuronic acid, a key solubilizer, are eliminated in the urine or in the small intestine. It has been observed that piperine inhibits glucuronic acid, allowing more chemicals to enter the cell [20].

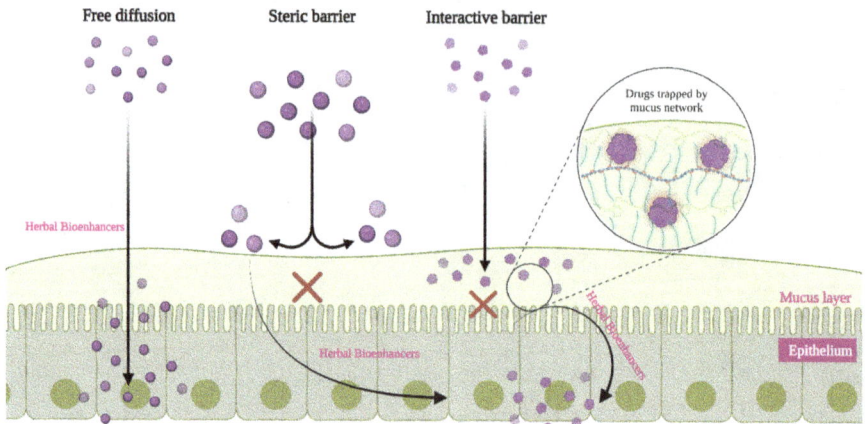

Figure 2.2: Predominant role of herbal bioenhancers during oral absorption.

2.2.2 Different mechanisms related to herbal bioenhancement

Multitudinous mechanisms are available to multiply the bioavailability using bioenhancers. The dominant mechanisms are listed below and shown in Figure 2.3.

– Magnifying medication absorption from the alimentary canal
– Modulating the biotransformation of therapeutics agent within the liver or intestine.
– Altering the system in such a way that the overall requirement of drugs shows a considerable decline.
– Increasing pathogen penetration or entry even when the pathogens become persistors within the macrophages. This ensures improved destruction of those organisms secured within the regions where the active medicine cannot reach.

- Inhibiting pathogens or the ability of the aberrant tissue to reject drugs such as efflux mechanisms, commonly found in antimalarial, anticancer, and antibacterial medications.
- Strengthening the accessibility of medicines to infections by altering the signaling pathway between the host and the pathogen.
- Increasing the drugs' binding to receptors at locations within the pathogen, thus potentiating and increasing its impact, ultimately showing increased antibiotic efficacy against infections.
- In addition to the aforementioned mechanisms, bioenhancers can also bolster the movement of medications across the difficult barrier.

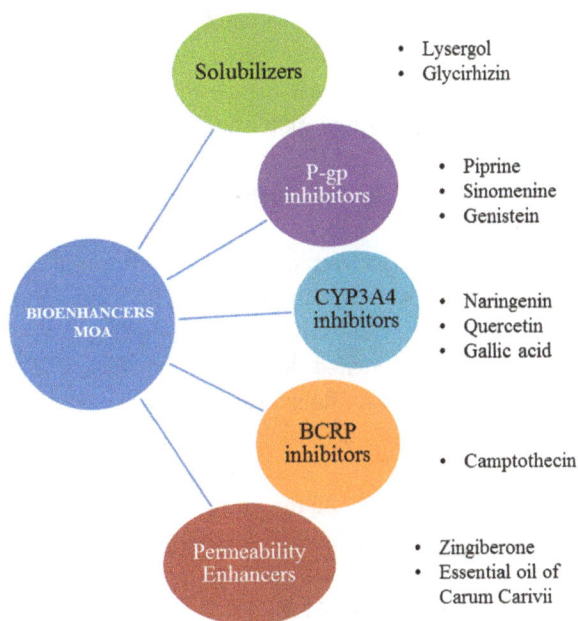

Figure 2.3: Mechanism of action of herbal bioenhancers.

2.3 Bioenhancers in formulation development

2.3.1 Piperine

Piperine (1-piperoyl piperidine) belongs to the Piperaceae family of plants, such as long pepper (long pepper) and pepper (black pepper), and is an amide alkaloid. Studies on the metabolism of piperine proved that the maximum concentration of piperine in the stomach could be a major reason for its less availability at other sites such as the intestine, spleen, kidney, and serum within a time period of ½ h to 24 h [21]. This evidence displayed the utilization of piperine as a sustained bioenhancer.

The therapeutic use of piperine as a bioenhancer commenced long back in the treatment of human tuberculosis. Piperine was discovered to spice up rifampicin, the first-line drug, bioavailability by roughly 60%. This resulted in a significant decrease in dosage from 450 to 200 mg [21, 22].

Subsequently, when piperine was used with nevirapine, a strong nonnucleoside inhibitor of Human immunodeficiency virus (HIV)-1 polymerase that is utilized in combination with other antiretroviral medicines to treat HIV-1 infection, it illustrated revamped bioavailability [23]. The ethnomedical properties of piperine such as anti-inflammatory, antipyretic, antifungal, antidiarrheal, and anticancer were supplementary to provide synergistic effect along with other therapeutic agents apart from bioenhancement [24]. Remodeling in membrane dynamics, clampdown of P-gp efflux, gastrointestinal metabolism, and hepatic metabolism are probable pathways for piperine's bioenhancing effects. The biopotentiation for resveratrol (3,5,4'-trihydroxystilbene) was remarkable by increasing area under the curve and C_{max} (maximum concentration) by 229% and 1,544%, respectively, when piperine was coadministered to C57BL mice orally [25].

The alteration in the pharmacokinetics of oxytetracycline, following oral administration of long pepper, in hens was reported by Singh et al. The poultry birds were fed with *P. longum* prior to the administration of oxytetracycline, which increased the entire duration of antimicrobial action and enriched the therapeutic efficacy [26]. Pattanaik et al. administered piperine bis in die (twice daily), simultaneously with carbamazepine, in epileptic patients, who were on carbamazepine monotherapy. There was an interesting enhancement of activity that might be attributed to retardation in elimination or increased absorption. Piperine accelerated the mean plasma concentration of carbamazepine in two different dose groups [27]. When piperine was coadministered, biopotentiation of antidepressant activities of curcumin was reported [28]. The reversal of beryllium-induced biochemical alteration and the oxidative stress of tiferron and piperine were evaluated individually and in combination. A mixture of tiferron with piperine could reverse all the variables significantly when compared to the control [29]. A patent reveals the effectiveness of piperine in inflating the bioavailability of bedaquiline and delaminid for antitubercular actions [27].

Amoxicillin, when studied by in vitro studies using noneverted gut sac method along with 99% pure piperine 4.36% w/w of gingerol resulted in a dose-dependent increase in bioavailability, whereas gingerol revealed no change [30].

Piperine puts forth significant changes, even at small doses, in the metabolic pathways of the drug, irrespective of their administration route.

2.3.2 Rutin

Kumar et al. developed a particulate system using polymer (Eudragit L100), lipoid (glycerol monostearate: soya lecithin), and rutin as bioenhancer for a partial dose reduction of isradipine (ISR). This resulted in 3.2–4.7-fold enhancement in oral bioavailability of coated solid lipid nanoparticle of ISR, when compared to formulation without rutin and suspension formulation [31].

2.3.3 Capsaicin

Capsaicin produces a burning sensation due to its irritant action on the tissue. It is the active component of red chili. Several marketed formulations such as Volini gel are available for its biopotentiation effect along with diclofenac sodium. Cruz et al. reported that the oral administration of *Capsicum annum* with aspirin toned down the bioavailability [32]. In another study on fluoroquinolone antibacterial (ciprofloxacin), capsaicin exhibited negligible or no impact on bioavailability. An investigation related to absorption and bioavailability of theophylline from a sustained-release gelatin capsule with and without capsicum fruit suspension was carried out revealing the potential of capsaicin as a bioenhancer [33].

2.3.4 Quercetin

It shows dual action by inhibiting cytochrome P3A4 and P-glycoprotein modulation [34], consequently achieving enhanced bioavailability of therapeutic agents.

When tamoxifen (10 mg/kg) with quercetin (2.5 mg/kg) was administered and its pharmacokinetic profile was compared to tamoxifen (10 mg/kg) alone, there was a significant boost in the parameters. Absolute bioavailability of tamoxifen with quercetin and also the relative bioavailability of tamoxifen showed remarkable increase as compared to the control group. This could be associated with the reduction of first-pass metabolism of tamoxifen and the promotion of intestinal absorption [35].

2.3.5 *Z. officinale*

Gingerols are one of the most potent actives present in the *Z. officinale* rhizome. High percentage of volatile oil ([6]-gingerol) constituents impart strong odor to ginger and restrict its use. Several therapeutic activities revealed by Ginger during its scientific screening, which includes anti-inflammatory, antiulcer [36], antithrombotic, antimicrobial, antifungal, and anticancer [37]. Ginger acts as a bioenhancer by regulating intestinal function, i.e., by promoting drug absorption [38].

Atazanavir (ATV) is usually coadministered with ritonavir as an inhibitor of protease and other host enzymes. Ginger attenuated the plasma concentration of the drug in blood, comparable to the ATV–RTV combination. The ideal range for oleoresin from ginger is 10–150 mg from the bioenhancement report [39].

A prior administration of 4 mL of ginger extract revealed better bioavailability and enhanced antimicrobial action of pefloxacin [40].

The database revealed ginger powder is preferred more than ginger extract in producing the particulate drug delivery. Sumatriptan granules with ginger powder show better compressibility and rapid disintegration [41].

The nanodelivery of oleanolic acid with the polyphenolic bioenhancers (*Z. officinale* extract) and oleanolic acid resulted in enhanced antitumor effect, preserving female fertility [42].

2.3.6 Niaziridin

Drumstick leaves, pods, and bark have been identified to contain niaziridin, a nitrile glycoside. *M. oleifera* has been useful by displaying antifertility properties, diuretic, antimicrobial, anti-inflammatory, anticancer, spasmolytic and hypotensive activity, etc. [43]. It boosts the effectiveness of antibiotics such as ampicillin, rifampicin, nalidixic acids, and tetracycline against Gram-positive and Gram-negative bacteria. The increase in action was about 1.2–19-fold; however, the action of clotrimazole against *Candida albicans* was modified only by 5–6 times. It also facilitates the uptake of Vitamin B12 via the intestinal gut membrane, in combination, revealing the bioenhancement potential [24]. A 16-fold reduction in the MIC of tetracycline was witnessed when investigated in combination with Niazirin and niaziridin. Both phytochemicals proved their drug resistance reversal activity through the inhibition of efflux pump and modulating the drug resistant genes expression pattern [44].

2.3.7 Glycyrrhizin

Glycyrrhizin is a glycoside located in liquorice roots and stolons (*Glycyrrhiza glabra*). It expresses an expectorant effect and is favorable in bronchitis, allergies, asthma, gastritis, peptic ulcers, rheumatism, and sore throat. It aids in the detoxification of medicines by the liver and is used to treat hepatic diseases. It is a diuretic and laxative, while also strengthening the immune system and stimulating the adrenal gland. Sugar is 50 times sweeter than glycyrrhizin. Treatment of peptic ulcers and stomach diseases as well as respiratory and intestinal channels is the most common application. It boosts the bioactivity of popular antibiotics, including rifampicin, ampicillin, tetracycline, and nalidixic acids, against Gram-positive and Gram-negative bacteria such as *M. smegmatis* and *Bacillus subtilis* [45]. It amplifies the efficacy of antifungal azoles such as clotrimazole against *Candida albicans* [46, 47].

Ultra-small spheres of thymol with dipotassium glycyrrhizinate as bioenhancer for ocular delivery demonstrated improvement in ex vivo and in vivo membrane permeation, in vitro release, and antioxidant activity of thymol [48].

The glycyrrhizin-conjugated particulated drug delivery is a novel area for research wherein doxorubicin and lamivudine were studied for liver targeting [49, 50].

2.4 Application of herbal bioenhancer in microparticulates through different routes

2.4.1 Bioenhancers in oral microparticulates

An ideal strategy for alleviation of deficiencies due to poor bioavailability of minerals such as iron and zinc could be enhancing the bioavailability from complementary food mixes and Indian flatbread mixes. The addition of spices of different concentrations, and a mix of bioenhancers at 2% enhanced the bioaccessibility of iron and zinc 3–6-fold and 1.7–2.5-fold, respectively [51].

Joshi et al. developed microparticulated granules of ciprofloxacin, including ethanolic extracts of turmeric, cow urine distillate, pepper, and zinger as bioenhancers, and investigated the antimicrobial activity. The formulation with pepper – pepper and cow urine distillate – revealed the highest bioenhancing effect, followed by turmeric and ginger [52]. The altered metabolism and, thereby, pharmacokinetic interaction could be the reason for piperine-assisted improvement of bioavailability. The metabolic pathway of Nisoldipine is the metabolism by cytochrome P4503A4 enzymes. A research was undertaken to assess the influence of piperine on the pharmacokinetics and pharmacodynamics of nisoldipine particles in rats. After oral administration in rats, in vivo kinetic and dynamic studies were carried out. It resulted in a 4.9-fold increase in the bioavailability of nisoldipine nanoparticles due to the alteration in cytochrome P4503A4 enzyme metabolism [53]. Baspinar et al. developed curcumin and piperine-loaded zein-chitosan nanoparticles. The developed nanoparticle formulation comprehends natural compounds that showed admirable cytotoxic effects [54]. Pingale et al. prepared and optimized a sustained-release formulation of isoniazid microspheres and probed the effect of *Piper nigrum* (black pepper) on its in vitro release. The coadministration of bioenhancers enhanced in vitro release to the extent of 100–107% [55]. Ravindra and Pingale et al. also studied an in vitro release of rifampicin from the microspheres containing different concentrations of bioenhancers, prepared by the complex co-acervation method. Formulation containing bioenhancer revealed better in vitro release against the formulation, devoid of bioenhancer [56].

The inhibition of ergosterol trafficking in the plasma membrane by allicin plays an important role in amphotericin B-induced vacuole membrane damage [57].

A study was carried out to examine the impression of piperine and its derivatives (5a–5d) (10 mg/kg) on the bioavailability of ibuprofen. A significant upliftment in the absorption and decrease in the elimination of ibuprofen (10 mg/kg) was observed, suggesting a newer mechanism of action [58].

2.4.2 Bioenhancers in topical microparticulates

Topical patches of 18β-glycyrrhitic acid were prepared and evaluated using synthetic polymers by Alsaad AA, 2021. There was a dramatic uptake in the permeation of drug across membranes through this reservoir-type patch, formulated using 42% ethanol, 50 g carbopol-934 gel base, and 5% menthol with 0.5% piperine [59].

Experts studied the release behavior of dexibuprofen in the presence of capsaicin and menthol as permeation enhancers. The dexibuprofen-capsaicin emulgel showed release of 81.342% ± 1.21 (with 100 mg menthol), 72.321% ± 1.31 (75 mg menthol), and 52.462% ± 1.23 (without menthol) of dexibuprofen. The permeation of dexibuprofen shares a direct relation with the concentration of menthol with capsaicin [60].

2.4.3 Bioenhancers in nasal microparticulates

It is anticipated that the nasal drug delivery technology market will transcend by $64 B in 2021 [61]. Nasal route is the most preferred approach in the administration of macromolecules; however, thousands of small peptides are marketed in intranasal formulations, including buserelin (1.2 kDa), desmopressin (1 kDa), gonadorelin (1.2 kDa), oxytocin (1 kDa), sCT (3.8 kDa), nafarelin (1.4 kDa), and pritorelin (362 Da) [62]. The highest variability in the performance of these molecules is around 1–5% [63]. Penetration enhancers (PEs) play a crucial role in reducing the variability observed in their performance for larger peptides and proteins (e.g., interferon, insulin, and human growth hormone) [64]. Most formulations use CriticalSorb® which contains cationic surfactant that alters nasal permeability at selected concentrations. However, other PEs used could be nonionic surfactants [65], thiolated polymers [66] chitosan, cyclodextrins [67], and cell penetrating peptides [68]. Several attempts have been made by researchers to overcome the transmucosal barrier with the help of bioenhancers. In one of the studies, novasomes of zolmitriptan using nonionic surfactant, a free fatty acid, and cholesterol were prepared for acceleration in the bioavailability rate of zolmitriptan [69].

Gerber et al. studied the paracellular transport of fluorescein isothiocyanate-labeled dextran 4,400 (MW 4,400 Da) across the nasal epithelium in the presence of capsaicin and piperine. A substrate for P-glycoprotein-mediated efflux used in the study was rhodamine 123. The inhibition of rhodamine 123 transport signifies that capsaicin inhibits P-glycoprotein efflux to a large extent. It was pertinent to study the cell cytotoxicity; so, the studies were carried out, indicating the nontoxic nature of capsaicin and piperine up to concentrations of 200 and 500 µM, respectively [70].

2.5 Impact of herbal bioenhancers on the microparticulate formulation development

2.5.1 Impact on processing

The sticky and volatile nature of the herbal bioenhancers can pose a surfeit of complications in the processing of formulation on the expeditious machine due to the elevated product temperature [71]. The inadequate and improper mixing of herbal bioenhancers could be a critical activity. The formulation optimization considering the concentration of the herbal bioenhancer is one of the key independent variables that lays the needs for the most developmental activity.

2.5.2 Impact on stability

Although the bioenhancers may radically enhance the performance of these formulations, it may reduce the shelf life of the product. The moisture content, carbohydrate content, lipid content, and microbial burden are some of the climacteric attributes of the herbal bioenhancers that need thorough consideration [72].

2.5.3 Interaction in finished product

To ensure the compatibility of the herbal bioenhancers with other excipients and active pharmaceutical ingredients, with reference to contaminants and residues, the World Health Organization has developed a new guideline for assessing the quality of herbal medicines [72].

2.6 Regulatory requirements for formulations with herbal bioenhancers

As the bioenhancers embrace a chemical entity, they can have their reactions as well, and when administered alongside other drugs, may have ramifications such as interactions and other side effects [73]. However, a segment of the perplexities experienced has been resolved by just modifying the physicochemical properties of the materials. This could include change in the functional surface area, drug stabilization and protection from degradation, alterations in interactions with biological barriers, and site-specific targeting [74]. Another challenge in the development of herbal bioenhancers is the standardization and large-scale production. There is consistently a requirement to scale-up laboratory or pilot innovations for possible commercialization. Making the nanoparticles or microparticles and maintaining their size and compatibility with the medications is, likewise, a subject of concern [75]. Advances in natural bioenhancers engender modern difficulties to administrative control. There is a requirement to develop guidelines that would represent the physicochemical and pharmacokinetics of drug products, which are not quite the same as traditional medication items [76]. Although the bioenhancers are considered as inactives because they are used in modest concentrations, yet these concentrations may themselves infer some activity. Therefore, the selection of a suitable bioenhancer is paramount and exigent; else, it may precede adverse events, attributed to antagonistic or synergistic effects. The United States' Food and Drug Administration and The European Medicines Evaluation Agency have initiated the recognition of some possible scientific and regulatory challenges [77]. In Japan, where herbal drugs are indicated as Kampo medicine, they undergo stringent regulatory framework as the allopathic system of medicine, and are well-integrated. The manufacturer of a new herbal product needs to send in details of heavy metal content, aflatoxin details, mycotoxin, and pesticide details to the regulatory authority. Along with these documents, the manufacturer needs to submit chemistry, manufacture, and control documents prior to market authorization [78].

2.7 Prospects in drug development with bioenhancers

With the availability of luxurious flora and fauna and the abundance of Ayurveda heritage, neoterics need to explore more bioenhancers, ignoring the available ones. The phytoconstituents must be extensively standardized before their inclusion in the formulations. Rational optimization techniques should be adopted, including critical material attributes as independent parameters, for evaluating the product performance. Compatibility studies should be performed and risk assessment with respect to aflatoxins, pesticide content, heavy metals, etc., should be carried out.

2.8 Conclusion

In an era where not only delivery at the site of action but also the bioavailability of therapeutic agents is a censorious issue, bioenhancers such as piperine, quercetin, and glycyrrhizin offer remedy. Even though they indicate some vices, the proportion of their benefit is high. Hence, it would be appropriate to acknowledge that bioenhancers have the aptitude for efficient delivery that lays the ground for their concrete exploration.

List of abbreviations

GI	Gastrointestinal
P-gp	P-glycoprotein
HIV-1	Human immunodeficiency virus
SLN	Solid lipid nanoparticle
ISR	Isradipine
ATV	Atazanavir
RTV	Ritonavir
sCT	Miacalcin®, Novartis, Switzerland
PEs	Penetration enhancers

References

[1] Rescigno A, Bioavailability and bioequivalence, Foundations of Pharmacokinetics, 2003, 171–177.
[2] Raeissi SD, Li J, Hidalgo IJ, The role of an α-amino group on H+-dependent transepithelial transport of cephalosporins in Caco-2 cells, Journal of Pharmacy and Pharmacology, 1999, 51(1), 35–40.
[3] Adachi Y, Suzuki H, Sugiyama Y, Quantitative evaluation of the function of small intestinal P-glycoprotein: Comparative studies between in situ and in vitro, Pharmaceutical Research, 2003, 20(8), 1163–1169.
[4] Engel R, Riggi S, Fahrenbach M, Insulin: Intestinal absorption as water-in-oil-in-water emulsions, Nature, 1968, 219(5156), 856–857.
[5] Patel H, Ryman BE, Oral administration of insulin by encapsulation within liposomes, FEBS Letters, 1976, 62(1), 60–63.
[6] Liversidge GG, Cundy KC, Particle size reduction for improvement of oral bioavailability of hydrophobic drugs: I. Absolute oral bioavailability of nanocrystalline danazol in beagle dogs, International Journal of Pharmaceutics., 1995, 125(1), 91–97.
[7] Veiga F, Fernandes C, Teixeira F, Oral bioavailability and hypoglycaemic activity of tolbutamide/cyclodextrin inclusion complexes, International Journal of Pharmaceutics, 2000, 202(1–2), 165–171.
[8] Patel H, Gohel M, A Review on Enteric Coated Pellets Composed of Core Pellets Prepared by Extrusion-Spheronization, Recent Patents on Drug Delivery & Formulation, 2019, 13(2), 83–90.

[9] Bale S, Khurana A, Reddy ASS, Singh M, Godugu C, Overview on therapeutic applications of microparticulate drug delivery systems, Critical Reviews™ in Therapeutic Drug Carrier Systems, 2016, 33(4), 309–361.

[10] Singh A, Verma BK, Pandey S, Exploring natural bioenhancers to enhancing bioavailability: An Overview, International Journal of Pharmacy & Life Sciences, 2021, 12, 2.

[11] Kang+ MJ, Cho+ JY, Shim BH, Kim DK, Lee J, Bioavailability enhancing activities of natural compounds from medicinal plants, Journal of Medicinal Plants Research, 2009, 3(13), 1204–1211.

[12] Lundin S, Artursson P, Absorption of a vasopressin analogue, 1-deamino-8-D-arginine-vasopressin (dDAVP), in a human intestinal epithelial cell line, Caco-2, International Journal of Pharmaceutics, 1990, 64(2–3), 181–186.

[13] Aungst BJ, Blake JA, Hussain MA, An in vitro evaluation of metabolism and poor membrane permeation impeding intestinal absorption of leucine enkephalin, and methods to increase absorption, Journal of Pharmacology and Experimental Therapeutics, 1991, 259(1), 139–145.

[14] Schipper NG, Olsson S, Hoogstraate JA, deBoer AG, Vårum KM, Artursson P, Chitosans as absorption enhancers for poorly absorbable drugs 2: Mechanism of absorption enhancement, Pharmaceutical Research, 1997, 14(7), 923–929.

[15] Paolino D, Vero A, Cosco D, Pecora TM, Cianciolo S, Fresta M, et al., Improvement of oral bioavailability of curcumin upon microencapsulation with methacrylic copolymers, Frontiers in Pharmacology, 2016, 7, 485.

[16] Kesarwani K, Gupta R, Bioavailability enhancers of herbal origin: An overview, Asian Pacific Journal of Tropical Biomedicine, 2013, 3(4), 253–266.

[17] Chivte VK, Tiwari SV, Pratima A, Nikalge G, Bioenhancers: A brief review, Advance Journal of Life Science and Pharma Research, 2017, 2, 1–18.

[18] McClements DJ, Zou L, Zhang R, Salvia-Trujillo L, Kumosani T, Xiao H, Enhancing nutraceutical performance using excipient foods: Designing food structures and compositions to increase bioavailability, Comprehensive Reviews in Food Science and Food Safety, 2015, 14(6), 824–847.

[19] Han H-K, The effects of black pepper on the intestinal absorption and hepatic metabolism of drugs, Expert Opinion on Drug Metabolism & Toxicology, 2011, 7(6), 721–729.

[20] Lambert JD, Hong J, Kim DH, Mishin VM, Yang CS, Piperine enhances the bioavailability of the tea polyphenol (–)-epigallocatechin-3-gallate in mice, The Journal of Nutrition, 2004, 134(8), 1948–1952.

[21] Bhat BG, Chandrasekhara N, Studies on the metabolism of piperine: Absorption, tissue distribution and excretion of urinary conjugates in rats, Toxicology, 1986, 40(1), 83–92.

[22] Du Toit LC, Pillay V, Mp D, Penny C, Formulation and statistical optimization of a novel crosslinked polymeric anti-tuberculosis drug delivery system, Journal of Pharmaceutical Sciences, 2008, 97(6), 2176–2207.

[23] Kasibhatta R, Naidu M, Influence of piperine on the pharmacokinetics of nevirapine under fasting conditions, Drugs in R & D, 2007, 8(6), 383–391.

[24] Alexander A, Qureshi A, Kumari L, Vaishnav P, Sharma M, Saraf S, et al., Role of herbal bioactives as a potential bioavailability enhancer for active pharmaceutical ingredients, Fitoterapia, 2014, 97, 1–14.

[25] Chinta G, Syed B, Ms C, Periyasamy L, Piperine: A comprehensive review of pre-clinical and clinical investigations, Current Bioactive Compounds, 2015, 11(3), 156–169.

[26] Singh M, Varshneya C, Telang R, Srivastava A, Alteration of pharmacokinetics of oxytetracycline following oral administration of Piper longum in hens, Journal of Veterinary Science, 2005, 6(3), 197–200.

[27] Pattanaik S, Hota D, Prabhakar S, Kharbanda P, Pandhi P, Pharmacokinetic interaction of single dose of piperine with steady-state carbamazepine in epilepsy patients, Phytotherapy

Research: An International Journal Devoted to Pharmacological and Toxicological Evaluation of Natural Product Derivatives, 2009, 23(9), 1281–1286.

[28] Bhutani MK, Bishnoi M, Kulkarni SK, Anti-depressant like effect of curcumin and its combination with piperine in unpredictable chronic stress-induced behavioral, biochemical and neurochemical changes, Pharmacology, Biochemistry, and Behavior, 2009, 92(1), 39–43.

[29] Chopra B, Dhingra AK, Kapoor RP, Prasad DN, Piperine and its various physicochemical and biological aspects: A review, Open Chemistry Journal, 2016, 3, 1.

[30] Barve K, Ruparel K, Effect of bioenhancers on amoxicillin bioavailability, ADMET and DMPK, 2015, 3(1), 45–50.

[31] Kumar V, Chaudhary H, Kamboj A, Development and evaluation of isradipine via rutin-loaded coated solid–lipid nanoparticles, Interventional Medicine and Applied Science, 2018, 10(4), 236–246.

[32] Cruz L, Castañeda-Hernández G, Navarrete A, Ingestion of chilli pepper (Capsicum annuum) reduces salicylate bioavailability after oral aspirin administration in the rat, Canadian Journal of Physiology and Pharmacology, 1999, 77(6), 441–446.

[33] Sumano-López H, Gutiérrez-Olvera L, Aguilera-Jiménez R, Gutiérrez-Olvera C, Jiménez-Gómez F, Administration of ciprofloxacin and capsaicin in rats to achieve higher maximal serum concentrations, Arzneimittelforschung, 2007, 57(05), 286–290.

[34] Bhimanwar R, Kothapalli L, Khawshi A, Quercetin as Natural Bioavailability Modulator: An Overview, Research Journal of Pharmacy and Technology, 2020, 13(4), 2045–2052.

[35] Shin S-C, Choi J-S LX, Enhanced bioavailability of tamoxifen after oral administration of tamoxifen with quercetin in rats, International Journal of Pharmaceutics, 2006, 313(1–2), 144–149.

[36] Asnaashari S, Dastmalchi S, Javadzadeh Y, Gastroprotective effects of herbal medicines (roots), International Journal of Food Properties, 2018, 21(1), 902–920.

[37] Shahrajabian MH, Wenli S, Cheng Q, Pharmacological uses and health benefits of ginger (Zingiber officinale) in traditional Asian and ancient Chinese medicine, and modern practice, Notulae Scientia Biologicae, 2019, 11(3), 309–319.

[38] Javed S, Ahsan W, Kohli K, The concept of bioenhancers in bioavailability enhancement of drugs–a patent review, Journal of Science Letters, 2016, 1, 143–165.

[39] Prakash S, Kherde P, Rangari V, Bioenhancement effect of piperine and ginger oleo resin on the bioavailability of atazanvir, Int J Pharm Pharm Sci, 2015, 7, 241–245.

[40] Nduka S, Adonu L, Okonta E, Okonta J Original Research The Influence of Ginger (Zingiber Officinale) Extract on the Pharmacokinetic Profile of Pefloxacin. 2008.

[41] Sonawane Harshali R, Surana Santosh S, Maru Avish DR, Allopolyherbo tablet formulation and evaluation for migraines from direct compression method, Journal of Medical Pharmaceutical and Allied Sciences, 2019, 8(2), 2090–2095.

[42] Sharma M, Sharma S, Sharma V, Sharma K, Yadav SK, Dwivedi P, et al., Oleanolic–bioenhancer coloaded chitosan modified nanocarriers attenuate breast cancer cells by multimode mechanism and preserve female fertility, International Journal of Biological Macromolecules, 2017, 104, 1345–1358.

[43] Mishra G, Singh P, Verma R, Kumar S, Srivastav S, Jha K, et al., Traditional uses, phytochemistry and pharmacological properties of Moringa oleifera plant: An overview, Der Pharmacia Lettre, 2011, 3(2), 141–164.

[44] Dwivedi GR, Maurya A, Yadav DK, Khan F, Gupta MK, Gupta P, et al., Comparative drug resistance reversal potential of natural glycosides: Potential of synergy Niaziridin & Niazirin, Current Topics in Medicinal Chemistry, 2019, 19(10), 847–860.

[45] Dudhatra GB, Mody SK, Awale MM, Patel HB, Modi CM, Kumar A, et al., A comprehensive review on pharmacotherapeutics of herbal bioenhancers, The Scientific World Journal, 2012, 637953.

[46] Reen RK, Jamwal DS, Taneja SC, Koul JL, Dubey RK, Wiebel FJ, et al., Impairment of UDP-glucose dehydrogenase and glucuronidation activities in liver and small intestine of rat and guinea pig in vitro by piperine, Biochemical Pharmacology, 1993, 46(2), 229–238.

[47] Bhardwaj RK, Glaeser H, Becquemont L, Klotz U, Gupta SK, Fromm MF, Piperine, a major constituent of black pepper, inhibits human P-glycoprotein and CYP3A4, Journal of Pharmacology and Experimental Therapeutics, 2002, 302(2), 645–650.

[48] Song K, Yan M, Li M, Geng Y, Wu X, Preparation and in vitro–in vivo evaluation of novel ocular nanomicelle formulation of thymol based on glycyrrhizin, Colloids and Surfaces. B, Biointerfaces, 2020, 194, 111157.

[49] Mishra D, Jain N, Rajoriya V, Jain AK, Glycyrrhizin conjugated chitosan nanoparticles for hepatocyte-targeted delivery of lamivudine, Journal of Pharmacy and Pharmacology, 2014, 66(8), 1082–1093.

[50] Chopdey PK, Tekade RK, Mehra NK, Mody N, Jain NK, Glycyrrhizin conjugated dendrimer and multi-walled carbon nanotubes for liver specific delivery of doxorubicin, Journal of Nanoscience and Nanotechnology, 2015, 15(2), 1088–1100.

[51] Jaiswal A, Pathania V, An exploratory trial of food formulations with enhanced bioaccessibility of iron and zinc aided by spices, LWT, 2021, 143, 111122.

[52] Joshi VIP, Kowti R, Acharya A, Janadri S, Study of bioenhancer for efficacy improvement of a model antibiotic drug, Advanced Pharmaceutical Journal, 2016, 1(2), 31–37.

[53] Rathee P, Kamboj A, Sidhu S, Enhanced oral bioavailability of nisoldipine-piperine-loaded poly-lactic-co-glycolic acid nanoparticles, Nanotechnology Reviews, 2017, 6(6), 517–526.

[54] Baspinar Y, Üstündas M, Bayraktar O, Sezgin C, Curcumin and piperine loaded zein-chitosan nanoparticles: Development and in-vitro characterisation, Saudi Pharmaceutical Journal, 2018, 26(3), 323–334.

[55] Pingale PL, Ravindra R, Effect of Piper nigrum on in-vitro release of Isoniazid from oral microspheres, International Journal of Pharm and Bio Sciences, 2013, 4(1), 1027–1036.

[56] Rp R, Effect of Piper Nigrum on In-Vitro Release of Rifampicin Microspheres, Asian Journal of Pharmaceutical and Clinical Research, 2013, 1, 79–83.

[57] Ogita AFK, Tanaka T, Enhancement of the fungicidal activity of amphotericin B by allicin: Effects on intracellular ergosterol trafficking, Planta Medica, 2009, Feb 75(03), 222–226.

[58] Tiwari AGS, Mahadik KR, Comparative Study on the Pharmacokinetics of Ibuprofen Alone or in Combination with Piperine and its Synthetic Derivatives as a Potential Bioenhancer, International Journal of Pharmaceutical Sciences and Research, 2021, 12(1), 363–371.

[59] Alsaad AAA, Formulation & evaluation of β-glycyrrhetinic acid patches with natural bioenhancer, Materials Today: Proceedings, 2021.

[60] Ik B, Mk K, Ba K, Uzair B, Va B, Qa J, Formulation Development, Characterization, and Evaluation of a Novel Dexibuprofen-Capsaicin Skin Emulgel with Improved In Vivo Anti-inflammatory and Analgesic Effects, AAPS PharmSciTech, 2020, 21(6), 211.

[61] Maher S, Casettari L, Illum L, Transmucosal Absorption Enhancers in the Drug Delivery Field, Pharmaceutics, 2019, 11(7), 339.

[62] Rohrer J, Lupo N, Bernkop-Schnürch A, Advanced formulations for intranasal delivery of biologics, International Journal of Pharmaceutics, 2018, 553(1), 8–20.

[63] Hinchcliffe M, Jabbal-Gill I, Smith A, Effect of chitosan on the intranasal absorption of salmon calcitonin in sheep, Journal of Pharmacy and Pharmacology, 2005, 57(6), 681–687.

[64] Fabrizio B, Giulia BA, Fabio S, Paola R, Gaia C, In vitro permeation of desmopressin across rabbit nasal mucosa from liquid nasal sprays: The enhancing effect of potassium sorbate, European Journal of Pharmaceutical Sciences, 2009, 37(1), 36–42.

[65] Lewis AL, Jordan F, Illum L, CriticalSorb™: Enabling systemic delivery of macromolecules via the nasal route, Drug Delivery and Translational Research, 2013, 3(1), 26–32.

[66] Jain A, Hurkat P, Jain A, Jain A, Jain A, Jain SK, Thiolated Polymers: Pharmaceutical Tool in Nasal Drug Delivery of Proteins and Peptides, International Journal of Peptide Research and Therapeutics, 2019, 25(1), 15–26.

[67] Merkus FWHM, Verhoef JC, Marttin E, Romeijn SG, Phm VDK, Hermens WAJJ, et al., Cyclodextrins in nasal drug delivery, Advanced Drug Delivery Reviews, 1999, 36(1), 41–57.

[68] Kristensen M, Nielsen HM, Cell-penetrating peptides as tools to enhance non-injectable delivery of biopharmaceuticals, Tissue Barriers, 2016, 4(2), e1178369.

[69] Abd-Elal RM, Shamma RN, Rashed HM, Bendas ER, Trans-nasal zolmitriptan novasomes: In-vitro preparation, optimization and in-vivo evaluation of brain targeting efficiency, Drug Delivery, 2016, 23(9), 3374–3386.

[70] Gerber W, Steyn D, Kotzé A, Svitina H, Weldon C, Hamman J, Capsaicin and piperine as functional excipients for improved drug delivery across nasal epithelial models, Planta Medica, 2019, 85(13), 1114–1123.

[71] Pingale PL, Ravindra R, Comparative Study of Herbal Extract of Piper Nigrum, Piper Album and Piper Longum on Various Characteristics of Pyrazinamide and Ethambutol Microspheres, Journal of Drug Delivery and Therapeutics, 2019, 9, (4-A):72–8.

[72] Organization WH. WHO guidelines on good agricultural and collection practices (GACP) for medicinal plants: World Health Organization; 2003.

[73] Liu S-H, Chuang W-C, Lam W, Jiang Z, Cheng Y-C, Safety surveillance of traditional Chinese medicine: Current and future, Drug Safety, 2015, 38(2), 117–128.

[74] Choi YH, Han H-K, Nanomedicines: Current status and future perspectives in aspect of drug delivery and pharmacokinetics, Journal of Pharmaceutical Investigation, 2018, 48(1), 43–60.

[75] Viswanathan P, Muralidaran Y, Ragavan G, Challenges in oral drug delivery: A nano-based strategy to overcome, Nanostructures for oral medicine, Elsevier, 2017, 173–201.

[76] Fan T-P, Deal G, Koo H-L, Rees D, Sun H, Chen S, et al., Future development of global regulations of Chinese herbal products, Journal of Ethnopharmacology, 2012, 140(3), 568–586.

[77] Dwyer JT, Coates PM, Smith MJ, Dietary supplements: Regulatory challenges and research resources, Nutrients, 2018, 10(1), 41.

[78] Liang Y-Z, Xie P, Chan K, Quality control of herbal medicines, Journal of Chromatography B, 2004, 812(1–2), 53–70.

Amarjitsing Rajput, Satish Mandlik, Shipa Dawre, Deepa Mandlik,
Prashant Pingale, Shital Butani

Chapter 3
Herbal bioenhancers in nanoparticulate drug delivery system

Abstract: Herbal bioenhancer compounds are capable of promoting and improving biological activity – the bioavailability and bioefficacy of a particular material or nutrient at low doses. They could decrease the dose of the active component and adverse effects associated with it. However, the combined use of both drugs with bioenhancer served as an economical and feasible approach too. Nowadays, researchers are diverting their research in identifying novel bioenhancers from the natural origin for drug substances with low bioavailability or significant adverse effects. Current research has indicated that several herbal components like piperine, genistein, naringin, sinomenine, glycyrrhizin, quercetin, and nitrile glycoside were capable of improving the bioavailability of certain therapeutic compounds and phytopharmaceuticals through the alteration of metabolic and intestinal absorption. Bioenhancers are also used in several nanocarrier systems such as solid lipid nanoparticles, nanostructured lipid carriers, nanoemulsion, microemulsion, liposomes, transferosomes, ethosomes, methosomes, and sphingosomes to improve their properties, especially bioavailability. Thus, bioenhancers play a vital role in nanotechnology by enhancing the bioavailability of several components due to their inherent property.

Satish Mandlik, Deepa Mandlik, Department of Pharmaceutics, Poona College of Pharmacy, Bharti Vidyapeeth Deemed University, Erandwane, Pune 411038, Maharashtra, India
Amarjitsing Rajput, Department of Pharmaceutics, Poona College of Pharmacy, Bharti Vidyapeeth Deemed University, Erandwane, Pune 411038, Maharashtra, India; Department of Biosciences and Bioengineering, Indian Institute of Technology Bombay, Powai 400076, Maharashtra, India
Shipa Dawre, Department of Pharmaceutics, School of Pharmacy and Technology Management, SVKM's Narsee Monjee Institute of Management and Studies (NMIMS), Mukesh Patel Technology Park, Shirpur 425405, Maharashtra, India
Prashant Pingale, Department of Pharmaceutics, GES's Sir Dr. M. S. Gosavi College of Pharmaceutical Education and Research, Nashik 422005, Maharashtra, India
Shital Butani, Department of Pharmaceutics and Pharmaceutical Technology, Institute of Pharmacy, Nirma University, S.G. Highway, Ahmedabad 382 481, Gujarat, India, e-mail: shital_26@yahoo.com

https://doi.org/10.1515/9783110746808-003

3.1 Bioenhancer

Herbal bioenhancers, also known as **"biopotentiers,"** are phytomolecules capable of promoting and improving biological activity, which is described as bioefficacy and bioavailability of a material or nutrient (at low doses) [1].

The term herbal bioenhancer principally originated from Ayurvedic medicine [2]. The word "Yogvahi" is used in Ayurveda to describe herbs that can increase or potentiate a drug's plasma concentration. The first scientifically confirmed example of Yogvahi is piperine, which is derived from black pepper. Trikatu (a mixture of ginger, long pepper, and black pepper) is an Ayurvedic preparation to manage various ailments [3].

Bioenhancers, also called "bioavailability enhancers," are compounds that can improve the absorption and bioavailability of a particular drug. They are administered simultaneously and enhance their activity in the body [1, 4]. Bioavailability is the rate and extent to which a compound reaches the blood circulation and becomes available at the desired site of action [5]. **Higher bioavailability** is achieved by compounds administered via the intravenous route. However, drugs administered orally are poorly bioavailable due to first-pass metabolism and partial absorption. Drug substances that are unutilized in the body may result in side effects and drug resistance. Hence, there is a need for substances that do not possess any therapeutic activity but, when combined with other compounds, increase their bioavailability [6]. Several **natural substances** from medicinal plants have the potential to augment the bioavailability when coadministered with another drug. Hence, bioenhancers are chemical moieties that promote and enhance the bioavailability of the drugs mixed with them and do not show a synergistic effect with the drug [7].

They could decrease the dose of the active component and adverse effects associated with it. However, combining both drugs with bioenhancer served as an economical and feasible approach [8]. Researchers are now focusing on discovering novel bioenhancers derived from natural sources for drug substances with poor bioavailability or severe side effects. Current research has indicated that several herbal components like piperine, genistein, naringenin, sinomenine, glycyrrhizin, quercetin, and nitrile glycoside could improve the bioavailability of certain **therapeutic compounds** and phytopharmaceuticals through the alteration of metabolic and intestinal absorption [9].

For example, the bioavailability of curcumin was enhanced by 154% in rats, and 200% in humans post coadministration with the first and promising bioenhancer from the natural source, i.e., piperine [10]. There is no doubt that developing a suitable formulation improved the absorption of poorly soluble substances; however, bioenhancers could be of clinical significance because they function by inhibiting the action of drug-metabolizing enzymes or by mediating transcellular transport through transporters [1, 4]. The positive impact of improved bioavailability of therapeutic compounds could eventually lead to improved potential and reduced dosing frequency [11]. The benefits of bioenhancers that are useful in pharmaceutical industry are shown in Figure 3.1.

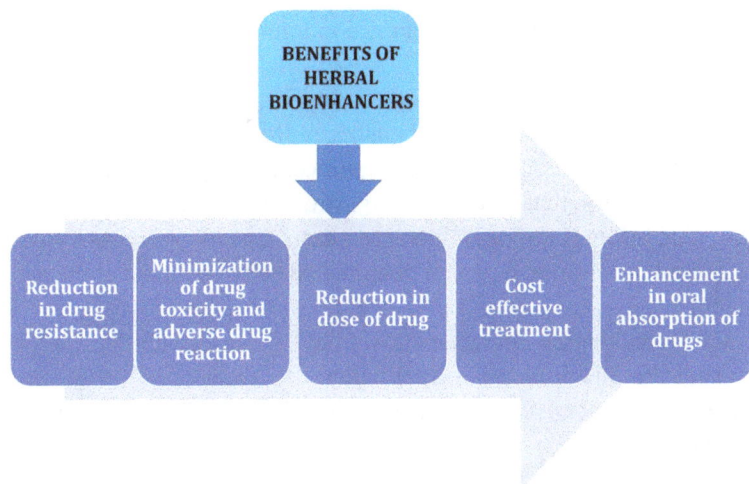

Figure 3.1: Benefits of bioenhancers useful in pharmaceutical industry.

3.1.1 Ideal characteristics of herbal bioenhancers

a) It should be nontoxic to humans and animals.
b) It should be effective in low concentrations.
c) It should be easy to fabricate.
d) It should be cost-effective.
e) It should improve the drug's uptake/**absorption and activity** [1].

3.2 Mechanism of bioenhancer

3.2.1 General mechanism

The main mechanisms involved in the bioenhancing activity of the plant extract and phytochemicals are summarized as follows:

i) Overcoming the enzymes (cytochrome P450) and isoenzymes (quercetin, gallic acid, piperine, naringin) or enhancement of gamma-glutamyl transpeptidase action are examples of changes in enzymatic potential (uptake of amino acids).

ii) For example, the **inhibition of the P glycoprotein** efflux pump causes changes in drug transporters (caraway, sinomenine, and genistein).

iii) **Cholagogues** have the same effect as liquorice in terms of bile promotion into the intestine.

iv) Garlic, ginger, and turmeric have thermogenic/bioenergetic activity, resulting in enhanced metabolic rate and improved **gastric motion**.

v) Reduced production of hydrochloric acid (liquorice, aloe, ginger, and niaziri-din), prevention of gastric time of emptying, gastrointestinal transit, and intestinal motility (tea, allum, liquorice), increased gastrointestinal mucosa blood supply and altered penetration behavior of epithelial cell membrane of the gastrointestinal tract.

vi) Alteration of physicochemical characteristics of phytosome formulation such as dissociation constant, lipophilicity, and solubility.

vii) Affect target drug receptors. The general mechanisms of herbal bioenhancers are shown in Figure 3.2.

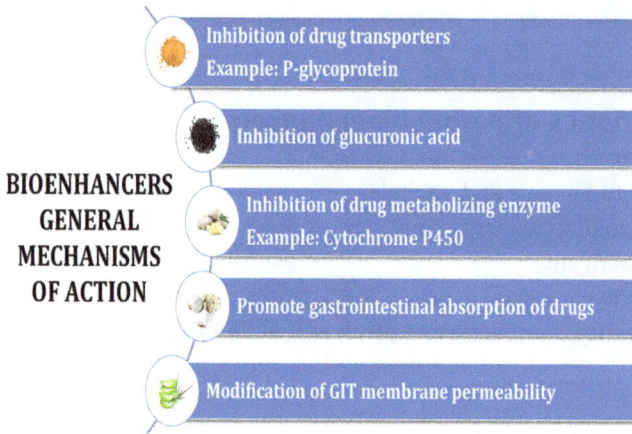

BIOENHANCERS GENERAL MECHANISMS OF ACTION

- Inhibition of drug transporters Example: P-glycoprotein
- Inhibition of glucuronic acid
- Inhibition of drug metabolizing enzyme Example: Cytochrome P450
- Promote gastrointestinal absorption of drugs
- Modification of GIT membrane permeability

Figure 3.2: General mechanisms of herbal bioenhancers.

3.2.2 Specific mechanism

Herbal bioenhancer also acts as a bioavailability enhancing agent via a specific mechanism. These include:

3.2.2.1 Enzymatic alterations

Herbal compounds are mixtures of biologically active components. The metabolism of these different substances occurs by the same mechanism as the administered drug, resulting in association and, ultimately, inhibition or enhancement of passengers or drug-metabolizing enzymes. The modification of the protein expression or physical/chemical/pharmacological competition affects the amount of compound or its metabolite pharmacokinetics. A different herb was found to react with enzyme (**cytochrome p450**), which exists in many polymorphic forms in humans and animals. It plays a crucial role in the metabolism or detoxification of herbs [12]. The CYP and UDP enzymes are essential for phase I metabolism of a variety of medications, endogenous compounds, nutrients, and environmental contaminants, and changes in their expression or **functional properties** have a significant impact on the effectiveness of treatment or the progression of toxic effect [13]. As a result of the inhibition, fewer drug substances will be metabolized, while an increased volume of unmetabolized drugs will move from the stomach into the bloodstream. The CYP3A4, 2D6, and 2C9 families are the essential isoenzymes involved in drug metabolism in humans [14].

Piperine was the first purified bioenhancer discovered to inhibit various types of enzymes like cytochrome P450 (CYP) (especially CYP3A4) [15] and enzyme **glucuronosyl transferases** present in the hepatic or intestinal region. This inhibition relies on several parameters such as route of administration, dosage, and **exposure length** [16]. Piperine has a potentiating effect on the effectiveness and rate and the extent to which drug reaches the systemic circulation in mice, rats, rabbits (laboratory animals), and human beings [15]. Still, there are few studies available for veterinary medicine. Along similar lines, animals (hens) treated with oxytetracycline and oral *Piper longum* 7 days before treatment showed increased bioavailability and pharmacokinetic improvements [17].

St. John's wort, which is anxiolytic, was found to cause CYP3A4, CYP2C19, CYP2C9, CYP1A2, CYP2B6, and CYP2E, resulting in a reduction in levels (plasma) of the substrate of several available medications like central nervous compounds, statins, immunosuppressants, antihistamines, antivirals, and chemotherapeutics, with or without pretreatment [18]. In dogs treated with cyclosporine, St. John's wort's repeated administration altered the pharmacokinetic profile, resulting in overall total blood amount and area under the curve (AUC) [19].

Ginkgo biloba (homeopathic tincture) is used in all food-processing species to improve their growth and in pets to treat ailments such as neurological problems and hormonal imbalances [20]. In rats, *G. biloba* was discovered to potentiate hepatic CYP enzymes (especially CYP2B). It also shows a significant impact on CYP2C19 and is found to enhance hydroxylation of omeprazole in a CYP2C19 genotype, thus reducing its effectiveness in human beings [21].

Grapefruit juice irreversibly inhibits CYP3A due to furanocoumarin's presence and increases the bioavailability of multiple drugs such as benzodiazepines, calcium channel blockers, and statins. Only the enterocyte cells lining the small intestine are inhibited by regular ingestion, although hepatic CYP is unaffected until high-risk concentrations are reached [22]. Grapefruit juice can improve the effectiveness and lower the cost of drugs destroyed by CYP3A post oral administration due to the first-pass metabolism, such as cyclosporine [22].

3.2.2.2 Using phytosomes formulation

Phytosomes are considered bioenhancing phytomolecules that were recently introduced among the all-available herbal formulations. Phosphatidylcholine and a plant portion formed a 1:1 and 2:1 complex (molecular) consisting of hydrogen bonds in phytosomes, natural products, and natural phospholipids. This complex's dual solubility properties result in better absorption and bioavailability than liposomes. The lipid solubility of phytosomes containing plant extracts (standardized) or water-soluble principal constituents (tannins, terpenoids, flavonoids, xanthones) has increased. They also protect active compounds from gastric degradation (by hydrochloric acid) and bacteria in the gut, resulting in improved bioavailability. Phytosomes increase absorption, biological activity, and transmission to the target tissue [23, 24].

Silybum martanum (milk thistle) contains hepatoprotective flavonoids (silymarin, silchristin, and silybin) but shows poor absorption via the oral route. On the other side, silybin formulated as phytosome (silybin-phospholipid complex) is more quickly absorbed, resulting in lower silymarin doses and improved biological effects, with fewer side or adverse effects. This drug is also being researched for cancer prevention and treatment (prostate) and chronic iron overload management [25, 26]. *Centella* (*Centella asiatica* L.) is another example where the triterpene type forms a phytosome with soy phospholipids, enhancing oral bioavailability and reducing bile salt interaction. In a phthalic anhydride-induced atopic dermatitis mouse model, *Centella* phytosomes were also found to increase anti-inflammatory effects by inhibiting NF-kB signaling, indicating that they could be used to treat atopic dermatitis [27].

3.2.2.3 Transporter protein alteration

P-glycoprotein (P-gp) is a transporter protein that is located on the apical surface of epithelial cells, the adrenal gland, blood–brain barrier endothelial cells, and the exterior of numerous neoplastic cells, where it facilitates the transfer of medicines from the blood via the intestinal lumen. Interfering with protein binding sites, ATP hydrolysis, or cell membrane lipid integrity serve as a replacement therapeutic approach that enhances therapeutic compound distribution and drug amount at the intracellular level [28]. Since currently available synthetic P-gp inhibitors have a higher occurrence of side effects, plant-based alternatives are considered possible drug compounds. In vitro and in vivo experiments have shown that a few active components from plants have an inhibitory effect. Since current P-gp inhibitors can't determine the difference between normal and cancerous tissues, intrinsic toxicity (cytotoxicity) is widespread, putting patients at risk. As a result, monitoring systems are needed to develop a preventative strategy for such severe interactions [29].

3.2.2.4 Cholagogue/choleretic effect

Cholagogues aid in the digestion of fats and the absorption of medicines from the gastrointestinal system by promoting the secretion and release of bile from the gallbladder. Choleretics **stimulate** gallbladder motility, and cholagogues enhance bile flow. They are widely utilized to treat cholecystitis and cholelithiasis, as well as situations requiring intestine spasmolytic activity. The substances with limited water solubility are dissolved in bile salt-phospholipid micelles and transported through the intestinal wall with improved bioavailability. Bile acids also act as signaling compounds for the control of hepatocyte metabolism by binding nuclear receptors, which would affect bioavailability [30]. The compounds of medicinal plant category with cholagogic/choleretic properties consist of St. John's wort (*Hyperium perforatum*), chamomile (*Chamomilla recutita*), rosemary (*Rosmarinus officinalis*), and *Chelidonium* (*Chelidonium majus*). Along with this, choleretic qualities can be found in essential oils from plants such as coriander, black pepper, turmeric, red chili, cumin, licorice, and peppermint. The bioenhancers, with their mechanism of action on different compounds for improved drug delivery, are as shown in Table 3.1.

3.3 Selected bioenhancers

3.3.1 *Long pepper (Piper longum)* and *black pepper (Piper nigrum)*

Piperine is a bioactive alkaloid produced from *Piper longum* Linn and *Piper nigrum* Linn [58]. Piperine is commonly used in India as a spice and flavoring agent for a variety of sauces. In the pharmaceutical area, it is identified to have several properties such as anti-inflammatory [59], antioxidant [60], antitumor [61], antipyretic [62], antithyroid [63], antidiarrheal [64], antimutagenic [65], antidepressant [66], analgesic, antihypertensive [67], hepatoprotective [68], and antifungal [69]. According to scientific evidence, piperine is the world's first **bioavailability** enhancer [70].

Table 3.1: Natural bioenhancers with their mechanism of action on different compounds for improved drug delivery [31].

Bioenhancer compound	Biological source	Mechanism of action	Study design model	Route of administration	References
Chitosan	Deacetylated chitin from crustaceans and fungi	Mucoadhesion; changes in lipid organization and loosening of intercellular filaments	In vitro (T146 cells 1)	Buccal	[32]
Aloe vera	Plant (*Aloe vera*)	Intercellular modulation	In vitro (Franz diffusion cells)	Buccal	[33]
Menthol	Plant (corn mint, peppermint, or other mint oils)	No mechanism specified	Ex vivo (porcine buccal mucosa)	Buccal	[34]
Oleic acid, eicosapentaenoic acid, docosahexaenoic acid	Animal (cod fish)	No mechanism specified	In vitro (membraneless dissolution test), in vivo (rat)	Buccal	[35]
Chitosan	Chemically modified chitosan (crustaceans, fungi)	Tight junction modulation	In vivo (sheep)	Nasal	[36]
Chitosan	Deacetylated chitin from crustaceans and fungi	Increased mucoadhesion; tight junction modulation	In vivo (sheep, human)	Nasal	[37]
Chitosan–TBA	Deacetylated chitin from crustaceans and fungi	Increased mucoadhesion; tight junction modulation	In vivo (rat)	Nasal	[38]
(–)-Epicatechin	Plant (woody plants)	Metabolism (glucuronidation) inhibition	Ex vivo (rat small intestine)	Oral	[39]
Caraway	Plant (meridian fennel/ Persian cumin: *Carum carvi*)	Local mucosal tissue modulation	In vivo (human)	Oral	[40]

Curcumin	Plant (turmeric: *Curcuma longa*)	Efflux transporter inhibition; metabolism inhibition	In vivo (rabbit)	Oral	[41]
Fulvic acid	Plant (decomposed material)	Metabolism enhancement (enhanced drug water solubility)	In vivo (rat)	Oral	[42]
Genistein	Plant (soybean: glycine max, kudzu: *Pueraria lobata*)	Efflux transporter (P-gp, BCRP, MRP2) inhibition; metabolism (CYP3A4) inhibition	In vivo (rat)	Oral	[43]
Lycopene	Lycopene (carotenoid)	Dual carotenoid/LDL receptor mechanism for targeted hepatic delivery	In vivo (human)	Oral	[44]
Lysergol	Plant (morning glory plant: *Ipomoea* spp.)	Metabolism inhibition	In vivo (rat)	Oral	[45]
Naringin	Plant (grapefruit, apple, onion, tea)	Metabolism (CYP3A4) inhibition	In vivo (rat)	Oral	[46]
Peppermint oil	Plant (peppermint: *Mentha pipertita*)	Metabolism (CYP3A) inhibition	Ex vivo (rat intestinal tissue)	Oral	[47]
Piperine	Plant (*Piper longum* and *Piper nigrum*)	Metabolism (CYP450) inhibition	In vivo (rat)	Oral	[48]
Quercetin	Plant (citrus fruits, vegetables, leaves, grains)	Efflux transporter (P-gp) inhibition	In vivo (rat), Ex vivo (rat and chick everted intestinal sac)	Oral	[49]
Quercetin	Plant (citrus fruits, vegetables, leaves, grains)	Efflux transporter (P-gp) inhibition	In vivo (human)	Oral	[50]
Resveratrol	Plant (berries, grape skins, red wine)	Metabolism (CYP2C9, CYP2E1) inhibition	In vivo (human)	Oral	[51]

(continued)

Table 3.1 (continued)

Bioenhancer compound	Biological source	Mechanism of action	Study design model	Route of administration	References
Aprotinin, bestatin	Animal (bovine lung tissue), bacteria (*Streptomyces olivoreticuli*)	Metabolism inhibition	In vivo (rat)	Pulmonary	[52]
Citric acid	Plant (citrus fruits and vegetables), fungi (*Aspergillus niger*)	Local mucosal tissue modulation; metabolism inhibition	In vivo (rat)	Pulmonary	[53]
HPBCD, Crysmeb	Plant (starch)	Tight junction modulation	In vitro (Calu-3 cells 5)	Pulmonary	[54]
Sodium glycocholate	Intestinal bacterial by-product	Tight junction modulation	Ex vivo (rabbit trachea and jejunum)	Pulmonary	[55]
Sodium taurocholate	Intestinal bacterial by-product	Metabolism enhancement (dissociation of insulin hexamers); tight junction modulation; metabolism (enzymatic degradation) inhibition	In vitro (Caco-2 cells 2), In vivo (dog)	Pulmonary	[56]
Dideoxycytidine	Plant (starch)	Tight junction modulation	In vitro (Calu-3 cells 5), in vivo (rat)	Pulmonary	[57]

Piperine can function as bioenhancer through two mechanisms: stimulating the absorption of drugs rapidly [71] and inhibiting enzymes involved in drug biotransformation [72]. Piperine is a potent blocker of the P-gp efflux transporter found in the intestinal wall and the main metabolizing enzyme, Cytochrome P450 3A4. Curcumin with piperine has a higher antigenotoxic activity due to curcumin's improved bioavailability when combined with piperine. By raising the residence time, altering the membrane lipid dynamics, and altering the structure of intestinal enzymes, piperine decreases the rate of metabolism of curcumin and increases its intestinal absorption. As a result, piperine is a possible bioavailability enhancer for hydrophobic drugs such as curcumin [73]. Piperine increased the oral bioavailability of norfloxacin and ampicillin in rabbits, according to Janakiraman et al., ampicillin and norfloxacin are antibiotics that are used to treat a variety of diseases, but they have poor oral bioavailability. Piperine supplementation at a dose of 20 mg/kg enhanced the bioavailability of poorly absorbed medicines. **Norfloxacin** oral bioavailability is increased by piperine to a lesser extent than ampicillin. The improved **pharmacokinetic** characteristics could be due to the inhibition of liver metabolic enzymes and increased penetrability through GIT cells [74]. Piperine is a commonly used bioenhancer that has been shown to improve the bioactivity of antibiotics, antituberculosis drugs, antifungals, antivirals, cardiovascular drugs, anti-inflammatory drugs and nutraceuticals. Acyclovir's bioavailability increased by 70% in the presence of piperine alone, while its bioactivity increased by 98% when combined with *Cuminum cyminum* extract. As a result, piperine can be combined with several bioenhancers including cumin and curcumin to improve drug bioavailability.

3.3.2 Ginger (*Zingiber officinale*)

The active gingerols contained in the *Z. officinale* rhizome extract changed to zingerone, shogaols, and paradol [75]. Ginger's odor is primarily due to its volatile oil, which yields 1-3% [76]. Gingerols improve gastrointestinal motility in laboratory animals and have sedative, analgesic, antipyretic, and antibacterial activities [77]. The main pungent concept in ginger is [6]-gingerol. [6]-Gingerol's chemopreventive properties make it a promising possible alternative to costly and harmful therapeutic agents [78]. **Ginger** exhibits various properties like antifungal [79], antimicrobial [80], anti-ulcer [81], antidiabetic [82], antithrombotic [83], anti-inflammatory [84], antiemetic [85], and anticancer [86]. Ginger has a strong effect on the mucous membrane of the gastrointestinal tract. Ginger's job is to control the activity of the intestine and make absorption easier. *Z. officinale* alone provide bioenhancing activity (30–75% range), and piperine (4–12 mg/kg) and *Z. officinale* (10–30 mg/kg) together provide bioenhancing activity (10–85% range). Extremely selective **bioavailability** enhancing activity is shown by the extracts or fraction of *Z. officinale* alone or in the presence of piperine [87].

3.3.3 Turmeric (*Curcuma longa*)

Turmeric is a popular condiment commonly used in Indian dishes. It is also used in ointments and creams as a coloring agent. Curcumin is the only curcuminoid found in *Curcuma longa* [88]. **Curcumin** is an anti-inflammatory agent, in addition to its conventional uses. It is used in antimicrobial and anticancer medicine as a bioenhancer. Curcumin works by destroying drug-metabolizing enzymes (CYP3A4) in the liver and causing alterations in the P-gp acting as a drug transporter [89, 90]. Curcumin potentiated the anticancer activity of docetaxel in rats; according to the report by Yan et al., curcumin (100 mg/kg) was given to a group of rats, 30 min before docetaxel was given. When compared to the control group, the rats that were given curcumin before receiving docetaxel had a higher AUC (approximately 8-fold increase) and a higher serum concentration (C_{max}) (about 10-fold increase). The bioavailability of docetaxel increased by around 8-fold in the treatment group than in the control group. Variations in the protein expressions involved in the metabolism and transport of docetaxel may explain the improved bioavailability [91]. Pavithra et al. looked into the effect of curcumin on the norfloxacin disposition profile in rabbits. **Norfloxacin** is a broad-spectrum antibiotic commonly used to treat urinary tract infections, gastrointestinal tract, and lungs. It has a poor absorption profile, according to studies. A total of 16 rabbits were used in the study, with each group consisting of eight rabbits. Curcumin was noted to double the activity of norfloxacin [92]. Curcumin has been discovered to have anticancer properties, suppressing carcinogens in the body. As a result, we can use curcumin in conjunction with an anticancer medication like docetaxel to boost the drug's anticancer effect.

3.3.4 Black cumin (*Nigella sativa*)

Nigella sativa, also recognized as black cumin/caraway, was earlier tested as an amoxicillin bioenhancing agent. The seeds were used to make *N. sativa* extracts, which were then extracted for 6 h with methanol and hexane. The transfer of amoxicillin (6 mg/mL) with or without *N. sativa* seed methanolic extract (3 mg) and hexane (6 mg) extract was studied using everted rat intestinal sacs. The volume of amoxicillin conveyed through the intestine was measured using spectrophotometry. Rats were given amoxicillin (25 mg/kg) and *N. sativa* hexane extract (25 mg/kg) orally for in vivo studies. After collecting blood samples at various time intervals after dosing, the amount of amoxicillin in rat plasma was quantified using UPLC-MS/MS. Both the extracts of *N. sativa* (methanol and hexane) were effective in an in vitro analysis. The permeation of amoxicillin was greatly increased by *N. sativa*, with the latter showing the most remarkable rise. As a result, hexane extract was chosen for preclinical testing. The preclinical research also revealed that rats had a massive surge in plasma amoxicillin levels. *N. sativa* extract improved the rate and degree of amoxicillin absorption with increased C_{max} and AUC. This bioenhancing effect of *N. sativa* was suggested due to the occurrence of fatty acids [93]. Fatty acids have previously been shown to improve low drug permeation by enhancing the flexibility of the apical and basolateral membranes [94]. Rabbits were used in a related in vivo analysis that obtained the opposite findings. After pretreatment with *Nigella* (200 mg/kg), the cyclosporine at a dose of 30 mg/kg decreased C_{max} and AUC in that analysis. However, there was a 2-fold increase in cyclosporine clearance, implying that intestinal P-gp and CYP3A4 are triggered in the existence of *N. sativa* [95].

3.3.5 Cumin (*Cuminum cyminum*)

Cumin aldehyde, γ-terpinene, and β-pinenes are the major components of *C. cyminum* oil [96]. *C. cyminum* has several properties like estrogenic [97], antinociceptive, anticonvulsant, anti-inflammatory, anticancer [98], hypolipidemic [99], antimicrobial [100], antitussive, and antioxidant [101, 102]. Its components, such as luteolin, volatile oils, and flavonoids, are thought to have bioenhancer properties. Luteolin works by interfering with the **P-gp transporter** [103]. Qazi et al. looked into the function of *C. cyminum* by increasing the systemic bioavailability of several medications like nutrients and ayurvedic preparations. To investigate the synergistic bioenhancer activity of *C. cyminum*, it is used singly or mixed with other pharmaceutical excipients or piperine. The dose range of a bioactive fraction of *C. cyminum* is 2–20 mg/kg, whereas the extract is 10–30 mg/kg. Bioavailability is increased by 25-335% in formulations containing *C. cyminum* extract or fractions [104].

3.3.6 Caraway (*Carum carvi*)

The dried and crushed seeds of *C. carvi* are used to make caraway oil. Limonene and carvone are the oil's most powerful constituents. The seeds are widely used in cakes, pickles, biscuits, sweets, and confectioneries as a spice [87, 104]. Diuretic activity [105], hypoglycemic effects [106], antimicrobial activity [107], antioxidant activity [108], anti-ulcer effects [109], anti-aflatoxigenic activity, and antifungal activity are among the properties of *C. carvi*. Qazi et al. investigated the impact of *C. carvi* extract, singly or along with piperine or *Z. officinale*, on the bioavailability and pharmacokinetic summary of various medications, nutraceuticals, and herbal formulations. *C. carvi*'s bioenhancing activity may be due to its ability to improve mucosal penetration or impact on the P-gp transporter. The bioactive part of the bioenhancer has a concentration of 1–55 mg/kg, whereas the bioenhancing activity of *C. carvi* extract has a concentration of 5–100 mg/kg body weight. It improves the bioavailability of antibiotics, anticancer medicines, antifungals, and antivirals. The addition of

Z. officinale and *C. carvi* has a more significant **synergism** effect on the above-mentioned products, raising bioavailability by 20–110%. *C. carvi* is more effective when combined with piperine at a dose of 3–15 mg/kg. *C. carvi* exhibited a 25-95% improvement in the bioavailability of a variety of powerful medications. Rifampicin's antitubercular function was also improved by *C. carvi* [87].

3.3.7 Morning glory plant (*Ipomoea* spp.)

The alkaloid **lysergol** is produced from morning glory plants and is an ergoline alkaloid. It was discovered that it increased curcumin bioavailability. Curcumin's activity on human P-gp and **breast cancer resistance protein** (BCRP) as primary efflux transporters was investigated by in situ penetration and in vitro pharmacokinetic analysis. For this purpose, specific substrates such as sulfasalazine for BCRP probes and digoxin for P-gp probes are used to determine the mechanism of increased bioavailability (pantoprazole for BCRP and verapamil for P-gp). The findings suggest that the BCRP efflux transport mechanism is responsible for lysergol's increased **bioavailability** [110]. The increment in the oral bioavailability of berberine in rats was observed using lysergol. However, the mechanism of lysergol has yet to be determined, even though it is thought to be linked to P-gp inhibition [111].

3.3.8 Garlic (*Allium sativum*)

Garlic contains an allyl sulfur-containing compound called allicin. **Allicin** has antiplatelet [112], antiparasitic [113], antibacterial [114], antioxidant [115], antifungal [116], anticancer [117], immunomodulating [118], antidiabetic [119], anti-inflammatory [120], and antiviral activity [121]. **Copper** has dose-dependent fungicidal activity against *Saccharomyces cerevisiae* cells, and with allicin, an allyl sulfur compound present in garlic, its lethal action significantly increased. *N*-Acetyl-cysteine or dithiothreitol are compounds containing sulfur that had no impact on the fungicidal function of copper. Copper displayed no defense against the fungicidal action of copper recently produced along with allicin, but it showed protection against the toxic effect of copper alone [122]. As an example, the impact of allicin on the fungicidal action of amphotericin B against the *Saccharomyces cerevisiae* yeast was studied. Amphotericin B is an antibiotic used in the treatment of severe fungal infections, and allicin significantly increased its antifungal efficacy. Amphotericin B caused damage to the vacuole membrane, making the organelles recognizable as small distinct particles, in addition to changing the permeability of the plasma membrane. A nonlethal dose of amphotericin B, allicin-increased amphotericin B prompted structural changes to the vacuole membrane. Allicin can also improve amphotericin B antifungal activity against *Candida albicans* and *Aspergillus fumigatus* fungus [123]. In *S. cerevisiae*, allicin inhibited the transfer of ergosterol from the cell membrane to the cytoplasm that is thought to be a cell defensive reaction to amphotericin B-encouraged vacuolar breakdown. The findings indicate that lethal action of amphotericin B against *C. albicans* is due to its vacuole-disrupting behavior under physical conditions favorable to the fungus's invasive development [116]. Allicin promotes amphotericin B-induced vacuole membrane degradation by preventing ergosterol trafficking from the plasma membrane to the vacuole membrane, according to Ogita et al. [124].

3.3.9 Liquorice (*Glycyrrhiza glabra*)

Glycyrrhizin is a glycoside found in liquorice stolons and roots (*Glycyrrhiza glabra*). It has an expectorant effect and can help with allergies, sore throat, bronchitis, gastritis, asthma, peptic ulcers, and rheumatism. It aids in the detoxification of drugs in the liver and is used to treat liver disease. It acts as a diuretic and laxative, as well as strengthens the immune system. Glycyrrhizin is 50 times sweeter than sugar. Several pharmacological activities are exhibited by Glycyrrhizin like anti-inflammatory [125], anticancer [126], antihepatotoxic [127], antiviral [128], peptic ulcer, and stomach treatment. It makes antimicrobials like tetracycline, rifampicin, and nalidixic acids more effective against gram-positive bacteria. Glycyrrhizin can work by inhibiting the transporter glycoprotein, which is found in the intestine [129]. The ability of glycyrrhizin to improve absorption is dependent on its conversion to glycyrrhetic acid by the intestinal bacterial enzyme β-glucuronidase [130]. Chen et al. used oral, tail vein, and hepatic portal vein administration to study the effects of diammonium glycyrrhizinate on the bioavailability and pharmacokinetic properties of aconitine in rats. When glycyrrhizinate was provided orally with aconitine, the peak plasma concentration area under the curve, the absolute bioavailability of aconitine was enhanced drastically. However, the half-life and clearance did not improve significantly. After delivery of glycyrrhizinate via tail vein and hepatic vein, the aconitine pharmacokinetics parameter did not change considerably. The suppression of the P-gp transporter in the intestine could be the cause of this increase in absorption [129]. Khanuja et al. also revealed the usefulness of glycyrrhizin in improving the bioavailability of several antibiotics, anticancer, and anti-infective drugs. **Glycyrrhizin** aids absorption of antibiotic and other compounds across biological membranes, increasing their plasma levels, and, consequently, bioavailability. When antibiotics like ampicillin, rifampicin, tetracycline, and nalidixic acid are coupled with Glycyrrhizin, they are more effective against gram-negative bacteria. Glycyrrhizin also makes antifungal drugs like clotrimazole more effective against *Candida albicans*. Taxol (paclitaxel) was also tested for its enhanced anticancer activity in terms of cell division inhibition. Taxol works by preventing MCF-74 cancer cells from proliferating and multiplying. Glycyrrhizin increases the activity of taxol by a factor of five. Taxol combined with glycyrrhizin has higher activity than taxol given alone. Because of its efficacy and as it is devoid of toxicity at low concentrations, glycyrrhizin has a lot of potential as a bioenhancing agent for several drugs like anticancer, antifungal, antibacterial, and nutraceuticals. It also helps lessen the side effects of anticancer drugs and the production of antimicrobial resistance [131].

3.3.10 Aloe vera

Aloe vera exhibits several pharmacological activities like a wound, burn-healing, hypoglycemic, anti-inflammatory, gastroprotective, immunomodulatory, antifungal, and anticancer, commonly used in human and veterinary medicine [132]. *Aloe vera* gel and leaf extracts enhanced vitamin C and E plasma concentrations and enhanced absorption. *Aloe vera* gel extract was very effective in improving ascorbate absorption as well as reducing it. Its plasma concentration was substantially prolonged, even after an overnight fast of 24 h. **Vitamin E** absorption and plasma concentration were improved by both the leaf and gel extracts of aloe vera, particularly after 8 h. Aloe vera is found to improve the bioavailability of these vitamins and would be seen as a potential herbal bioenhancing agent in the future [133]. In streptozotocin-induced diabetic rats, an ethanol extract of aloe vera was shown to augment the hypoglycemic action of **glipizide** [134]. It was discovered that combining aloe vera and pantoprazole to treat gastroesophageal reflux symptoms in mustard gas sufferers was more successful than using them separately. This is attributed to its cytoprotective effect on the gastrointestinal tract mucosa via the production of endogenous prostaglandins [135]. In a rat model, treatment of aloe was shown to stimulate the roles of P-gp and **CYP3A**, lowering the bioavailability of cyclosporine [136].

3.3.11 Drumstick pods

Drumstick (*Moringa oleifera*) leaves, pods, and bark have all been found to contain niaziridin, a nitrile glycoside [137]. *M. oleifera* has found to display actions like antimicrobial [138], antifertility [139], hepatoprotective [140], antiteratogenic [141], diuretic [142], anticancer [143], antioxidant [144], hypolipidemic [145], anti-inflammatory [146], hypotensive and spasmolytic [147], antifungal [148], antiulcer [149], and anti-arthritic [150]. It improves the bioactivity of commonly used antibiotics such as tetracycline, ampicillin, rifampicin, and nalidixic acid against gram-positive and gram-negative bacteria such as *Bacillus subtilis*, *Mycobacterium smegmatis*, and *E. coli*. It increases the activity of ampicillin, nalidixic acids, rifampicin, and tetracycline against gram-positive bacteria by 1.2- to 19-fold. It enhances the efficacy of azole antifungals, such as clotrimazole, against *Candida albicans* by 5 to 6 times. However, at a greater quantity (10 g/mL), the compound's antifungal potency was only slightly improved.

It also works as a **bioavailability enhancer** by facilitating the ingestion of vitamin B_{12} via the intestinal membrane [151].

3.3.12 Genistein

It is an isoflavone flavonoid derived from *Pueraria lobata* and *Glycine max*. It is a strong phytoestrogen [152]. It has anti-inflammatory and anticancer properties. It acts by preventing P-gp from performing its efflux functions [153, 154] and BCRP2 [155]. Li et al. investigated the effect of orally administered genistein on the pharmacokinetic profile of paclitaxel, an anticancer medication that can be taken orally or intravenously. Thirty minutes after being given genistein (10 mg/kg orally), rats were given paclitaxel (30 mg/kg orally) or (3 mg/kg intravenously). In oral administration of paclitaxel, the existence of 10 mg/mL genistein resulted in a significant rise in AUC and C_{max}. The addition of genistein (10 mg/kg) to intravenous paclitaxel resulted in a substantial increase in AUC by 40.5%. The dual inhibition of **CYP3A** and **P-gp** by genistein explains the improved bioavailability of paclitaxel [43]. The combination treatment of genistein and epigallocatechin-3-gallate (EGCG) in HT 29 human colon cancer cells results in a 2- to 5-fold rise in cytosolic EGCG concentration relative to treatment with EGCG alone. When mice were given genistein at 200 mg/kg with EGCG at 75 mg/kg, they improved AUC and half-life significantly; thus, **genistein** can enhance the bioavailability of EGCG. However, in mice, combination of EGCG (0.01%) in drinking water and genistein (0.2%) in the diet increased the tumorigenesis process in the intestine. As a result, careful administration of this combination is needed [156]. Genistein is a dietary supplement that is used in the daily diets of many people. Genistein could be used to improve the bioavailability of anticancer medications by inhibiting CYP3A and P-gp.

3.3.13 Naringin

It is the main **flavonoid** present in tomatoes, grapes, and apples [157]. It has pharmacological effects such as anti-ulcer, anticancer, antioxidant, anti-allergic, and blood lipid-lowering [158]. It is shown to inhibit CYP 3A4 [159], P-gp [160], and reduce the metabolism of drugs such as verapamil, diltiazem, and paclitaxel, causing a lower drug dosage and higher plasma concentration. In rats, Choi et al. examined how naringin affected the oral bioavailability of the anticancer medication, paclitaxel. In this investigation, rats were administered paclitaxel at a dose of 40 mg/kg, or a combination of paclitaxel and naringin at a dose of 3 mg/kg orally. As a result of this, the AUC and C_{max} of paclitaxel increased, and when combined with naringin, the absolute bioavailability of paclitaxel was significantly higher than in the control group. In comparison to the control group, the bioavailability also improved. Increased bioavailability of paclitaxel may be due to the **dominance** of CYP-450 and P-gp in the gastric mucosa [161]. The absorptive behavior of diltiazem was assessed, following oral administration of naringin (5 mg/kg) and diltiazem (15 mg/kg) in rats. The diltiazem C_{max} and AUC improved 2-fold, but the plasma **half-life** did not improve significantly. The relative and absolute bioavailability of diltiazem also improved significantly. As a result, it can be concluded that naringin reduces the dosage and adverse effects of diltiazem by inhibiting P-gp and metabolism of the medication in the intestine [162]. Naringin, a **bioflavonoid**, has been discovered to have an antiproliferative activity on cancerous cells and inhibits the P-gp efflux pump and the CYP3A enzyme. As a result, it might be used in conjunction with anticancer medications to treat a variety of cancers.

3.3.14 Sinomenium acutum

Sinomenine is an **alkaloid** that can be found in the *Sinomenium acutum* plant. Rheumatoid disorders are often treated with Sinomenine. Sinomenine is used to increase paeoniflorin bioavailability. The increase in bioavailability was thought to be due to sinomenine's ability to reduce paeoniflorin efflux transport by P-gp. Chan et al. studied the impacts of sinomenine on paeoniflorin absorption activities, in an everted rat gut model [163]. The tissue culture experiments with sinomenine (16 and 136 μM) and paeoniflorin (20 μM) were used to assess the activity of sinomenine on the improvement of paeoniflorin intestinal absorption. There was a strong linear association between absorption of paeoniflorin and incubation time, when paeoniflorin was given alone. The connection became nonlinear when combined with sinomenine. After 45 min of incubation, paeoniflorin absorption increased significantly. When verapamil and quinidine (P-gp inhibitors) were given along with paeoniflorin, their absorption rate increased, suggesting that they act similarly to sinomenine. When **paeoniflorin** and **sinomenine** HCl are given together, the plasma sinomenine concentration is elevated, relative to paeoniflorin alone. The results of the AUC findings displayed that the paeoniflorin oral bioavailability was increased by 12 times in rats that were given sinomenine concurrently. Unrestrained conscious rats were administered a single dosage of paeoniflorin alone or in combination with sinomenine hydrochloride by gastric gavage. Paeoniflorin C_{max} increased, AUC increased, and volume of distribution reduced after coadministration of sinomenine. Sinomenine hydrochloride considerably enhanced the bioavailability of paeoniflorin in rats [164]. The absorption profile of **digoxin** (P-gp substrate) increased 2.5-fold when treated with sinomenine, suggesting that paeoniflorin is possibly a P-gp substrate in the intestine. As a result, the sinomenine **enhanced the bioavailability** of paeoniflorin due to the P-gp efflux transport inhibition [163]. Hence, sinomenine can be added to any P-gp substrate to increase its bioactivity. Sinomenine has anti-inflammatory properties, so it has a better anti-rheumatoid arthritis effect when combined with a medication like paeoniflorin. The bioavailability enhancement mechanism of herbal bioactives is as shown in Figure 3.3.

3.4 Role of bioenhancers in nanoparticulate drug delivery systems

Natural bioenhancers are very attractive in nano-based drug delivery [4]. They demonstrated a critical role in the bioavailability enhancement of nanocarriers, avoiding synthetic bioenhancers, resulting in increased **safety** and reduction of **adverse effects**. Thus, several researchers are showing interest in the development of nanocarriers in combination with natural bioenhancers [166]. This section systematically discusses the inclusion of natural bioenhancers in nanoparticulate drug delivery.

3.4.1 Nanoparticles

Nanoparticles are colloidal drug delivery systems that display sizes in the range of less than 1 μm. A variety of materials can produce them, viz., proteins, polymers, and solid lipids. Nanoparticles can be designed via two strategies: **capsule-type** and **matrix-type** named nanocapsules and nanospheres [167]. In nanocapsules, drugs are condensed in the interior area, enclosed by the polymeric membrane; however, the drug either physically adheres or uniformly disperses within the matrix in nanospheres. Commonly, nanoparticles are classified based on the substance used for its construction.

3.4.1.1 Solid lipid nanoparticles (SLNs)

Solid lipid nanoparticles (SLNs) are colloidal systems whose matrix is made of biodegradable solid lipids. Generally, solid lipids used to fabricate SLNs are waxes, steroids, fatty acids, and triglycerides. They are comparatively more stable than polymeric nanoparticles and revealed size ranges from 10 to 1,000 nm. There are several benefits of SLNs, such as more excellent permeability, enhanced bioavailability, decreased cytochrome P 450 metabolism, less P-gp efflux, prolonged-release, reduced first-pass metabolism, and superior tissue distribution as well as lymphatic uptake. Kumar et al. developed SLNs loaded with isradipine in combination with bioenhancing agent rutin, against hypertension. Pharmacokinetic studies revealed 4.7 times higher oral bioavailability as compared to drug-loaded nanoparticles without rutin. Thus, a study confirmed significantly higher **bioavailability** observed in the presence of bioenhancer [168].

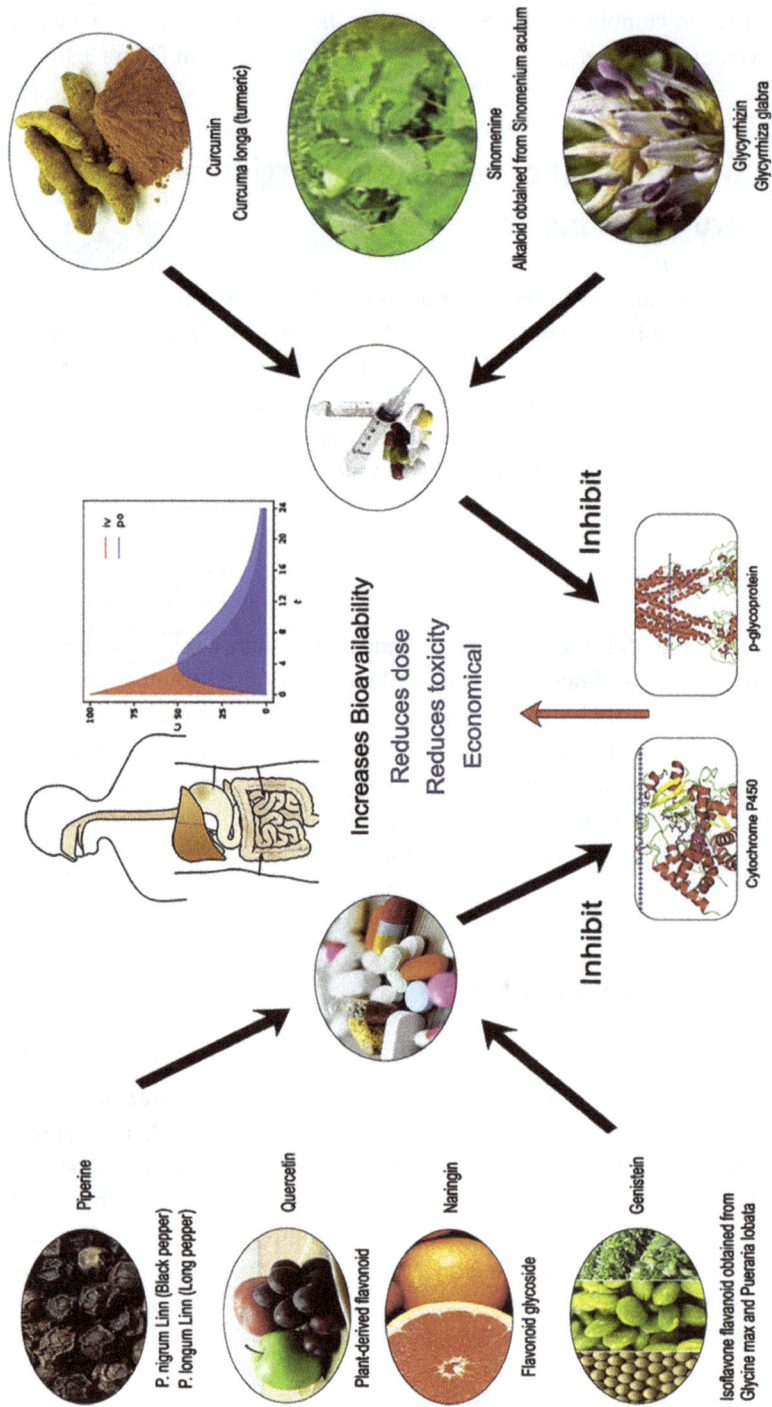

Figure 3.3: Bioavailability enhancement mechanism of herbal bioactives. Reproduced with permission from Alexander et al. [165].

3.4.1.2 Nanostructured lipid carriers (NLCs)

Muller first developed nanostructured lipid carriers (NLCs) in 1999 to overcome some observed shortcomings of SLNs, like less drug encapsulating capacity and drug discharge at the time of storage [169]. NLCs were designed to replace the little amount of compact lipids with liquefied lipids to formulate a drug-dispersed template. They are developed to increase the plasma concentration through the oral route of many lipophilic molecules [170]. Currently, NLCs are considered a potential drug delivery system and are incorporated in many commercial dermal creams and cosmetic preparations. Furthermore, NLCs have been explored for drug targeting to numerous systems like brain tissues, targeting cancerous cells, pulmonary, and investigating to deal with multidrug resistance. **Curcumin**-loaded NLCs were investigated with partially hydrolyzed ginsenoside. In vitro release showed that in the presence of hydrolyzed ginsenoside, the release rate of curcumin was higher. Therefore, the study concludes that it may help increase the bioavailability of curcumin [171]. In another study, NLCs-loaded curcumin delivered intranasally was studied in astrocytoma-glioblastoma cancer cell line (**U373MG**) and revealed higher cytotoxicity of curcumin-loaded NLC than curcumin alone [172]. Aditya et al. developed NLCs coloaded with curcumin and genistein against prostate cancer [173].

3.4.1.3 Polymeric nanoparticles

In contrast to SLNs, polymeric nanoparticles exhibit particle sizes less than 500 nm and are designed by using biocompatible and biodegradable polymers from natural or synthetic sources. Polymeric nanoparticles have been reported for a long time to increase the bioavailability of lipophilic molecules. Additionally, to enhance the efficacy of nanoparticles, they were developed in combination with herbal bioenhancers. **Zein–carrageenan** nanoparticles loaded with curcumin were formulated and assembled with piperine. Nanoparticles encapsulating curcumin and piperine demonstrated higher in vitro antioxidant activity than curcumin solution [174]. In another work on zein–carrageenan, nanoparticles containing curcumin and piperine revealed good cytotoxic effects in neuroblastoma cells [175]. In another exciting study, folic acid-conjugated chitosan nanoparticles-encapsulated curcumin were fabricated with a combination of bioenhancers, piperine and gingerol, to decrease hormone-refractory prostate cancer. The pharmacokinetics study confirmed 7.7-fold enhancement of bioavailability in comparison to curcumin solution by oral route [176]. Table 3.2 disclosed that this association of herbal bioenhancers with nanoparticles exhibits the remarkable potential to enhance the bioavailability of several lipophilic drugs.

Table 3.2: Natural bioenhancers laden in nanoparticulate drug delivery systems.

Nanoparticulate DDS	Drug	Herbal bioenhancers	Findings	References
SLN	Isradipine	Rutin	Enhanced oral bioavailability (4.7 times) with rutin	[177]
NLCs	Curcumin	Hydrolyzed ginsenoside	Higher drug release flux with bioenhancer	[171]
Chitosan nanoparticles	–	Piperine	High cytotoxicity on human brain cancer cell line (Hs683)	[178]
Guar gum nanoparticles	Amphotericin B	Piperine	Nanoparticles revealed good efficacy in golden Hamster *L. donovani* model, high bioavailability, and antileishmanial activity with up to 96% inhibition of the parasite	[179]
PLGA nanoparticles	Nislodipine	Piperine	Pharmacokinetic studies showed a 4.9-fold increase in oral bioavailability and a >28.376 ± 1.32% reduction in systemic blood pressure by using nanoparticles as compared to control (nisoldipine suspension)	[180]
Chitosan nanoparticles	–	Piperine	Brain-targeted therapy in Alzheimer's disease revealed significantly improved cognitive functions as efficient as standard drug (donpezil injection)	[181]
Polyethylene glycol-polylactide-co-glycolide 39 nanoparticles	Paclitaxel	Piperine	Reduction in IC50 value of paclitaxel was obtained with NPs and piperine, and cytotoxicity was higher in breast cancer MCF-7 128 cells	[182]
Human serum albumin nanoparticles	Curcumin	Piperine	Nanoparticles demonstrated higher cytotoxicity in breast cancer MCF-7 128 cells	[183]

Formulation	Drug	Bioenhancer	Observation	Ref.
SNEDDS	Raloxifen	Piperine and quercetin	No significant bioavailability enhancement was observed	[175]
SMEDDS	Curcumin	Curcumin oleoresins	29 times higher bioavailability was obtained with bioenhancers	[177]
SMEDDS	Curcumin	Piperine	Combination showed improved anti-inflammatory activity in ulcerative colitis	[178]
SMEDDS	–	Resveratrol	Higher cell vitality and less efflux of drug in Caco2 cells	[184]
SMEDDS	–	Resveratrol	Improved solubility and drug loading	[185]
SMEDDS	–	Resveratrol	Higher solubility reported and antioxidant activity	[185]
Liposomes	Curcumin	Resveratrol	Enhanced antioxidant activity	[179]
Transferrin-liposomes		Resveratrol	Liposomes showed tumor inhibition and high survival rate	[180]
Liposomes	Curcumin	Piperine	Enhanced bioavailability	[181]
Liposomes	–	Quercetin	Reduced dose, high penetration in blood–brain barrier	[186]
Niosomes	Methotrexate	Piperine and curcumin	Improved bioavailability	[181]
Transferosomes	Capsaicin		Enhanced skin permeation	[187]
Nano-transferosomes	Naringenin	Pravastatin	Reduced side effects and improved bioavailability	[189]
Phytosomes	Curcumin	Piperine	Improved glycemic function, hepatic function, and serum cortisol	[213]
Phytosomes	Domperidone	Piperine	Increased bioavailability of domperidone	[210]
Phytosomes	Dexibuprofen	Curcumin and piperine	Better efficacy in decreasing neuropathic pain	[187]

(continued)

Table 3.2 (continued)

Nanoparticulate DDS	Drug	Herbal bioenhancers	Findings	References
Nano-ethosomes	Nano-ZnO and TiO$_2$	Naringin	Good retention in skin for sun protection	[217]
Ethosomes	Sulforaphane	Nano-curcumin	Improved therapy for skin cancer	[219]
Ethosomes	Curcumin	–	Greater anti-inflammatory activity	[188]
Transethosomes	Resveratrol	–	High permeation through skin	[189]
Transethosomes	Resveratrol	–	Enhanced permeation	[222]
Transethosomes	Quercetin	–	Improved permeation	[190]

3.4.2 Emulsion

Emulsions are drug delivery systems with macroscopic globules (1–10 μm) consisting of oil, emulsifier, and water. Types of emulsions are multiple emulsions, oil-in-water (o/w), and water-in-oil (w/o). This drug delivery system has gone through several modifications to increase its efficiency.

3.4.2.1 Nanoemulsions

Nanoemulsions are primarily comprised of surfactants, oil, water, and co-surfactants. They are one-phase systems **and thermodynamically unstable**, while kinetically stable with a globule dimension in the range of 20–200 nm. Compared to other drug delivery systems, nanoemulsions showed several benefits like high drug-carrying ability, improved solubility, capture by the lymphatic system, increased residence duration in the gastrointestinal tract resulting in higher absorption, and bioavailability [191]. Self-nano-emulsifying drug delivery systems (SNEDDS) are discrete types of **lipid-based** drug delivery system. They have the property of converting instantaneously into emulsions in an aqueous phase with the assistance of cosolvents, surfactants, and cosurfactants with moderate stirring with globule size <100 nm [192]. Hence, nanoemulsions and SNEDDS are always an attractive drug delivery system for lipophilic drugs. In a study, raloxifene BCS II drug was dispersed in SNEDDS with bioenhancers (piperine and quercetin). Pharmacokinetic studies suggested that SNEDDS showed 4 times greater oral bioavailability than drug solution. However, no additional improvement in bioavailability was obtained in the presence of bioenhancers [193].

3.4.2.2 Microemulsions

Microemulsions are precise, transparent, thermodynamically stable, optically isotropic liquid dispersions comprising oil, water, surfactant, and cosurfactants. In comparison to nanoemulsions, they are self-assembled and do not require high-energy input. A self-micro-emulsifying drug delivery system (SMEDDS) is a blend of oil, cosolvents, and surfactants that rapidly produces fine globule size (<250 nm) upon mild agitation [194]. Vijayan et al. reported SMEDDS loaded with curcumin in combination with Curcuma oleoresins as bioenhancers. The study elucidates that significant bioavailability enhancement (29 times) was obtained, compared to curcumin suspension [195]. In another interesting work, curcumin and piperine were co-encapsulated in SMEDDS against ulcerative colitis. The anti-inflammatory activity of **curcumin** and **piperine** loaded in SMEDDS showed high anti colitis activity of SMEDDS by targeting this formulation to inflammatory colon tissue [196].

3.4.3 Lipid-based vesicular carriers

3.4.3.1 Liposomes

Liposomes are unilamellar or multilamellar vesicles that consist of cholesterol combined with phospholipids, demonstrating a globule size of 20–1,000 nm. The aqueous core surrounded by phospholipid bilayer is able to encapsulate the drug within. **Liposomes** exhibit the potential to capture both hydrophilic and lipophilic drugs. Presently, several commercial liposomal formulations are available for the delivery of lipophilic molecules. Huang et al. formulated liposomes loaded with **curcumin** and **resveratrol** to enhance antioxidant activity [197]. Jhaveri et al. developed **transferrin**-conjugated liposomes encapsulating resveratrol for the treatment of glioblastoma. Liposomes revealed good antitumor activity and improved survival in mice compared to other therapies [198]. Verma et al. designed liposomes loaded with curcumin and enhanced the bioavailability with piperine. The pharmacokinetic study suggests that a combination of curcumin and piperine in liposomes demonstrated **maximum bioavailability** compared to only curcumin [199].

3.4.3.2 Niosomes

Niosomes are vesicular systems that precisely consist of amphiphilic molecules with a neutral charge and neutral surfactants. These neutral surfactants are safer and inexpensive compared to ionic surfactants [200]. They can encapsulate both lipophilic and hydrophilic molecules. Schlich et al. designed resveratrol proniosomes and suggested improved drug bioavailability [201]. In one more study, curcumin-loaded proniosomes were fabricated and studied for antiviral activity [202]. Charyulu et al. developed niosomes loaded with methotrexate and studied them with bioenhancers, piperine and curcumin. The results demonstrate that the release rate of methotrexate significantly increased with bioenhancers [203].

3.4.3.3 Transferosomes

Transferosomes are modified liposomes specifically designed to transfer drugs across transdermal tissue. They consist of one aqueous core surrounded by a lipid layer and an edge activator. The unique properties of transferosomes are that they are **self-deformable**, self-regulating, stress-responsive, and **highly adaptable** [204]. In comparison to liposomes, they are very elastic and deformable as they can squeeze themselves as integral vesicles through narrow apertures of the skin that are considerably smaller for the vesicle size. Thus, transferosomes are advanced drug delivery systems that are explored for ferrying the high and low molecular weight drugs via the transdermal route. Long et al. developed capsaicin transferosomes and proved enhanced skin permeation of these transferosomes through rat abdomen skin epidermal membrane [205]. Vinod et al. tailored **transferosomes** encapsulating piperine and showed that more drug could reach through the skin [206]. In another work, Verma et al. designed mannosylated transfersomes loaded with naringenin for skin carcinoma. Ex vivo skin permeation and deposition study revealed that the marketed formulation and drug suspension transferosomes showed higher retention and permeation [199]. Hosny et al. investigated nano-transfersomes encapsulating **pravastatin–naringenin** to diminish the hepatic side effects and enhance bioavailability. The in vivo study revealed that naringenin reduced the hepatic toxicity caused by pravastatin and improved the bioavailability of the drug [207].

3.4.3.4 Phytosomes

A novel promising formulation strategy emerged to **improve the absorption** of pharmacological active phytoconstituents known as phytosomes. Complexing active phytoconstituents produce phytosomes with phospholipids under controlled conditions [208]. These phytosomes consist of an amphipathic phospholipid, which acts as carriers of active constituents helping them in crossing the gastrointestinal membranes, ultimately reaching blood circulation. Currently, phytosomes are gaining a lot of attention from researchers and are reviewed in detail, elsewhere [209]. Islam and coworkers fabricated phytosomes containing domperidone with piperine. Pharmacokinetic studies suggest that improved bioavailability was obtained [210]. In another study, **quercetin** was complexed with phospholipid, and its bioavailability was boosted with a reduction in liver damage. In addition, numerous studies have demonstrated that the formation of curcumin-loaded phytosomes improve its efficacy and safety against various human ailments such as inflammatory diseases [211], schizophrenia [212], liver cirrhosis [213], cancer, retinopathy, diabetes, and osteoarthritis [214]. Teng et al. reported **curcumin-loaded phytosomes** were found to be chemoprotective against hepatitis B virus-induced hepatocellular carcinoma [215].

3.4.3.5 Ethosomes

In addition to other lipidic drug delivery systems, ethosomes are modified liposomes composed of phospholipids, water, and an extraordinarily high ethanol concentration (45% w/w). Several reports suggest that ethosomes are superior to liposomes because they exhibit high drug-loading capacity, are smaller in size with negative zeta potential, more skin penetration, and stable [216]. Thus, these nanocarriers have successfully been applied for transdermal drug delivery. Gollavilli et al. prepared naringin nano-ethosomal sunscreen creams. Ethosomes plus nano-ZnO and -TiO$_2$ dispersed in cream demonstrated superior skin permeation and deposition of naringin [217]. Furthermore, Li et al. constructed curcumin-loaded ethosomes to enhance transdermal delivery [218]. Kumar et al. designed ethosomal cream loaded with piperine for the management of atopic dermatitis. Ethosomal cream showed significantly higher efficacy as compared to conventional cream (tacrolimus 0.1%) and is safer. In another study, Soni et al. fabricated **ethosomal nanogel** co-loaded with **nano-curcumin** and **sulforaphane** to treat skin cancer. The developed formulation demonstrated better anticancer activity in the B16-F10 murine tumor cell line [219].

3.4.3.6 Transethosomes

Transethosomes were first designed by Song et al., and they are a new-fangled ethosomal carrier [220]. They consist of basic ingredients of traditional ethosomes with an added substance, i.e., **edge activator** or penetration enhancer. This innovative vesicular system has been designed to integrate the benefits of classical ethosomes and transferosomes named Transethosomes [221]. Several researchers reported superior features of transethosomes over ethosomes. Wu et al. prepared transethosomes loaded with resveratrol and found enhanced permeation via transethosomes [222]. Transethosomes-incorporated quercetin was developed and showed higher permeation [223].

3.4.3.7 Menthosomes

Menthosomes are the unique ultra-deformable nanocarriers that comprise phospholipids, edge activators, and menthol. Yasuko Obata et al. have developed and characterized meloxicam-loaded menthosomes using a composite experimental design. In menthosomes, cetylpyridinium chloride and menthol are used as a penetration enhancer, and cholesterol is used as a membrane stabilizer [224]. In menthosomes, cationic surfactant, viz., **cetrimide** is used as an **edge activator** with a high radius of curvature that weakens the lipidic bilayers of the menthosomes, leading to enhanced deformation of the bilayers [225]. The comparative studies of menthosomes with conventional liposomes performed by Yasuko Obata et al. showed significant improvement in skin permeability of meloxicam stabilizer [226]. Zaky and Tawfick have studied the antifungal efficacy of terbinafine hydrochloride-loaded menthosome, transferosome, and ethosome formulations under nonocclusive conditions [225]. These ultra-deformable vesicles are proven as the potential carriers for transdermal drug delivery for various drugs.

3.4.3.8 Sphingosomes

Sphingosomes are bilayered lipidic vesicles composed of sphingolipid and cholesterol [227]. The pH of the polar interior environment is less than the exterior environment. Sphingolipids are obtained from natural sources, namely soybeans, eggs,
milk, and other animal sources. Sphingosomes are advantageous over liposomes
because of their **high stability** to acid hydrolysis and improved **drug retention capacity** [228]. Sphingosomes may demonstrate an effective carrier for drug targeting
to the effector tissue because of being biofriendly, biodegradable, and resembling
cell membrane. Sphingosome acts as a vehicle for managing malignant tumors and
various immunological, infectious, vascular, rheumatoid, and inflammatory disorders [227, 229]. Different lipophilic drug moieties, namely vincristine, vinblastine,
swainsonine, or etoposide, have been incorporated into the Sphingosomes and
shown significant output [230]. The list of natural bioenhancers laden in nanoparticulate drug delivery systems is shown in Table 3.2.

3.5 Future perspectives

Herbal bioenhancers have been reported in Ayurveda as Trikatu and have, ever
since, been directly as well as indirectly used for bioavailability enhancement of
drugs. Currently, the use of rich medicinal herbs occurring in diverse ecosystems
has increased tremendously. The underlying mechanisms of **herbal bioenhancers**
with their clinical outcomes can also be explored and researched in future. The
filed patents and ongoing clinical trials have proven the importance of bioenhancers in **bioavailability enhancement**. Novel herbal bioenhancers could be
identified from the endangered forest and plants across the globe. Several phytoconstituents are being explored and investigated for effective and improved drug
delivery, with enhanced bioavailability with minimum effective dose. Many pharmaceutical and food industries are working on applications of bioenhancers. Future
research can also be focused on marine herbal bioenhancers isolated from sea
plants and weeds and in exploring their novel principles and mechanisms. In the
near future, **combinational bioenhancers,** along with their **bioenhancing ability,**
mechanism, and toxicity can be explored. In the coming years, attention can be
given to plants such as *Bacopa monnierie, Rauwolfia serpentine, Catharanths roseus,
Taxus baccata,* and *Artemisia anuua* for their bioenhancing ability.

Conflict of interest

Authors have no conflict of interest.

List of abbreviations

AUC	Area under the curve
P-gp	P-glycoprotein
BCRP	Breast cancer resistance protein
SLNs	Solid lipid nanoparticles
NLCs	Nanostructured lipid carriers
SMEDDS	Self-micro-emulsifying drug delivery system

References

[1] Dudhatra GB, Mody SK, Awale MM, Patel HB, Modi CM, Kumar A, Kamani DR, Chauhan BN. A comprehensive review on pharmacotherapeutics of herbal bioenhancers, The Scientific World Journal, 2012, 2012.

[2] Singh R, Devi S, Patel J, Patel U, Bhavsar S, Thaker A. Indian herbal bioenhancers: A review, Pharmacognosy Reviews, 2009, 3, 90.

[3] Jhanwar B, Gupta S. Biopotentiation using herbs: Novel technique for poor bioavailable drugs, International Journal of Pharmtech Research, 2014, 6, 443–454.

[4] Kesarwani K, Gupta R. Bioavailability enhancers of herbal origin: An overview, Asian Pacific Journal of Tropical Biomedicine, 2013, 3, 253–266.

[5] Brahmankar D, Jaiswal SB. Biopharmaceutics and Pharmacokinetics-A Treatise, Vallabh Prakashan, New Delhi, 1995.

[6] Drabu S, Khatri S, Babu S, Lohani P. Use of herbal bioenhancers to increase the bioavailability of drugs, Research Journal of Pharmaceutical, Biological and Chemical Sciences, 2011, 2, 107–119.

[7] Patil UK, Singh A, Chakraborty AK. Role of piperine as a bioavailability enhancer, International Journal of Recent Advances in Pharmaceutical Research, 2011, 4, 16–23.

[8] Atal N, Bedi K. Bioenhancers: Revolutionary concept to market, Journal of Ayurveda and Integrative Medicine, 2010, 1, 96.

[9] Murugaiyah V, Mattson MP. Neurohormetic phytochemicals: An evolutionary–bioenergetic perspective, Neurochemistry International, 2015, 89, 271–280.

[10] Shoba G, Joy D, Joseph T, Majeed M, Rajendran R, Srinivas P. Influence of piperine on the pharmacokinetics of curcumin in animals and human volunteers, Planta Medica, 1998, 64, 353–356.

[11] Alexander A, Qureshi A, Kumari L, Vaishnav P, Sharma M, Saraf S, Saraf S. Role of herbal bioactives as a potential bioavailability enhancer for active pharmaceutical ingredients, Fitoterapia, 2014, 97, 1–14.

[12] Liu M-Z, Zhang Y-L, Zeng M-Z, He F-Z, Luo Z-Y, Luo J-Q, Wen J-G, Chen X-P, Zhou -H-H, Zhang W. Pharmacogenomics and herb-drug interactions: Merge of future and tradition, Evidence-Based Complementary and Alternative Medicine, 2015, 2015, 321091.

[13] Teo YL, Ho HK, Chan A. Metabolism-related pharmacokinetic drug–drug interactions with tyrosine kinase inhibitors: Current understanding, challenges and recommendations, British Journal of Clinical Pharmacology, 2015, 79, 241–253.

[14] Sanjay K, Rajiv S, Abhijit R. Modulation of cytochrome-P450 Inhibition (CYP) in drug discovery: A medicinal chemistry perspective, Current Medicinal Chemistry, 2012, 19, 3605–3621.

[15] Mhaske D, Sreedharan S, Mahadik K. Role of piperine as an effective bioenhancer in drug absorption, Pharmaceutica Analytica Acta, 2018, 9, 591.

[16] Lee SH, Kim HY, Back SY, Han H-K. Piperine-mediated drug interactions and formulation strategy for piperine: Recent advances and future perspectives, Expert Opinion on Drug Metabolism & Toxicology, 2018, 14, 43–57.

[17] Singh M, Varshneya C, Telang R, Srivastava A. Alteration of pharmacokinetics of oxytetracycline following oral administration of Piper longum in hens, Journal of Veterinary Science, 2005, 6.

[18] Grimstein M, Huang S-M. A regulatory science viewpoint on botanical–drug interactions, Journal of Food and Drug Analysis, 2018, 26, S12–S25.

[19] Fukunaga K, Orito K. Time-course effects of St John's wort on the pharmacokinetics of cyclosporine in dogs, Journal of Veterinary Pharmacology and Therapeutics, 2012, 35, 446–451.

[20] Araujo JA, Landsberg GM, Milgram NW, Miolo A. Improvement of short-term memory performance in aged beagles by a nutraceutical supplement containing phosphatidylserine, Ginkgo biloba, vitamin E, and pyridoxine, The Canadian Veterinary Journal, 2008, 49, 379.

[21] Yin OQP, Tomlinson B, Waye MMY, Chow AHL, Chow MSS. Pharmacogenetics and herb–drug interactions: Experience with: Ginkgo biloba: And omeprazole, Pharmacogenetics and Genomics, 2004, 14.

[22] Hanley MJ, Cancalon P, Widmer WW, Greenblatt DJ. The effect of grapefruit juice on drug disposition, Expert Opinion on Drug Metabolism & Toxicology, 2011, 7, 267–286.

[23] Galanakis CM. Nutraceutical and Functional Food Components: Effects of Innovative Processing Techniques, Academic Press, 2016.

[24] Hetal T, Bindesh P, Sneha T. A review on techniques for oral bioavailability enhancement of drugs, International Journal of Pharmaceutical Sciences Review and Research, 2010, 4, 203–223.

[25] Bergman Å, Heindel JJ, Jobling S, Kidd KA, Zoeller RT, State of the Science of Endocrine Disrupting Chemicals 2012 Summary for Decision-Makers, 2013.

[26] El-Gazayerly ON, Makhlouf AIA, Soelm AMA, Mohmoud MA. Antioxidant and hepatoprotective effects of silymarin phytosomes compared to milk thistle extract in CCl4 induced hepatotoxicity in rats, Journal of Microencapsulation, 2014, 31, 23–30.

[27] Ju Ho P, Jun Sung J, Cheon KK, Jin Tae H. Anti-inflammatory effect of Centella asiatica phytosome in a mouse model of phthalic anhydride-induced atopic dermatitis, Phytomedicine, 2018, 43, 110–119.

[28] Amin ML. P-glycoprotein inhibition for optimal drug delivery, Drug Target Insights, 2013, 7, DTI. S12519.

[29] Mealey KL, Fidel J. P-glycoprotein mediated drug interactions in animals and humans with cancer, Journal of Veterinary Internal Medicine, 2015, 29, 1–6.

[30] Holm R, Müllertz A, Mu H. Bile salts and their importance for drug absorption, International Journal of Pharmaceutics, 2013, 453, 44–55.

[31] Peterson B, Weyers M, Steenekamp JH, Steyn JD, Gouws C, Hamman JH. Drug bioavailability enhancing agents of natural origin (Bioenhancers) that modulate drug membrane permeation and pre-systemic metabolism, Pharmaceutics, 2019, 11, 33.

[32] Portero A, Remuñán-López C, Nielsen HM. The potential of chitosan in enhancing peptide and protein absorption across the TR146 cell culture model – an in vitro model of the buccal epithelium, Pharmaceutical Research, 2002, 19, 169–174.

[33] Ojewole E, Mackraj I, Akhundov K, Hamman J, Viljoen A, Olivier E, Wesley-Smith J, Govender T. Investigating the effect of aloe vera gel on the buccal permeability of didanosine, Planta Medica, 2012, 78, 354–361.

[34] Shojaei AH, Khan M, Lim G, Khosravan R. Transbuccal permeation of a nucleoside analog, dideoxycytidine: Effects of menthol as a permeation enhancer, International Journal of Pharmaceutics, 1999, 192, 139–146.

[35] Morishita M, Barichello JM, Takayama K, Chiba Y, Tokiwa S, Nagai T. Pluronic® F-127 gels incorporating highly purified unsaturated fatty acids for buccal delivery of insulin, International Journal of Pharmaceutics, 2001, 212, 289–293.

[36] Hinchcliffe M, Jabbal-Gill I, Smith A. Effect of chitosan on the intranasal absorption of salmon calcitonin in sheep, Journal of Pharmacy and Pharmacology, 2005, 57, 681–687.

[37] Illum L, Watts P, Fisher AN, Hinchcliffe M, Norbury H, Jabbal-Gill I, Nankervis R, Davis SS. Intranasal delivery of morphine, The Journal of Pharmacology and Experimental Therapeutics, 2002, 301, 391–400.

[38] Krauland AH, Guggi D, Bernkop-Schnürch A. Thiolated chitosan microparticles: A vehicle for nasal peptide drug delivery, International Journal of Pharmaceutics, 2006, 307, 270–277.

[39] Mizuma T, Awazu S. Dietary polyphenols (–)-epicatechin and chrysin inhibit intestinal glucuronidation metabolism to increase drug absorption, Journal of Pharmaceutical Sciences, 2004, 93, 2407–2410.

[40] Choudhary N, Khajuria V, Gillani Z, Tandon V, Arora E. Effect of *Carum carvi*, a herbal bioenhancer on pharmacokinetics of antitubercular drugs: A study in healthy human volunteers, Perspectives in Clinical Research, 2014, 5, 80–84.

[41] Pavithra BH, Prakash N, Jayakumar K. Modification of pharmacokinetics of norfloxacin following oral administration of curcumin in rabbits, Journal of Veterinary Science, 2009, 10, 293–297.

[42] Ghosal S. Delivery system for pharmaceutical, nutritional and cosmetic ingredients, in, Google Patents, 2003.

[43] Li X, Choi J-S. Effect of genistein on the pharmacokinetics of paclitaxel administered orally or intravenously in rats, International Journal of Pharmaceutics, 2007, 337, 188–193.

[44] Petyaev IM. State of the art paper
Improvement of hepatic bioavailability as a new step for the future of statin, Archives of Medical Science, 2015, 11, 406–410.

[45] Patil S, Dash RP, Anandjiwala S, Nivsarkar M. Simultaneous quantification of berberine and lysergol by HPLC-UV: Evidence that lysergol enhances the oral bioavailability of berberine in rats, Biomedical Chromatography, 2012, 26, 1170–1175.

[46] Choi J-S, Kang KW. Enhanced tamoxifen bioavailability after oral administration of tamoxifen in rats pretreated with naringin, Archives of Pharmacal Research, 2008, 31, 1631–1636.

[47] Wacher VJ, Wong S, Wong HT. Peppermint oil enhances cyclosporine oral bioavailability in rats: Comparison with d-α-Tocopheryl Poly(ethylene glycol 1000) Succinate (TPGS) and Ketoconazole, Journal of Pharmaceutical Sciences, 2002, 91, 77–90.

[48] Sama V, Nadipelli M, Yenumula P, Bommineni MR, Mullangi R. Effect of piperine on antihyperglycemic activity and pharmacokinetic profile of nateglinide, Arzneimittelforschung, 2012, 62, 384–388.

[49] Babu PR, Babu KN, Peter PLH, Rajesh K, Babu PJ. Influence of quercetin on the pharmacokinetics of ranolazine in rats and in vitro models, Drug Development and Industrial Pharmacy, 2013, 39, 873–879.

[50] Kim K-A, Park P-W, Park J-Y. Short-term effect of quercetin on the pharmacokinetics of fexofenadine, a substrate of P-glycoprotein, in healthy volunteers, European Journal of Clinical Pharmacology, 2009, 65, 609–614.

[51] Bedada SK, Yellu NR, Neerati P. Effect of resveratrol treatment on the pharmacokinetics of diclofenac in healthy human volunteers, Phytotherapy Research, 2016, 30, 397–401.

[52] Machida M, Hayashi M, Awazu S. The effects of absorption enhancers on the pulmonary absorption of recombinant human granulocyte colony-stimulating factor (rhG-CSF) in Rats, Biological & Pharmaceutical Bulletin, 2000, 23, 84–86.

[53] Todo H, Okamoto H, Iida K, Danjo K. Effect of additives on insulin absorption from intratracheally administered dry powders in rats, International Journal of Pharmaceutics, 2001, 220, 101–110.

[54] Salem LB, Bosquillon C, Dailey LA, Delattre L, Martin GP, Evrard B, Forbes B. Sparing methylation of β-cyclodextrin mitigates cytotoxicity and permeability induction in respiratory epithelial cell layers in vitro, Journal of Controlled Release, 2009, 136, 110–116.

[55] Morimoto K, Uehara Y, Iwanaga K, Kakemi M. Effects of sodium glycocholate and protease inhibitors on permeability of TRH and insulin across rabbit trachea, Pharmaceutica Acta Helvetiae, 2000, 74, 411–415.

[56] Johansson F, Hjertberg E, Eirefelt S, Tronde A, Hultkvist Bengtsson U. Mechanisms for absorption enhancement of inhaled insulin by sodium taurocholate, European Journal of Pharmaceutical Sciences, 2002, 17, 63–71.

[57] Yang T, Mustafa F, Bai S, Ahsan F. Pulmonary delivery of low molecular weight heparins, Pharmaceutical Research, 2004, 21, 2009–2016.

[58] Badmaev V, Majeed M, Norkus EP. Piperine, an alkaloid derived from black pepper increases serum response of beta-carotene during 14-days of oral beta-carotene supplementation, Nutrition Research, 1999, 19, 381–388.

[59] Kumar S, Singhal V, Roshan R, Sharma A, Rembhotkar GW, Ghosh B. Piperine inhibits TNF-α induced adhesion of neutrophils to endothelial monolayer through suppression of NF-κB and IκB kinase activation, European Journal of Pharmacology, 2007, 575, 177–186.

[60] Gülçin I. The antioxidant and radical scavenging activities of black pepper (Piper nigrum) seeds, International Journal of Food Sciences and Nutrition, 2005, 56, 491–499.

[61] Sunila E, Kuttan G. Immunomodulatory and antitumor activity of Piper longum Linn. and piperine, Journal of Ethnopharmacology, 2004, 90, 339–346.

[62] Parmar VS, Jain SC, Bisht KS, Jain R, Taneja P, Jha A, Tyagi OD, Prasad AK, Wengel J, Olsen CE. Phytochemistry of the genus Piper, Phytochemistry, 1997, 46, 597–673.

[63] Vijayakumar RS, Nalini N. Piperine, an active principle from Piper nigrum, modulates hormonal and apolipoprotein profiles in hyperlipidemic rats, Journal of Basic and Clinical Physiology and Pharmacology, 2006, 17, 71–86.

[64] Bajad S, Bedi K, Singla A, Johri R. Antidiarrhoeal activity of piperine in mice, Planta Medica, 2001, 67, 284–287.

[65] Wongpa S, Himakoun L, Soontornchai S, Temcharoen P. Antimutagenic effects of piperine on cyclophosphamide-induced chromosome aberrations in rat bone marrow cells, Asian Pacific Journal of Cancer Prevention, 2007, 8, 623–627.

[66] Wattanathorn J, Chonpathompikunlert P, Muchimapura S, Priprem A, Tankamnerdthai O. Piperine, the potential functional food for mood and cognitive disorders, Food and Chemical Toxicology, 2008, 46, 3106–3110.

[67] Taqvi SIH, Shah AJ, Gilani AH. Blood pressure lowering and vasomodulator effects of piperine, Journal of cardiovascular pharmacology, 2008, 52, 452–458.

[68] Matsuda H, Ninomiya K, Morikawa T, Yasuda D, Yamaguchi I, Yoshikawa M. Protective effects of amide constituents from the fruit of Piper chaba on d-galactosamine/TNF-α-induced cell death in mouse hepatocytes, Bioorganic & Medicinal Chemistry Letters, 2008, 18, 2038–2042.

[69] Navickiene HMD, Alécio AC, Kato MJ, Bolzani VDS, Young MCM, Cavalheiro AJ, Furlan M. Antifungal amides from Piper hispidum and Piper tuberculatum, Phytochemistry, 2000, 55, 621–626.

[70] Atal C, Zutshi U, Rao P. Scientific evidence on the role of Ayurvedic herbals on bioavailability of drugs, Journal of Ethnopharmacology, 1981, 4, 229–232.

[71] Saraf S. Applications of novel drug delivery system for herbal formulations, Fitoterapia, 2010, 81, 680–689.

[72] Khan J, Alexander A, Saraf S, Saraf S. Recent advances and future prospects of phyto-phospholipid complexation technique for improving pharmacokinetic profile of plant actives, Journal of Controlled Release, 2013, 168, 50–60.

[73] Sehgal A, Kumar M, Jain M, Dhawan D. Combined effects of curcumin and piperine in ameliorating benzo (a) pyrene induced DNA damage, Food and Chemical Toxicology, 2011, 49, 3002–3006.

[74] Janakiraman K, Manavalan R. Compatibility and stability studies of ampicillin trihydrate and piperine mixture, International Journal of Pharmaceutical Sciences and Research, 2011, 2, 1176.

[75] Jolad SD, Lantz RC, Solyom AM, Chen GJ, Bates RB, Timmermann BN. Fresh organically grown ginger (Zingiber officinale): Composition and effects on LPS-induced PGE2 production, Phytochemistry, 2004, 65, 1937–1954.

[76] Evans W. Trease and Evans Pharmacognosy, WB Sauders Company Ltd, London, 2002.

[77] O'Hara M, Kiefer D, Farrell K, Kemper K. A review of 12 commonly used medicinal herbs, Archives of Family Medicine, 1998, 7, 523.

[78] Oyagbemi AA, Saba AB, Azeez OI. Molecular targets of [6]-gingerol: Its potential roles in cancer chemoprevention, Biofactors, 2010, 36, 169–178.

[79] Ficker C, Smith M, Akpagana K, Gbeassor M, Zhang J, Durst T, Assabgui R, Arnason J. Bioassay-guided isolation and identification of antifungal compounds from ginger, Phytotherapy Research: An International Journal Devoted to Pharmacological and Toxicological Evaluation of Natural Product Derivatives, 2003, 17, 897–902.

[80] Jagetia GC, Baliga MS, Venkatesh P, Ulloor JN. Influence of ginger rhizome (Zingiber officinale Rosc) on survival, glutathione and lipid peroxidation in mice after whole-body exposure to gamma radiation, Radiation Research, 2003, 160, 584–592.

[81] Johji Y, Michihiko M, Rong HQ, Hisashi M, Hajime F. The anti-ulcer effect in rats of ginger constituents, Journal of Ethnopharmacology, 1988, 23, 299–304.

[82] Al-Amin ZM, Thomson M, Al-Qattan KK, Peltonen-Shalaby R, Ali M. Anti-diabetic and hypolipidaemic properties of ginger (Zingiber officinale) in streptozotocin-induced diabetic rats, British Journal of Nutrition, 2006, 96, 660–666.

[83] Thomson M, Al-Qattan K, Al-Sawan S, Alnaqeeb M, Khan I, Ali M. The use of ginger (Zingiber officinale Rosc.) as a potential anti-inflammatory and antithrombotic agent, Prostaglandins, Leukotrienes, and Essential Fatty Acids, 2002, 67, 475–478.

[84] Ojewole JA. Analgesic, anti-inflammatory and hypoglycaemic effects of ethanol extract of Zingiber officinale (Roscoe) rhizomes (Zingiberaceae) in mice and rats, Phytotherapy Research: An International Journal Devoted to Pharmacological and Toxicological Evaluation of Natural Product Derivatives, 2006, 20, 764–772.

[85] Chaiyakunapruk N, Kitikannakorn N, Nathisuwan S, Leeprakobboon K, Leelasettagool C. The efficacy of ginger for the prevention of postoperative nausea and vomiting: A meta-analysis, American Journal of Obstetrics and Gynecology, 2006, 194, 95–99.

[86] Shukla Y, Singh M. Cancer preventive properties of ginger: A brief review, Food and Chemical Toxicology, 2007, 45, 683–690.

[87] Qazi G, Bedi K, Johri R, Tikoo M, Tikoo A, Sharma S, Abdullah S, Suri O, Gupta B, Suri K, Bioavailability enhancing activity of Zingiber officinale Linn and its extracts/fractions thereof, in, Google Patents, 2003.

[88]   Ahshawat M, Saraf S, Saraf S. Preparation and characterization of herbal creams for improvement of skin viscoelastic properties, International Journal of Cosmetic Science, 2008, 30, 183–193.

[89]   Zhang W, Lim L-Y. Effects of spice constituents on P-glycoprotein-mediated transport and CYP3A4-mediated metabolism in vitro, Drug Metabolism and Disposition, 2008, 36, 1283–1290.

[90]   Zhang W, Tan TMC, Lim L-Y. Impact of curcumin-induced changes in P-glycoprotein and CYP3A expression on the pharmacokinetics of peroral celiprolol and midazolam in rats, Drug Metabolism and Disposition, 2007, 35, 110–115.

[91]   Yan Y-D, Kim DH, Sung JH, Yong CS, Choi HG. Enhanced oral bioavailability of docetaxel in rats by four consecutive days of pre-treatment with curcumin, International Journal of Pharmaceutics, 2010, 399, 116–120.

[92]   Pavithra B, Prakash N, Jayakumar K. Modification of pharmacokinetics of norfloxacin following oral administration of curcumin in rabbits, Journal of Veterinary Science, 2009, 10, 293.

[93]   Ali B, Amin S, Ahmad J, Ali A, Ali M, Mir SR. Bioavailability enhancement studies of amoxicillin with Nigella, The Indian Journal of Medical Research, 2012, 135, 555.

[94]   Sinha VR, Kaur MP. Permeation enhancers for transdermal drug delivery, Drug Development and Industrial Pharmacy, 2000, 26, 1131–1140.

[95]   Al-Jenoobi F, Al-Suwayeh S, Muzaffar I, Alam MA, Al-Kharfy KM, Korashy HM, Al-Mohizea AM, Ahad A, Raish M. Effects of Nigella sativa and Lepidium sativum on cyclosporine pharmacokinetics, BioMed Research International, 2013, 2013.

[96]   Iacobellis NS, Lo Cantore P, Capasso F, Senatore F. Antibacterial activity of Cuminum cyminum L. and Carum carvi L. essential oils, Journal of Agricultural and Food Chemistry, 2005, 53, 57–61.

[97]   Malini T, Vanithakumari G. Estrogenic activity of Cuminum cyminum in rats, Indian Journal of Experimental Biology, 1987, 25, 442–444.

[98]   Gagandeep SD, Mendiz E, Rao AR, Kale RK. Chemopreventive effects of Cuminum cyminum in chemically induced forestomach and uterine cervix tumors in murine model systems, Nutrition and Cancer, 2003, 47, 171–180.

[99]   Dhandapani S, Subramanian VR, Rajagopal S, Namasivayam N. Hypolipidemic effect of Cuminum cyminum L. on alloxan-induced diabetic rats, Pharmacological Research, 2002, 46, 251–255.

[100] Gachkar L, Yadegari D, Rezaei MB, Taghizadeh M, Astaneh SA, Rasooli I. Chemical and biological characteristics of Cuminum cyminum and Rosmarinus officinalis essential oils, Food Chemistry, 2007, 102, 898–904.

[101] El-Ghorab AH, Nauman M, Anjum FM, Hussain S, Nadeem M. A comparative study on chemical composition and antioxidant activity of ginger (Zingiber officinale) and cumin (Cuminum cyminum), Journal of Agricultural and Food Chemistry, 2010, 58, 8231–8237.

[102] Pai MB, Prashant G, Murlikrishna K, Shivakumar K, Chandu G. Antifungal efficacy of Punica granatum, Acacia nilotica, Cuminum cyminum and Foeniculum vulgare on Candida albicans: An in vitro study, Indian Journal of Dental Research, 2010, 21, 334.

[103] Boumendjel A, Di Pietro A, Dumontet C, Barron D. Recent advances in the discovery of flavonoids and analogs with high-affinity binding to P-glycoprotein responsible for cancer cell multidrug resistance, Medicinal Research Reviews, 2002, 22, 512–529.

[104] Qazi GN, Bedi KL, Johri RK, Tikoo MK, Tikoo AK, Sharma SC, Absullah ST, Suri OP, Gupta BD, Suri KA, Bioavailability/bioefficacy enhancing activity of Cuminum cyminum and extracts and fractions thereof, in, Google Patents, 2006.

[105] Lahlou S, Tahraoui A, Israili Z, Lyoussi B. Diuretic activity of the aqueous extracts of Carum carvi and Tanacetum vulgare in normal rats, Journal of Ethnopharmacology, 2007, 110, 458–463.

[106] Eddouks M, Lemhadri A, Michel J-B. Caraway and caper: Potential anti-hyperglycaemic plants in diabetic rats, Journal of Ethnopharmacology, 2004, 94, 143–148.

[107] De Martino L, De Feo V, Fratianni F, Nazzaro F. Chemistry, antioxidant, antibacterial and antifungal activities of volatile oils and their components, Natural Product Communications, 2009, 4, 1934578X0900401226.

[108] Najda A, Dyduch J, Brzozowski N. Flavonoid content and antioxidant activity of caraway roots (Carum carvi L.), Vegetable Crops Research Bulletin, 2008, 68, 127.

[109] Khayyal MT, El-Ghazaly MA, Kenawy SA, Seif-El-Nasr M, Mahran LG, Kafafi YA, Okpanyi SN. Antiulcerogenic effect of some gastrointestinally acting plant extracts and their combination, Arzneimittelforschung, 2001, 51, 545–553.

[110] Shukla M, Malik M, Jaiswal S, Sharma A, Tanpula D, Goyani R, Lal J. A mechanistic investigation of the bioavailability enhancing potential of lysergol, a novel bioenhancer, using curcumin, RSC Advances, 2016, 6, 58933–58942.

[111] Patil S, Dash RP, Anandjiwala S, Nivsarkar M. Simultaneous quantification of berberine and lysergol by HPLC-UV: Evidence that lysergol enhances the oral bioavailability of berberine in rats, Biomedical Chromatography, 2012, 26, 1170–1175.

[112] Makheja A, Bailey J. Antiplatelet constituents of garlic and onion, Agents and Actions, 1990, 29, 360–363.

[113] Anthony J-P, Fyfe L, Smith H. Plant active components–a resource for antiparasitic agents?, Trends in Parasitology, 2005, 21, 462–468.

[114] Cai Y, Wang R, Pei F, Liang -B-B. Antibacterial activity of allicin alone and in combination with β-lactams against Staphylococcus spp. and Pseudomonas aeruginosa, The Journal of Antibiotics, 2007, 60, 335–338.

[115] Chung LY. The antioxidant properties of garlic compounds: Allyl cysteine, alliin, allicin, and allyl disulfide, Journal of Medicinal Food, 2006, 9, 205–213.

[116] Borjihan H, Ogita A, Fujita KI, Hirasawa E, Tanaka T. The vacuole-targeting fungicidal activity of amphotericin B against the pathogenic fungus Candida albicans and its enhancement by allicin, The Journal of Antibiotics, 2009, 62, 691–697.

[117] Hirsch K, Danilenko M, Giat J, Miron T, Rabinkov A, Wilchek M, Mirelman D, Levy J, Sharoni Y. Effect of purified allicin, the major ingredient of freshly crushed garlic, on cancer cell proliferation, Nutrition and Cancer, 2000, 38, 245–254.

[118] Kang N, Moon E, Cho C, Pyo S. Immunomodulating effect of garlic component, allicin, on murine peritoneal macrophages, Nutrition Research, 2001, 21, 617–626.

[119] Eidi A, Eidi M, Esmaeili E. Antidiabetic effect of garlic (Allium sativum L.) in normal and streptozotocin-induced diabetic rats, Phytomedicine, 2006, 13, 624–629.

[120] Wilson EA, Demmig-Adams B. Antioxidant, anti-inflammatory, and antimicrobial properties of garlic and onions, Nutrition & Food Science, 2007.

[121] Naithani R, Huma LC, Holland LE, Shukla D, McCormick DL, Mehta RG, Moriarty RM. Antiviral activity of phytochemicals: A comprehensive review, Mini Reviews in Medicinal Chemistry, 2008, 8, 1106–1133.

[122] Ogita A, Hirooka K, Yamamoto Y, Tsutsui N, Fujita K-I, Taniguchi M, Tanaka T. Synergistic fungicidal activity of Cu2+ and allicin, an allyl sulfur compound from garlic, and its relation to the role of alkyl hydroperoxide reductase 1 as a cell surface defense in Saccharomyces cerevisiae, Toxicology, 2005, 215, 205–213.

[123] Ogita A, Fujita K-I, Taniguchi M, Tanaka T. Enhancement of the fungicidal activity of amphotericin B by allicin, an allyl-sulfur compound from garlic, against the yeast Saccharomyces cerevisiae as a model system, Planta Medica, 2006, 72, 1247–1250.

[124] Ogita A, Fujita K-I, Tanaka T. Enhancement of the fungicidal activity of amphotericin B by allicin: Effects on intracellular ergosterol trafficking, Planta Medica, 2009, 75, 222–226.

[125] Fujisawa Y, Sakamoto M, Matsushita M, Fujita T, Nishioka K. Glycyrrhizin inhibits the lytic pathway of complement – possible mechanism of its anti-inflammatory effect on liver cells in viral hepatitis, Microbiology and Immunology, 2000, 44, 799–804.

[126] Shibata S, Antitumor-promoting and anti-inflammatory activities of licorice principles and their modified compounds, in, ACS Publications, 1994.

[127] Nose M, Ito M, Kamimura K, Shimizu M, Ogihara Y. A comparison of the antihepatotoxic activity between glycyrrhizin and glycyrrhetinic acid, Planta Medica, 1994, 60, 136–139.

[128] Crance JM, Scaramozzino N, Jouan A, Garin D. Interferon, ribavirin, 6-azauridine and glycyrrhizin: Antiviral compounds active against pathogenic flaviviruses, Antiviral Research, 2003, 58, 73–79.

[129] Chen L, Yang J, Davey A, Chen Y-X, Wang J-P, Liu X-Q. Effects of diammonium glycyrrhizinate on the pharmacokinetics of aconitine in rats and the potential mechanism, Xenobiotica, 2009, 39, 955–963.

[130] Imai T, Sakai M, Ohtake H, Azuma H, Otagiri M. Absorption-enhancing effect of glycyrrhizin induced in the presence of capric acid, International Journal of Pharmaceutics, 2005, 294, 11–21.

[131] Khanuja S, Arya J, Srivastava S, Shasany A, Kumar TS, Darokar M, Kumar S, Antibiotic pharmaceutical composition with lysergol as bio-enhancer and method of treatment, in, Google Patents, 2007.

[132] Maan AA, Nazir A, Khan MKI, Ahmad T, Zia R, Murid M, Abrar M. The therapeutic properties and applications of Aloe vera: A review, Journal of Herbal Medicine, 2018, 12, 1–10.

[133] Vinson JA, Al Kharrat H, Andreoli L. Effect of Aloe vera preparations on the human bioavailability of vitamins C and E, Phytomedicine, 2005, 12, 760–765.

[134] Naveen P, Padma J, Vasudha B, Gouda T. Herb-drug interaction between ethanolic extract of aloe vera with glipizide in streptozotacin induced diabetic rats, Indo American Journal of Pharmaceutical Research, 2016, 6, 4265–4269.

[135] Panahi Y, Aslani J, Hajihashemi A, Kalkhorani M, Ghanei M, Sahebkar A, Effect of aloe vera and pantoprazole on gastroesophageal reflux symptoms in mustard gas victims: a randomized controlled trial, 2016.

[136] Yang M-S, Yu C-P, Huang C-Y, Chao P-DL, Lin S-P, Hou Y-C. Aloe activated P-glycoprotein and CYP 3A: A study on the serum kinetics of aloe and its interaction with cyclosporine in rats, Food & Function, 2017, 8, 315–322.

[137] Shanker K, Gupta MM, Srivastava SK, Bawankule DU, Pal A, Khanuja SP. Determination of bioactive nitrile glycoside (s) in drumstick (Moringa oleifera) by reverse phase HPLC, Food Chemistry, 2007, 105, 376–382.

[138] Caceres A, Cabrera O, Morales O, Mollinedo P, Mendia P. Pharmacological properties of Moringa oleifera. 1: Preliminary screening for antimicrobial activity, Journal of Ethnopharmacology, 1991, 33, 213–216.

[139] Shukla S, Mathur R, Prakash AO. Antifertility profile of the aqueous extract of Moringa oleifera roots, Journal of Ethnopharmacology, 1988, 22, 51–62.

[140] Pari L, Kumar NA. Hepatoprotective activity of Moringa oleifera on antitubercular drug-induced liver damage in rats, Journal of Medicinal Food, 2002, 5, 171–177.

[141] Saravillo K, Herrera A, Biological activity of Moringa oleifera Lam. (Malunggay) crude seed extract, Philippine Agricultural Scientist (Philippines), 2004.

[142] Cáceres A, Saravia A, Rizzo S, Zabala L, De Leon E, Nave F. Pharmacologie properties of Moringa oleifera. 2: Screening for antispasmodic, antiinflammatory and diuretic activity, Journal of Ethnopharmacology, 1992, 36, 233–237.

[143] Aruna K, Sivaramakrishnan V. Anticarcinogenic effects of some Indian plant products, Food and Chemical Toxicology, 1992, 30, 953–956.

[144] Neveda O, Asna U, Preetham Paul P, Narayan Prasad N. Effect of dietary lipids and drumstick leaves (Moringa oleifera) on lipid profile & antioxidant parameters in rats, Food and Nutrition Sciences, 2012, 2012.

[145] Mehta K, Balaraman R, Amin A, Bafna P, Gulati O. Effect of fruits of Moringa oleifera on the lipid profile of normal and hypercholesterolaemic rabbits, Journal of Ethnopharmacology, 2003, 86, 191–195.

[146] Guevara A, Vargas C, Uy M, Anti-inflammatory and antitumor activities of seed extracts of malunggay, Moringa oleifera L. (Moringaceae), Philippine Journal of Science (Philippines), 1996.

[147] Gilani AH, Aftab K, Suria A, Siddiqui S, Salem R, Siddiqui BS, Faizi S. Pharmacological studies on hypotensive and spasmolytic activities of pure compounds from Moringa oleifera, Phytotherapy Research, 1994, 8, 87–91.

[148] Nwosu MO, Okafor JI. Preliminary studies of the antifungal activities of some medicinal plants against Basidiobolus and some other pathogenic fungi: Vorläufige Studien zur antimyzetischen Aktivität einiger offizineller Pflanzen auf Basidiobolus und andere pathogene Pilze, Mycoses, 1995, 38, 191–195.

[149] Pal SK, Mukherjee PK, Saha B. Studies on the antiulcer activity of Moringa oleifera leaf extract on gastric ulcer models in rats, Phytotherapy Research, 1995, 9, 463–465.

[150] Mahajan SG, Mali RG, Mehta AA. Protective effect of ethanolic extract of seeds of Moringa oleifera Lam. against inflammation associated with development of arthritis in rats, Journal of Immunotoxicology, 2007, 4, 39–47.

[151] Khanuja SPS, Arya JS, Tiruppadiripuliyur RSK, Saikia D, Kaur H, Singh M, Gupta SC, Shasany AK, Darokar MP, Srivastava SK, Nitrile glycoside useful as a bioenhancer of drugs and nutrients, process of its isolation from Moringa oleifera, in, Google Patents, 2005.

[152] Kurzer MS, Xu X. Dietary phytoestrogens, Annual Review of Nutrition, 1997, 17, 353–381.

[153] Huisman MT, Chhatta AA, van Tellingen O, Beijnen JH, Schinkel AH. MRP2 (ABCC2) transports taxanes and confers paclitaxel resistance and both processes are stimulated by probenecid, International Journal of Cancer, 2005, 116, 824–829.

[154] Sparreboom A, Van Asperen J, Mayer U, Schinkel AH, Smit JW, Meijer DK, Borst P, Nooijen WJ, Beijnen JH, Van Tellingen O. Limited oral bioavailability and active epithelial excretion of paclitaxel (Taxol) caused by P-glycoprotein in the intestine, Proceedings of the National Academy of Sciences, 1997, 94, 2031–2035.

[155] Doyle LA, Ross DD. Multidrug resistance mediated by the breast cancer resistance protein BCRP (ABCG2), Oncogene, 2003, 22, 7340–7358.

[156] Lambert JD, Kwon S-J, Ju J, Bose M, Lee M-J, Hong J, Hao X, Yang CS. Effect of genistein on the bioavailability and intestinal cancer chemopreventive activity of (-)-epigallocatechin-3-gallate, Carcinogenesis, 2008, 29, 2019–2024.

[157] Dixon RA, Steele CL. Flavonoids and isoflavonoids–a gold mine for metabolic engineering, Trends in Plant Science, 1999, 4, 394–400.

[158] Nijveldt RJ, Van Nood E, Van Hoorn DE, Boelens PG, Van Norren K, Van Leeuwen PA. Flavonoids: A review of probable mechanisms of action and potential applications, The American Journal of Clinical Nutrition, 2001, 74, 418–425.

[159] Zhang H, Wong C, Coville P, Wanwimolruk S. Effect of the grapefruit flavonoid naringin on pharmacokinetics of quinine in rats, Drug Metabolism and Drug Interactions, 2000, 17, 351–364.

[160] Lim SC, Choi JS. Effects of naringin on the pharmacokinetics of intravenous paclitaxel in rats, Biopharmaceutics & Drug Disposition, 2006, 27, 443–447.

[161] Choi J-S, Shin S-C. Enhanced paclitaxel bioavailability after oral coadministration of paclitaxel prodrug with naringin to rats, International Journal of Pharmaceutics, 2005, 292, 149–156.

[162] Choi J-S, Han H-K. Enhanced oral exposure of diltiazem by the concomitant use of naringin in rats, International Journal of Pharmaceutics, 2005, 305, 122–128.

[163] Chan K, Liu ZQ, Jiang ZH, Zhou H, Wong YF, Xu H-X, Liu L. The effects of sinomenine on intestinal absorption of paeoniflorin by the everted rat gut sac model, Journal of Ethnopharmacology, 2006, 103, 425–432.

[164] Liu ZQ, Zhou H, Liu L, Jiang ZH, Wong YF, Xie Y, Cai X, Xu HX, Chan K. Influence of co-administrated sinomenine on pharmacokinetic fate of paeoniflorin in unrestrained conscious rats, Journal of Ethnopharmacology, 2005, 99, 61–67.

[165] Alexander A, Qureshi A, Kumari L, Vaishnav P, Sharma M, Saraf S, Saraf S. Role of herbal bioactives as a potential bioavailability enhancer for active pharmaceutical ingredients. Fitoterapia, 2014, 97, 1–14. (Original work published 2014).

[166] Chavhan S, Shinde S, Gupta I. Current trends on natural bioenhancers; A review, Internal Journal of Pharmacognosy and Chinese Medicine, 2018, 2, 2576–4772.

[167] Khan I, Saeed K, Khan I. Nanoparticles: Properties, applications and toxicities, Arabian Journal of Chemistry, 2017, 12, 908.

[168] Sivakumar M, Ruckmani K. Microwave-assisted extraction of polysaccharides from Cyphomandra betacea and its biological activities, International Journal of Biological Macromolecules, 2016, 92, 682–693.

[169] Müller RH, Alexiev U, Sinambela P, Keck CM. Nanostructured lipid carriers (NLC): The second generation of solid lipid nanoparticles, In: Percutaneous Penetration Enhancers Chemical Methods in Penetration Enhancement, Springer, 2016, 161–185.

[170] Salvi VR, Pawar P. Nanostructured lipid carriers (NLC) system: A novel drug targeting carrier, Journal of Drug Delivery Science and Technology, 2019, 51, 255–267.

[171] Selvaraj K, YOO B.-K. Curcumin-loaded nanostructured lipid carrier modified with partially hydrolyzed ginsenoside. AAPS PharmSciTech, 2019, 20, 1–9.

[172] Madane RG, Mahajan HS. Curcumin-loaded nanostructured lipid carriers (NLCs) for nasal administration: Design, characterization, and in vivo study, Drug Delivery, 2016, 23, 1326–1334.

[173] Aditya NP, Shim M, Lee I, Lee Y, Im MH, Ko S. Curcumin and genistein coloaded nanostructured lipid carriers: In vitro digestion and antiprostate cancer activity, Journal of Agricultural and Food Chemistry, 2013, 61, 1878–1883.

[174] Chen S, Li Q, McClements DJ, Han Y, Dai L, Mao L, Gao Y. Co-delivery of curcumin and piperine in zein-carrageenan core-shell nanoparticles: Formation, structure, stability and in vitro gastrointestinal digestion, Food Hydrocolloids, 2020, 99, 105334.

[175] Baspinar Y, Üstündas M, Bayraktar O, Sezgin C. Curcumin and piperine loaded zein-chitosan nanoparticles: Development and in-vitro characterisation. Saudi Pharmaceutical Journal, 2018, 26, 323–334.

[176] Sharma M, Sharma S, Sharma V, Sharma K, Yadav SK, Dwivedi P, Agrawal S, Paliwal SK, Dwivedi AK, Maikhuri JP. Oleanolic–bioenhancer coloaded chitosan modified nanocarriers attenuate breast cancer cells by multimode mechanism and preserve female fertility, International Journal of Biological Macromolecules, 2017, 104, 1345–1358.

[177] Kumar V, Kharb R, Chaudhary H. Optimization & design of isradipine loaded solid lipid nanobioparticles using rutin by Taguchi methodology. International Journal of Biological Macromolecules, 2016, 92, 338–346.

[178] Sedeky AS, Khalil IA, Hefnawy A, El-Sherbiny IM. Development of core-shell nanocarrier system for augmenting piperine cytotoxic activity against human brain cancer cell line. European Journal of Pharmaceutical Sciences, 2018, 118, 103–112.

[179] Ray L, Karthik R, Srivastava V, Singh SP, Pant A, Goyal N, Gupta KC. Efficient antileishmanial activity of amphotericin B and piperine entrapped in enteric coated guar gum nanoparticles. Drug delivery and translational research, 2021, 11, 118–130.

[180] Rathee P, Kamboj A, Sidhu S. Enhanced oral bioavailability of nisoldipine-piperine-loaded poly-lactic-co-glycolic acid nanoparticles. Nanotechnology Reviews, 6, 517–526.

[181] Elnaggar YS, Etman SM, Abdelmonsif DA, Abdallah OY. Intranasal piperine-loaded chitosan nanoparticles as brain-targeted therapy in Alzheimer's disease: optimization, biological efficacy, and potential toxicity. Journal of pharmaceutical sciences, 2015, 104, 3544–3556.

[182] Pachauri M, Gupta ED, Ghosh PC. Piperine loaded PEG-PLGA nanoparticles: Preparation, characterization and targeted delivery for adjuvant breast cancer chemotherapy. Journal of drug delivery science and technology, 2015, 29, 269–282.

[183] Abolhassani H, Shojaosadati A. A comparative and systematic approach to desolvation and self-assembly methods for synthesis of piperine-loaded human serum albumin nanoparticles. Colloids and Surfaces B: Biointerfaces, 2019, 184, 110534.

[184] Seljak KB, Berginc K, Trontelj J, Zvonar A, Kristl A, Gašperlin M. A self-microemulsifying drug delivery system to overcome intestinal resveratrol toxicity and presystemic metabolism. Journal of Pharmaceutical Sciences, 2014, 103, 3491–3500.

[185] Tang H, Xiang S, Li X, Zhou J, Kuang C. Preparation and in vitro performance evaluation of resveratrol for oral self-microemulsion. PLos One, 2019, 14, e0214544.

[186] Priprem A, Watanatorn J, Sutthiparinyanont S, Phachonpai W, Muchimapura, S. Anxiety and cognitive effects of quercetin liposomes in rats. Nanomedicine, 2008, 4, 70–78.

[187] Di Pierro F, Settembre R. Safety and efficacy of an add-on therapy with curcumin phytosome and piperine and/or lipoic acid in subjects with a diagnosis of peripheral neuropathy treated with dexibuprofen. J Pain Res, 2013, 6, 497–503.

[188] Zhao Y-Z, Lu C-T, Zhang Y, Xiao J, Zhao Y-P, Tian J-L, Xu Y-Y, Feng Z-G, Xu C-Y. Selection of high efficient transdermal lipid vesicle for curcumin skin delivery. International Journal of Pharmaceutics, 2013, 454, 302–309.

[189] Scognamiglio I, De Stefano D, Campani V, Mayol L, Carnuccio R, Fabbrocini G, Ayala LA, Rotonda MI, De Rosa G. Nanocarriers for topical administration of resveratrol: A comparative study. International Journal of Pharmaceutics, 2013, 440, 179–187.

[190] Ramadon D, Pramesti SS, Anwar E. Formulation, stability test and in vitro penetration study of transethosomal gel containing green tea (Camellia sinensis L. Kuntze) leaves extract. Int J Appl Pharm, 2017, 9, 91–96.

[191] Anton N, Vandamme TF. Nano-emulsions and micro-emulsions: Clarifications of the critical differences, Pharmaceutical Research, 2011, 28, 978–985.

[192] Dokania S, Joshi AK. Self-microemulsifying drug delivery system (SMEDDS)–challenges and road ahead, Drug Delivery, 2015, 22, 675–690.

[193] Thakur PS, Singh N, Sangamwar AT, Bansal AK. Investigation of Need of Natural Bioenhancer for a Metabolism Susceptible Drug – Raloxifene, in a Designed Self-Emulsifying Drug Delivery System, AAPS Pharmaceutical Scientists, 2017, 18, 2529–2540.

[194] Kale SN, Deore SL. Emulsion micro emulsion and nano emulsion: A review, Systematic Reviews in Pharmacy, 2017, 8, 39.

[195] Vijayan UK, Varakumar S, Sole S, Singhal RS. Enhancement of loading and oral bioavailability of curcumin loaded self-microemulsifying lipid carriers using Curcuma oleoresins, Drug Development and Industrial Pharmacy, 2020, 46, 889–898.

[196] Li Q, Zhai W, Jiang Q, Huang R, Liu L, Dai J, Gong W, Du S, Wu Q. Curcumin–piperine mixtures in self-microemulsifying drug delivery system for ulcerative colitis therapy, International Journal of Pharmaceutics, 2015, 490, 22–31.
[197] Huang M, Liang C, Tan C, Huang S, Ying R, Wang Y, Wang Z, Zhang Y. Liposome co-encapsulation as a strategy for the delivery of curcumin and resveratrol, Food & Function, 2019, 10, 6447–6458.
[198] Jhaveri A, Deshpande P, Pattni B, Torchilin V. Transferrin-targeted, resveratrol-loaded liposomes for the treatment of glioblastoma, Journal of Controlled Release, 2018, 277, 89–101.
[199] Verma N, Saraf S. Development and optimization of mannosylated naringenin loaded transfersomes using response surface methodology for skin carcinoma, International Journal of Applied Pharmaceutics, 2021, 235–241.
[200] Bartelds R, Nematollahi MH, Pols T, Stuart MC, Pardakhty A, Asadikaram G, Poolman B. Niosomes, an alternative for liposomal delivery, PLoS One, 2018, 13, e0194179.
[201] Schlich M, Lai F, Pireddu R, Pini E, Ailuno G, Fadda A, Valenti D, Sinico C. Resveratrol proniosomes as a convenient nanoingredient for functional food, Food Chemistry, 2020, 310, 125950.
[202] Badria FA, Abdelaziz AE, Hassan AH, Elgazar AA, Mazyed EA. Development of provesicular nanodelivery system of curcumin as a safe and effective antiviral agent: Statistical optimization, vitro characterization, and antiviral effectiveness, Molecules, 2020, 25, 5668.
[203] Jose J, Priya S, Shastry C. Influence of bioenhancers on the release pattern of niosomes containing methotrexate, Journal of Health and Allied Sciences NU, 2012, 2, 36–40.
[204] Bajaj KJ, Parab BS, Shidhaye SS. Nano-transethosomes: A Novel Tool for Drug Delivery through Skin, Indian Journal of Pharmaceutical Education and Research, 2021, 55, S1–S10.
[205] Benson HA. Transfersomes for transdermal drug delivery, Expert Opinion on Drug Delivery, 2006, 3, 727–737.
[206] Vinod K, Anbazhagan S, Kumar MS, Sandhya S, Banji D, Rani AP. Developing ultra deformable vesicular transportation of a bioactive alkaloid in pursuit of vitiligo therapy, Asian Pacific Journal of Tropical Disease, 2012, 2, 301–306.
[207] Hosny KM, Alharbi WS, Almehmady AM, Bakhaidar RB, Alkhalidi HM, Sindi AM, Hariri AH, Shadab MD, Zaki RM. Preparation and optimization of pravastatin-naringenin nanotransfersomes to enhance bioavailability and reduce hepatic side effects, Journal of Drug Delivery Science and Technology, 2020, 57, 101746.
[208] Lu M, Qiu Q, Luo X, Liu X, Sun J, Wang C, Lin X, Deng Y, Song Y. Phyto-phospholipid complexes (phytosomes): A novel strategy to improve the bioavailability of active constituents, Asian Journal of Pharmaceutical Sciences, 2019, 14, 265–274.
[209] Azeez NA, Deepa VS, Sivapriya V. Phytosomes: Emergent promising nano vesicular drug delivery system for targeted tumor therapy, Advances in Natural Sciences: Nanoscience and Nanotechnology, 2018, 9, 033001.
[210] Islam N, Irfan M, Hussain T, Mushtaq M, Khan IU, Yousaf AM, Ghori MU, Shahzad Y. Piperine phytosomes for bioavailability enhancement of domperidone, Journal of Liposome Research, 2021, 1–9.
[211] Baradaran S, Moghaddam AH, Jelodar SK, Moradi-Kor N. Protective effects of curcumin and its nano-phytosome on carrageenan-induced inflammation in mice model: Behavioral and biochemical responses, Journal of Inflammation Research, 2020, 13, 45.
[212] Moghaddam AH, Maboudi K, Bavaghar B, Sangdehi SRM, Zare M. Neuroprotective effects of curcumin-loaded nanophytosome on ketamine-induced schizophrenia-like behaviors and oxidative damage in male mice. Neuroscience Letters, 2021, 765, 136249.
[213] Cicero AF, Sahebkar A, Fogacci F, Bove M, Giovannini M, Borghi C. Effects of phytosomal curcumin on anthropometric parameters, insulin resistance, cortisolemia and non-alcoholic

fatty liver disease indices: A double-blind, placebo-controlled clinical trial, European Journal of Nutrition, 2020, 59, 477–483.

[214] Mirzaei H, Shakeri A, Rashidi B, Jalili A, Banikazemi Z, Sahebkar A. Phytosomal curcumin: A review of pharmacokinetic, experimental and clinical studies. Biomedicine & Pharmacotherapy, 2017, 85, 102–112.

[215] Teng C-F, Yu C-H, Chang H-Y, Hsieh W-C, Wu T-H, Lin J-H, Wu H-C, Jeng L-B, Su I-J. Chemopreventive effect of phytosomal curcumin on hepatitis b virus-related hepatocellular carcinoma in a transgenic mouse model, Scientific Reports, 2019, 9, 1–13.

[216] Abdulbaqi IM, Darwis Y, Khan NAK, Abou Assi R, Khan AA. Ethosomal nanocarriers: The impact of constituents and formulation techniques on ethosomal properties, in vivo studies, and clinical trials, International Journal of Nanomedicine, 2016, 11, 2279.

[217] Gollavilli H, Hegde AR, Managuli RS, Bhaskar KV, Dengale SJ, Reddy MS, Kalthur G, Mutalik S. Naringin nano-ethosomal novel sunscreen creams: Development and performance evaluation. Colloids and Surfaces B: Biointerfaces, 2020, 193, 111122.

[218] Li Y, Xu F, Li X, Chen S-Y, Huang L-Y, Bian -Y-Y, Wang J, Shu Y-T, Yan G-J, Dong J. Development of curcumin-loaded composite phospholipid ethosomes for enhanced skin permeability and vesicle stability, International Journal of Pharmaceutics, 2021, 592, 119936.

[219] Soni K, Mujtaba A, Akhter MH, Zafar A, Kohli K. Optimisation of ethosomal nanogel for topical nano-CUR and sulphoraphane delivery in effective skin cancer therapy, Journal of Microencapsulation, 2020, 37, 91–108.

[220] Song CK, Balakrishnan P, Shim C-K, Chung S-J, Chong S, Kim -D-D. A novel vesicular carrier, transethosome, for enhanced skin delivery of voriconazole: Characterization and in vitro/ in vivo evaluation, Colloids and Surfaces. B, Biointerfaces, 2012, 92, 299–304.

[221] Nainwal N, Jawla S, Singh R, Saharan VA. Transdermal applications of ethosomes–a detailed review, Journal of Liposome Research, 2019, 29, 103–113.

[222] Wu P-S, Li Y-S, Kuo Y-C, Tsai S-J-J, Lin -C-C. Preparation and evaluation of novel transfersomes combined with the natural antioxidant resveratrol, Molecules, 2019, 24, 600.

[223] Ramadon D, Anwar E, Harahap Y. In vitro penetration and bioavailability of novel transdermal quercetin-loaded ethosomal gel, Indian Journal of Pharmaceutical Sciences, 2018, 79, 948–956.

[224] Duangjit S, Obata Y, Sano H, Kikuchi S, Onuki Y, Opanasopit P, Ngawhirunpat T, Maitani Y, Takayama K. Menthosomes, novel ultradeformable vesicles for transdermal drug delivery: Optimization and characterization, Biological & Pharmaceutical Bulletin, 2012, 35, 1720–1728.

[225] Zaky A. Comparative study of terbinafine hydrochloride transfersome, menthosome and ethosome nanovesicle formulations via skin permeation and antifungal efficacy, Al-Azhar Journal of Pharmaceutical Sciences, 2016, 54, 18–36.

[226] Duangjit S, Obata Y, Sano H, Onuki Y, Opanasopit P, Ngawhirunpat T, Miyoshi T, Kato S, Takayama K. Comparative study of novel ultradeformable liposomes: Menthosomes, transfersomes and liposomes for enhancing skin permeation of meloxicam, Biological & Pharmaceutical Bulletin, 2014, b13–00576.

[227] Ashok K, Kumar AR, Nama S, Brahmaiah B, Desu P, Rao C. Sphingosomes: A novel vesicular drug delivery system, International Journal of Pharmaceutical Research and Bio-Science, 2013, 2, 305–312.

[228] Saraf S, Gupta D, Kaur CD, Saraf S. Sphingosomes a novel approach to vesicular drug delivery, International Journal of Current Research, 2011, 1, 63–68.

[229] Lankalapalli S, Damuluri M. Sphingosomes: Applications in targeted drug delivery, International Journal of Pharmaceutical and Chemical Sciences, 2012, 2, 507–516.

[230] Webb MS, Bally MB, Mayer LD, Miller JJ, Tardi PG, Sphingosomes for enhanced drug delivery, in, Google Patents, 1998.

Shashikant B. Bagade, Shivanee Vyas, Amit B. Page, Kiran D. Patil
Chapter 4
Role of herbal bioenhancers in tuberculosis and drug delivery thereof

Abstract: In recent years, there has been an increase in demand in the medical field for simple drug delivery matrices that have fast bioavailability, are easy to use, and cost-effective to produce. Through the use of novel approaches such as bioenhancers, solutions of herbal origin can meet these requirements.

Bioenhancers are compounds that may not be therapeutic, but when combined with an active drug, potentiate the drug's pharmacologic action. They work through a variety of pathways that can influence drug metabolism, absorption, and drug target action. Bioenhancers can be of herbal or animal origin. Herbal bioenhancers have been shown to enhance the bioavailability and bioefficiency of different classes of drugs such as antitubercular drugs, antibiotics, antiviral, antifungal, and anticancer drugs. These agents have been found to increase the bioavailability of some drugs, including the antitubercular drug (e.g., rifampicin), even when reduced doses of drugs are present in such formulations. Bioenhancers reduce the dose and toxicity of drugs; they may shorten the treatment period. Extensive research on these bioenhancers is the need of the hour so that they could be utilized in drug formulations for antitubercular treatment in the future.

4.1 Introduction

Drugs that might be administered orally undergo a dissolution process, after which, they permeate throughout the gastric membrane earlier than when they are seen inside the blood stream. Drug bioavailability refers to the amount of medicine that enters the bloodstream from the point of administration [1]. Low bioavailability is a problem from which several drugs suffer [2]. Orally administered drugs have a low bioavailability because of their vulnerability to first-pass metabolism and insufficient absorption, which is not an issue for intravenous administration. Thus, these

Shashikant B. Bagade, SVKM's NMIMS School of Pharmacy and Technology Management, Shirpur, Dist. Dhule 425405, Maharashtra, India, Phone No: +91 9637474753, e-mail: shashikant.bagade@nmims.edu
Shivanee Vyas, Amit B. Page, SVKM's NMIMS School of Pharmacy and Technology Management, Shirpur, Dist. Dhule 425405, Maharashtra, India
Kiran D. Patil, SVKM's Institute of Pharmacy, Dist. Dhule 424001, Maharashtra, India

https://doi.org/10.1515/9783110746808-004

unused medications in the body might have negative consequences as well as lead to drug resistance [3].

Despite these limitations, the oral route of drug administration has been the most popular due to ease of administration, high patient compliance, cost-effectiveness, low requirement for sterile conditions, and flexibility in dosage form design [4]. The usage of bioenhancers is a promising strategy for overcoming bioavailability issues [5]. A large percentage of the world's population uses plant-based medicines. Thousands of natural medications, including some for unusual conditions, are mentioned in our Ayurvedic literature. Plant-based medications are also found in over a quarter of the modern pharmacopoeias. The concept of herbal bioenhancers dates all the way back to Ayurveda medicine's ancient heritage.

In Ayurveda, this theory is known as "Yogvahi" and is used to increase the therapeutic effect of medicines by improving the oral bioavailability, increase tissue distribution, specifically for those drugs with poor oral bioavailability, decrease their dose and adverse effects, and circumvent the parenteral routes of drug administration. The idea of bioenhancers was first described by Bose in 1929, which showed an increase in the anti-asthmatic effects of Vasaka (*Adhatoda vasica*) leaves by the addition of long pepper to it [6]. Bioenhancers are chemical entities that stimulate the bioavailability of the drug. Bioenhancers are such agents, which by themselves are not therapeutic entities, but when combined with an active constituent proceed with potentiating the pharmacological effect of the drug [7]. Bioenhancement is the phenomenon of increasing the total availability of any chemical entity or drug molecule in the biological fluid or systemic circulation, and bioavailability enhancers or bioenhancers are the secondary agents responsible for this increase in plasma concentration of the principal ingredient [8, 9].

4.2 Bioenhancers

The most important rate-limiting step for therapeutic activity of oral and topical formulations is the bioavailability of the drugs. Bioavailability mainly depends on the hydrophilic and lipophilic characteristic of the drugs. Lipophilic drugs have more bioavailability compared to hydrophilic drugs because of the lipid bilayer transportation barrier [10]. The size of the molecule also plays an important role in the transportation of the drug across the biological membranes. Moreover, there are many other factors that affect the bioavailability of the drugs, such as salivary pH, gastric pH, enzymes, emptying time, protein binding, drug metabolism, excretion, and person's diseased condition. In addition to these, if the drug gets bioavailable in the cell, it gets rapidly excreted because of the efflux pump of the cells. Because of all these hurdles in the bioavailability of the drugs, oral and topical routes are less effective as compared to other drugs [11, 12]. To address these issues, it is critical to design certain approaches for increasing drug bioavailability. Bioenhancers are one of the most widely used techniques for increasing drug bioavailability [13, 14]. Bio-enhancers are the agents that enhance the bioavailability of the drugs without altering their pharmacological response. However, they may increase the therapeutic role of the drugs by increasing the concentration of the drug at the target site.

Bioenhancers offer the following benefits to drug development.
- They reduce the dosage of the drug.
- They reduce the cost of the drug.
- They prevent drug resistance.
- They decrease the percent level of adverse drug reactions or side effects.
- They increase the efficacy of the drug.
- Increased bioavailability.
- Raw material requirement for drug manufacture will be decreased.
- Economically, they are useful to the world economy.
- Treatment cost of patient decreases.

4.2.1 Need for bioenhancers

Aqueous solubility, drug permeability, dissolving rate, first-pass metabolism, pre-systemic metabolism, and sensitivity to efflux mechanisms are factors that affect drug absorption and bioavailability [12]. Bioenhancers modify one or more of these factors to improve absorption and decrease the metabolism, to reduce dose, toxicity, and cost of drugs. The need for bioenhancers especially arises for costly and chronic-administered drugs that are poorly bioavailable and have toxic and adverse drug effects. By enhancing the efficacy of drugs, bioenhancers can further reduce chances of microbial resistance.

As can be seen from the literature further, bioenhancers not only enhance bioavailability by affecting membrane permeability or solubility of drugs, but can reduce the metabolism of the drug by interfering enzymes. Bioenhancers also increase the permeability of drugs in microorganisms, leading to more antimicrobial actions of drugs [11].

4.2.2 Effect of bioenhancers on drugs

The drug dose is reduced and the potential of drug resistance is significantly reduced. The drug dose-dependent toxicity and cost will be decreased, particularly for antitubercular medications. Bioenhancers were found to improve the bioavailability and bioefficacy of a variety of medications, including antitubercular drugs, antibiotics, antiviral, antifungal, and anticancer agents at small doses [10].

Bioenhancers may act through one or more mechanisms as follows [10–14]:
1. Alteration of gastrointestinal (GI) permeability, e.g., glycerrhiza and aloe.
2. Inhibition of presystemic metabolism of drugs by inhibiting hepatic enzymes, e.g., quercetin and naringin.
3. Inhibiting cellular efflux, e.g., piperine, naringin, and quercetin.

4.2.3 Classification of bioenhancers

Bioenhancers are classified based on their origin such as plant origin and animal origin. They are also classified on the basis of their mechanism of action such as P-glycoprotein (P-gp), cytochrome-450 (CYP 450) inhibitors, and gastrointestinal function regulators (Figure 4.1) [11–14].

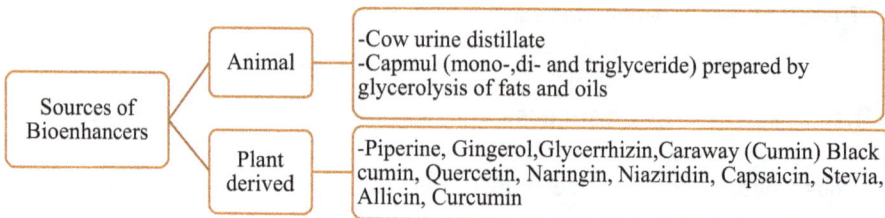

a)

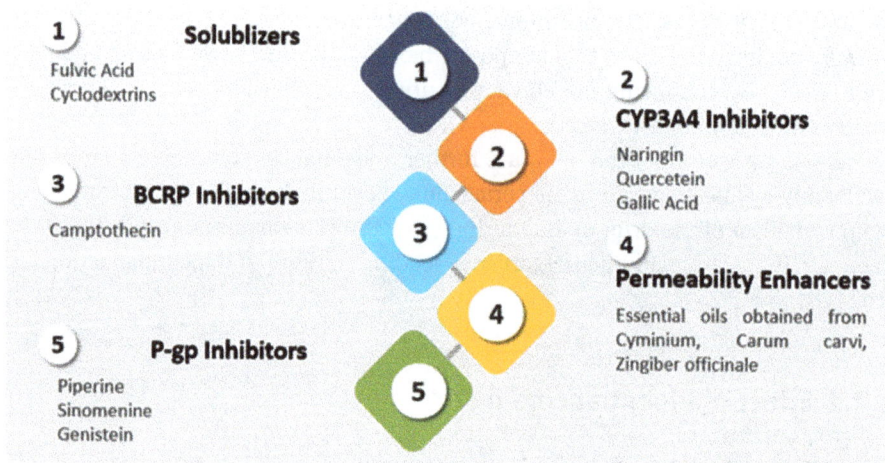

b)

Figure 4.1: Classification of bioenhancers: (a) sources and (b) mechanism of action.

4.3 Tuberculosis

Tuberculosis (TB), a ubiquitous, highly contagious, chronic, communicable, granulomatous bacterial infection caused by the bacterium *Mycobacterium tuberculosis* *(M. TB)* or to a much lesser extent by *Mycobacterium bovis*, *Mycobacterium africanum*, and *Mycobacterium microti*. TB, is an infectious disease caused by an *actinobacteria*, the *M. TB*. Though it is primarily a respiratory infection affecting the lungs and the alveoli (pulmonary TB), it can affect other tissues such as the intestine (intestinal TB), meninges (tuberculous meningitis), bones and joints (tuberculous arthritis), and kidneys (renal TB),, leading to a combined disease known as extra pulmonary TB. *M. TB* is a fast spreading, extremely infectious gram-positive aerobic rod-shaped acid-fast bacillus with a modest growth rate. Tubercle bacilli can live in macrophages because their cell walls contain a lot of lipid. M.TB is most commonly found in humans [15, 16]. The human race is the primary host for *M.TB*.

TB is one of the deadliest diseases as the number of casualties caused by it has crossed even the deaths caused by human immunodeficiency virus (HIV). According to the World Health Organization, near about 30% of the total world's population is suffering from the disease [17]. TB reduces human efficiency at work, thus reducing the GDP of a country. TB affects mainly the population in the slums and with low hygiene; thus, the poor have 5 times higher chance of getting this disease. Incomplete treatment, poor patient compliance, etc. lead to resistance in the strains of mycobacteria, making its treatment extremely difficult. These TB strains with multidrug resistance (MDR-TB) have low cure rates and significant fatality rates [18]. In addition to these, several cases of total drug-resistant TB have been reported in the clinics [19–21]. In addition to the threat of extensively drug-resistant (XDR)-TB, one more threat is TB with diabetes. According to the Centre for Disease Control and Prevention, TB can appear to be a life-threatening complication for people suffering from diabetes [22].

Diabetic patients are more prone to get the primary infection of TB more rapidly than a nondiabetic, while chances of development of latent TB to a full-blown disease are more in diabetics. Also, diabetics take a longer time to be cured than nondiabetic TB patients. Diabetics develop pulmonary, cavitary TB more as compared to nondiabetics [23].

4.3.1 Epidemiology

According to a recent World Health Organization research (2019), worldwide about 10.0 million new (incident) TB cases were found, of which 57% were men, 32% women, and 11% were children. HIV-positive people accounted for 8.6% of all new TB cases. The mortality in HIV-negative TB patients in 2018 was almost 1.2 million, while HIV-positive patients accounted for 251,000 deaths [24].

4.3.2 Pathogenesis

TB infection is spread primarily in the form of droplets containing viable infectious bacteria from an infected patient to an uninfected person during coughing or sneezing or even talking. Such droplets having viable tubercle bacilli are then inhaled and travel through lungs to the alveoli. Alveolar macrophages then engulf these bacilli. These macrophages then become the place for bacilli lodgment and multiplication.

The fibrosis around such a clump of affected macrophages form a hard shell called granuloma, keeping these bacteria under control. Multiplying bacteria in this granuloma then burst outside to develop the active disease, the threat of which is dependent on several factors, including patients' age, duration of latent infection and above all, patient's immunity [25]. The mortality of untreated smear-positive cases may be as high as 50–80%; while though inconsistent, the treatment can reduce it to 30%. If the disease is treated through DOTS (Directly Observed Treatment, possibly short course) or such other TB control programs, mortality rates can be reduced to 5% [26].

4.3.3 Conventional therapy of TB

Use of rifampicin (R), isoniazid (H), pyrazinamide (Z), ethambutol (E), and streptomycin (S) in a specific pattern constitute the conventional chemotherapy of TB. These drugs are called the first-line drugs. The second line treatment is with drugs such as ethionamide, amikacin, kanamycin, *para*-amino salicylate, cycloserine (CS), fluoroquinolones, thioacetazone, and capreomycin. First-line medications are very effective in the case of drug-susceptible TB, but second-line medications are needed to be used when the first-line regimen breaks due to the development of drug resistance. Recent therapy, i.e., DOTS for drug-susceptible TB consists of two phases (Table 4.1). The intensive phase consists of two months treatment with first-line drugs in which the majority of the viable bacilli are killed.

In order to sterilize the lesion and kill all slow-growing bacteria, the two major first-line drugs, viz. rifampicin and isoniazid, are continued for the next 4–6 months, either daily- or thrice-a-week regimen.

4.4 Challenges in conventional therapy of tuberculosis

One of the major reasons for the development of drug resistance in TB includes the use of drugs causing life-threatening adverse drug reactions. These ADRS also lead to noncompliance by patients. One more reason for noncompliance is the complicated regimen of the treatment. More than 480,000 cases of MDR-TB patients are observed

Table 4.1: Conventional chemotherapy for the treatment of drug-susceptible tuberculosis.

Category of patients	Phase-I (2 months) with rifampicin, isoniazid, pyrazinamide, and ethambutol	Phase-II (4 months)
Newly identified patients	All four drugs daily	Rifampicin + isoniazid daily
	All four drugs daily	Rifampicin + isoniazid thrice a week
	All four drugs thrice a week	
Patient with previous incomplete treatment	All four drugs + streptomycin daily for 2 months. Followed by standard treatment for next 1 month.	Rifampicin + isoniazid + ethambutol daily for 5 months

per year, of which XDR-TB accounts for 9% of the total cases [5]. Treatment for MDR-TB necessitates a minimum of two years of second-line treatment; however, treatment for XDR-TB is more difficult due to fluoroquinolone and injectable second-line drugs. Furthermore, XDR-TB treatment is time-consuming, expensive, and has a low cure rate. It also has life-threatening adverse effects [25, 26]. Success rate of MDR-TB accounts to 50%, while the same for XDR-TB is reported to be around 27% with less than 20% of the XDR-TB cases of surviving a good life [28–30]. Tubercle bacilli create drug resistance in a variety of ways, but the most common cause is a decrease in drug bioavailability at the target location due to increased efflux or inactivation.

4.5 Role of herbal bioenhancers in tuberculosis therapy

The main reason behind the treatment failure of TB is the high-dose multidrug therapy, which leads to toxic side effects and patient non-compliance. Moreover, it also results in the development of drug-resistant strains. In order to increase patient compliance, one must try to reduce the dose and dosing frequency of the toxic drugs. The dose of the drugs can be reduced by increasing the bioavailability of the drug at the target site. Bioenhancers can be the effective strategy to stimulate and enhance drug concentration at the site of action. Bioenhancers work by several mechanisms to increase the bioavailability. They can act as efflux pump inhibitors to avoid the exit of the drug from cells such as P-gp inhibitors. They also increase the concentration of drugs by inhibiting their enzymatic degradation such as cytochrome P-450 inhibitors. Moreover, they can also regulate the gastrointestinal functions to maximize the absorption of the drug, which leads to an increase in bioavailability.

Following are some of the important herbal bioenhancers that can be used to increase the bioavailability of antitubercular drugs and patient compliance.

4.5.1 Piperine from pepper

Piperine is an alkaloidal phytoconstituent found in the fruits of the Piperaceae family such as black pepper. Piperine's bioenhancing properties were first utilized in humans to treat TB. Because of the ability of piperine to increase bioavailability of rifampicin by about 60%, the dose of the latter could be reduced to 200 mg from 450 mg. Nevirapine's (an antiviral with nucleoside inhibitor of HIV1-reverse transcriptase) bioavailability was also increased when combined with piperine, It could therefore be used in combination with other antiretroviral drugs to treat HIV1 infection effectively. Piperine is also found to improve the absorption of curcumin.

Piperine is shown to increase bioavailability by 20× when combined with 20 mg of curcumin. Several animal studies on piperine have shown that it has interesting bioenhancing properties for a variety of medicines [31–33].

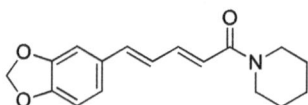

Piperine

4.5.2 Ginger

Gingerol is the chief chemical constituent in ginger. It helps to improve absorption of drugs by modulating GI functions. The bioenhancers extract's effective dose is around 10–30 mg/kg body weight. Gingerol increases the bioavailability of rifampicin by 65% while that of ethionamide is increased by 56%. Using gingerol, the bioavailability of antibiotic drugs (azithromycin – 78% bioavailability), antifungal drugs (ketoconazole – 125% bioavailability), antiviral drugs (zidovudine – 105% bioavailability), and anticancer drugs (5-fluorouracil – 110% bioavailability) are also improved [34–36].

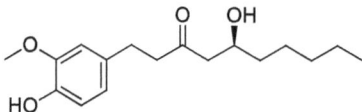

Gingerol

4.5.3 Liquorice

The dried roots and stolons of *Glycyrrhiza glabra* are shown to have antioxidant, contraceptive, anti-inflammatory, expectorant, and liver protecting properties. At a concentration of 1 g/ml, glycyrrhizin, its chief chemical constituent is shown to increase the bioavailability of rifampicin to about 6 times the original, while that of taxols is increased to about 5 times [37].

4.5.4 Caraway/cumin

It is a P-gp efflux pump inhibitor made from the dried ripe fruits of *Carum carvi*, a plant in the Umbelliferae family. Antioxidant, antimicrobial, diuretic, and carminative properties are all present. Carvone and limonene are the main components of caraway. The effective dose of the extract was between the range of 5–100 mg/kg body weight. Rifampicin has a 110% increase in bioavailability; CS has a 75% increase; and ethionamide has a 68% increase. Caraway improves the bioavailability of cloxacillin to 100%, antifungal amphotericin B to 78%, antiviral zidovudine to 92%, and anticancer drug 5-fluorouracil to 90%, with doses in the range of 1–55 mg/kg body weight [38, 39].

4.5.5 Black cumin

It is obtained from the plant *Cuminum cyminum*, and has carminative, estrogenic, antibacterial, anti-inflammatory, and antioxidant properties. Its 3′, 5-dihydroxyflavone -7-O–D-galactouronide-4′-O-D-glucopyranoside is the most important component of cumin. The bioenhancer extract's effective dose ranges from 0.5 to 25 mg/kg body weight. Rifampicin has a 250% increase in bioavailability; CS has an 89% increase; and ethionamide has a 78% increase [40, 41].

4.5.6 Quercetin

It is a flavonoid glycoside found in many plants such as onions and citrus fruits. Antioxidant, radical scavenging, anti-inflammatory, and anti-atherosclerotic properties are all present. It operates by blocking CYP3A4 and the P-gp efflux pump. Diltiazem, digoxin, verapamil, etoposide, and paclitaxel have all been shown to increase the bioavailability, blood levels, and efficacy of quercetin [42–44].

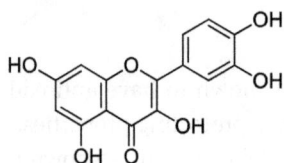

Quercetin

4.5.7 Naringin

It is obtained from grapefruit, apples, onions, and tea, and all of them contain this essential flavonoid glycoside. It has antiulcer, antioxidant, antiallergic, and antihyperlipidemic properties, among others. Naringin serves as a bioenhancer by inhibiting intestinal CYP3A4, CYP3A1, CYP3A2, and P-gp. In a dose-dependent manner, pretreatment with oral naringin at multiple levels enhances the area under curve for *I.V.* for intravenous dose of paclitaxel, around 3 mg/kg. Naringin at three doses of 3.0–10.0 mg/kg of body weight augments the bioavailability of paclitaxel. Diltiazem, verapamil, saquinavir, and cyclosporine A are some of the other biogreater capsules [45].

4.5.8 Niaziridine

It is found in drumstick pods and is nitrile glycoside, which is an effective bioenhancer. It is used for the regulation of gastrointestinal function to increase absorption. Furthermore, it enhances the bioavailability of rifampicin by 38.8-fold at 1.0 µg/mL [46].

4.5.9 Capsaicin

This is an active phytoconstituents from *Capsicum annum* and other species of the family. It increases the bioavailability of theophylline [47].

4.5.10 *Stevia* (Honey leaf)

The plant *Stevia rebaudiana* contains steviosides, which is bioenhancer phytoconstituent present in the leaf of this plant. It is found to enhance the bioavailability of various drugs from the therapeutic category of antitubercular, antileprotic, anticancer, antifungal, and antiviral drugs. The extract of this plant is effective in the range of 0.01–50 mg/kg body weight [48].

4.5.11 Gallic acid

Gallic acid is found to exert a synergistic bioenhancement effect with piperine in adverse conditions such as beryllium-induced hepatorenal dysfunction and of oxidative stress [49]. Most of the esters of gallic acid such as propyl, octyl and aluryl gallate have shown bioavailability enhancement of several drugs, e.g., nifedipine [50, 51].

4.5.12 Allicin

It is obtained from garlic, *Allium sativum*. Allicin was first isolated by Dr. Chester Cavallito and John Bailey in 1944. Since then, it is one of the most sought after chemical for the study of bioenhancement. This organophosphorus compound is shown to be effective in the enhancement of bioavailability of amphotericin B against a variety of pathogenic fungal strains [52].

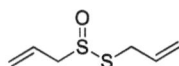

Allicin

4.5.13 Turmeric

It is a dry powder of turmeric rhizome obtained from *Curcuma longa* and is a common food ingredient of various forms, as a spice or suspension in milk or lozenge with jaggery. *Curcumin*, the chief chemical constituent in turmeric, is a flavonoid in nature. It is shown to decrease hepatic microsomal enzymes such as CYP3A4. It is reported to show change in protein transporter such as P-gp (efflux inhibition). It increases bioavailability of certain drugs in rats, e.g., celiprolol and midazolam [50–52]. It shows the bioenhancing effect similar to that of piperine [53]. It also suppresses the uridine diphosphate glucuronyl transferase level in the GI tract and in the hepatic cells demonstrating a mechanism of enhancing bioavailability [54]. It is also involved in the modification of gastric activities for better absorption of drugs.

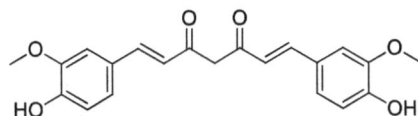

Curcumin

4.5.14 Distillate from cow's urine

It is well known in the Indian system as *Kamdhenu ark*. The distillate of cow's urine is a good bioenhancer. It improves the bioavailability of antibiotics many folds, e.g., ampicillin and tetracycline. In addition, it enhances the efficiency of the anti-cancer drug taxol against MCF-7 cell lines. It enhances the bioavailability of drugs such as rifampicin, ampicillin, and clotrimazole. It has antitoxic activity against cadmium chloride poisoning and it increases bioavailability of Zn. The bioenhancing capacity is by facilitating the absorption of drugs throughout the cell membrane [55, 56].

4.5.15 Capmul

Capmul is a glycerolysis product (MCM C10) of fatty acid and oils. In a study with ceftriaxone, the capmul showed 80% increased bioavailability of ceftriaxone [57].

4.6 Future perspectives

The bioenhancing concept is found to be very innovative in the modern era. In spite of the enormous beneficial effects of bioenhancers to mankind, the job of exploring new herbal bioenhancers is still in the nascent stage. There are still abundant bioenhancers of herbal origin that need to be explored in several vital areas. There is a huge range of untapped herbs that need to be investigated for their bioenhancing effectiveness [58].

Pharmaceutical companies all around the world have spent billions of dollars on new drug discovery programs in the hope of developing novel blockbuster medications for a variety of conditions. Pharmaceutical research seems to have had a lot of success as new compounds have been produced. However, the recently found compounds have a number of disadvantages, including low water solubility, poor bioavailability, and so on. The principal process responsible for decreasing bioavailability is the metabolism of API by cytochrome P450 in the gastrointestinal tract and the liver. EDTs such as P-gp and multidrug-resistant-associated protein are also responsible for the decreased bioavailability of therapeutically active medicines, notably those used to treat cancer.

As a result, a different approach is required to improve the bioavailability of pharmacological compounds. Coadministration of these drugs with bioenhancers can change the pharmacokinetic and pharmacodynamic characteristics of the main drugs [59].

4.7 Conclusion

Bioenhancers are an innovative, novel concept as this invention is based on ancient and conventional structure of the Indian medicine. One can expect complementary action of bioenhancers to lead to a drop in treatment cost, toxicity, adverse effects, and to have a valuable influence on the economy of the nation. Bioenhancers can be easily cultivated and procured; they are economical, nonaddictive, safe, and effective; and have a wide range of applications [60].

In developing countries like India, treatment expenditure is a key concern for modern medicine. Efficient novel means are the need of the hour for reducing the costs. The new drug development technique has been explored aiming the economics of treatment. As a result, treatments are now becoming more inexpensive for broad sections of the population, including the economically challenged [61].

When piperine is added to rifampicin, its bioavailability is improved by 60%. As a result of adding the bioenhancer "piperine," the rifampicin dose is reduced from 450 to 200 mg. The immunomodulatory and hepatoprotective properties of piperine may aid TB treatment. As a result, the dosage, cost, and toxicity of rifampicin are all reduced.

Researchers are now exploring methods aimed at reducing drug dosage and thus the drug treatment costs; making the treatment affordable to a wider section of the society. The economic benefits will help poor patients needing prolonged and expensive antituberculosis or anticancer treatment. The main objective is to target expensive, toxic, and scarce drugs or drugs that exhibit poor bioavailability [62].

Conflict of interests

All the authors declare no conflict of interests.

List of abbreviations

CS	Cycloserine
CYP 450	Cytochrome P450
CYP3A	Cytochrome P450 3A
DOTS	Directly observed treatment
GI	Gastrointestinal
HIV	*Human immunodeficiency virus*
M. TB	*Mycobacterium tuberculosis*
MDR-TB	Multidrug-resistant tuberculosis
P-gp	P-glycoprotein
TB	Tuberculosis
XDR-TB	Extensively drug-resistant tuberculosis

References

[1] Chow SC, Bioavailability and bioequivalence in drug development, Wiley Interdisciplinary Reviews. Computational Statistics, 2014, 6(4), 304–312.

[2] Brahmankar DM, Jaiswal SB, Biopharmaceutics and Pharmacokinetics - A Treatise, 3rd Edition, 2014, Vallabh Prakashan, Delhi, 24–26.

[3] Gopal V, Prakash Yoganandam G, Velvizhy Thilagam T, Bio-enhancer: A pharmacognostic perspective, European Journal of Molecular Biology and Biochemistry, 2016, 3(1), 33–38.

[4] Verma P, Thakur A, Deshmukh K, Jha A, Verma S, Routes of drug administration, International Journal of Pharmaceutical Studies and Research, 2010, 1(1), 54–59.

[5] Kesarwani K, Gupta R, Bioavailability enhancers of herbal origin: An overview, Asian Pacific Journal of Tropical Biomedicine, 2013, 3(4), 253–266.

[6] Drabu S, Khatri S, Babu S, Lohani P, Use of herbal bioenhancers to increase the bioavailability of drugs, Research Journal of Pharmaceutical, Biological and Chemical Sciences, 2011, 2(4), 107–119.

[7] Khanuja SPS, Kumar S, Shasany AK, Arya JS, Darokar MP, Singh M, Sinha P, Awasthi S, Gupta SC, Gupta VK, Gupta MM, Verma RK, Agarwal S, Mansinghka SB, Dawle SH Use of bioactive fraction from cow urine distillate ("Go-mutra") as a bioenhancers of anti-infective, anti-cancer agents and nutrients U.S. Patent US 7235262; 2007.

[8] Lim SC, Choi JS, Effects of naringin on the pharmacokinetics of paclitaxel in rats, Biopharmaceutics & Drug Disposition, 2006, 27, 443–447.

[9] Tatiraju DV, et al., Natural Bioenhancers: An overview, Journal of Pharmacognosy and Phytochemistry, 2(3), 2013, 55–60.

[10] Tatiraju DV, Bagade VB, Karambelkar PJ, Jadhav VM, Kadam V, Natural Bioenhancers: An overview, Journal of Pharmacognosy and Phytochemistry, 2013, 2, 55–60.

[11] Dudhatra GB, Mody SK, Awale MM, Patel HB, Modi CM, Kumar A, Kamani DR, Chauhan BN, A Comprehensive Review on Pharmacotherapeutics of Herbal Bioenhancers, The Scientific World Journal, 2012, 1, 1–33.

[12] Vaishnavi Chivte K, Bioenhancers: A brief review, Advanced Journal of Pharmacie and Life Science Research, 5(2), 2017, 1–18.

[13] Jain G, Umesh Patil K, Strategies for enhancement of bioavailability of medicinal agents with natural products, International Journal of Pharmaceutical Sciences and Research, 6(12), 2015, 5315–5325.

[14] Randhawa GK, Kullar JS R, Bioenhancers from Mother Nature and their applicability in modern medicine, International Journal of Applied and Basic Medical Research, 2011, 1(1), 5–10.

[15] Katzung BG, Basic & clinical pharmacology; antimycobacterial drugs, 14th ed., USA, McGraw-Hill Education, 2018.

[16] Lemke TL, Williams DA, (Ed.) Foye's Principles of Medicinal Chemistry, 7th Edition, 2013, Wolters Kluwer, Lippincott Williams & Wilkins, Philadelphia, 1274–1303.

[17] World Health Organization, Global tuberculosis report 2019, 2019.

[18] W.H. World Health Organization, Global tuberculosis report 2016, (2016).

[19] Zumla A, Nahid P, Cole ST, Advances in the development of new tuberculosis drugs and treatment regimens, Nature Reviews Drug Discovery, 12, 2013, 388–404.

[20] Klopper M, Warren RM, Hayes C, van Pittius NCG, Streicher EM, Müller B, Sirgel FA, Chabula-Nxiweni M, Hoosain E, Coetzee G, Emergence and spread of extensively and totally drug-resistant tuberculosis, South Africa, Emerging Infectious Diseases, 19, 2013, 449.

[21] Slomski A, South Africa warns of emergence of "totally" drug-resistant tuberculosis, Jama, 309, 2013, 1097–1098.

[22] Centers for Disease Control and Prevention (CDC), Tuberculosis, Tuberculosis and Diabetes, https://www.cdc.gov/tb/topic/basics/tb-and-diabetes.html (accessed on 27/102021)
[23] Restrepo BI, Diabetes and Tuberculosis, Microbiology Spectrum, 2016, 4(6), 01–19.
[24] WHO, WHO TB Report, WHO Libr. Cat. Data World. (2019) 7.
[25] Jilani TN, Avula A, Gondal AZ, Siddiqui AH, in Active Tuberculosis, StatPearls [Internet] publishing, Florida, (updated on October19, 2021), Active Tuberculosis.
[26] Singh S, Tuberculosis, Current Anaesthesia & Critical Care, 15, 2004, 165–171.
[27] Chan ED, Strand MJ, Iseman MD, Treatment outcomes in extensively resistant tuberculosis, The New England Journal of Medicine, 2008, 359, 657–659. DOI: 10.1056/NEJMc0706556.
[28] BMJ, 2015 Feb 26, 350, h882. doi: 10.1136/bmj.h882.
[29] W.H. World Health Organization, Global Tuberculosis Programme, WHO treatment guidelines for drug-resistant tuberculosis: 2016 update, Who. (2016) 56. https://doi.org/WHO/HTM/TB/2016.04.
[30] D'Ambrosio L, Centis R, Sotgiu G, Pontali E, Spanevello A, Migliori GB, New anti-tuberculosis drugs and regimens: 2015 update, ERJ Open Research, 1, 2015, 10–2015.
[31] Singh M, Varshneya C, Telang RS, Srivastava AK, Alteration of pharmacokinetics of oxytetracycline following oral administration of Piper longum in hens, Journal of Veterinary Science, 2005, 6(3), 197–220.
[32] Atal N, Bioenhancers: Revolutionary concept to market, Journal of Ayurveda and Integrative Medicine, 2010 Apr, 1(2), 96–99. doi: 10.4103/0975-9476.65073.
[33] Kashibhatta R, Naidu MU, Influence of piperine on the pharmacokinetics of nevirapine under fasting conditions: A randomized, crossover, placebo-controlled study, Drugs RD, 2007, 8(6), 383–391.
[34] Singh A, Pawar VK, Jakhmola V, Parabia MH, Awasthi R, Sharma G, et al., In vivo assessment of enhanced bioavailability of metronidazole with piperine in rabbits, Journal of Pharmaceutical, Biological and Chemical Sciences, 2010, 1(4), 27.
[35] Janakiraman K, Manavalan R, Studies on effect of piperine on oral bioavailability of ampicillin and norfloxacin, African Journal of Traditional, Complementary, and Alternative Medicines: AJTCAM / African Networks on Ethnomedicines, 2008, 5, 257–262.
[36] Qazi GN, Tikoo L, Gupta AK, Ganju K, Gupta DK, Jaggi BS et al. Bioavailability enhancing activity of Zingiber officinale and its extracts/ fractions thereof. European patent EP 1465646; 2002.
[37] Mukhopadhyay N, Khan S, Aswatha RH, Natural bioenhancers: Scope for newer combinations-a short review, JJoGPT, 2018, 10(11), 1–4.
[38] Jain G, Patil UK, Strategies for enhancement of bioavailability of medicinal agents with natural products, JIJPSR, 2015, 6(12), 5315–5324.
[39] Khanuja SPS, Kumar S, Arya JS, Shasany AK, Singh M, Awasthi S, Darokar MP, Rahman LU Composition comprising pharmaceutical/nutraceutical agent and a bio-enhancer obtained from Glycyrrhiza glabra. United States Patent, Number 6979471, 2002.
[40] Choudhary N, Khajuria V, Gillani ZH, Tandon VR, Arora E, Effect of Carum carvi, an herbal bioenhancer on pharmacokinetics of anti-tubercular drugs: A study in healthy human volunteers, Perspectives in Clinical Research, 2014, 5, 80–84.
[41] Qazi GN, Bedi KL, Johri RK, Tikoo MK, Tikoo AK, Sharma SC et al. Bioavailability enhancing activity of Carum carvi extracts and fractions thereof. US Patent US 20060257505; 2006.
[42] Quzi GN, Bedi KL, Johri RK, Tikoo MK, Tikoo AK, Sharma SC, Abdullah ST, Suri OP, Gupta BD, Suri KA, Satti NK, Khajuria RK, Singh S, Khajuria A, Kapahi BK. Bioavailability/bioefficacy enhancing activity of Cuminum cyminum and extracts and fractions there of United States Patent, Number 7070814, 2003.
[43] Bedi K, Gupta BD, Rakesh KJ, Khan IA, Qazi GN, Johri RK et al. Use of herbal agents for potentiation of bioefficacy of anti infectives. U.S. patent US 7119075 B1; 2006.

[44] Anup K, Sonia G, Swati K, Shrirang N, Waheed R, Vadim I, et al. The studies on bio enhancer effect of red onions and other nutrients on the absorption of epigallocatechin gallate from green tea extract in human volunteers. Boston: 2nd International Conference on Tumor Progression & Therapeutic Resistance Proceedings; 2005:89.

[45] Mekala P, Arivuchelvan A, Bioenhancer for animal health and production: A review, Agriculture, 2012, 6(11), 1–6.

[46] Dubey S, Valecha V, Prodrugs: A Review, World Journal of Pharmaceutical Research, 3(7), 2014, 277–297.

[47] Shanmugam S, Natural Bioenhancers: Current Outlook. Clinical Pharmacology & Biopharmaceutics, 4:e, 2015, 116. doi:10.4172/2167-065X.1000e116.

[48] Bouraoui A, Braziel ZL, Zougahi H, Effects of capsicum fruit on theophylline absorption and bioavailability in rabbits, Drug-Nutrient Interactions, 1988, 5(4), 345–350.

[49] Gokaraju GR, Gokaraju RR, D'Souza C, Frank E Bioavailability/Bio-efficacy enhancing activity of Stevia rebaudiana and extracts and fractions and compounds thereof. US Patent US 0112101; 2010.

[50] Zhao JQ, Du GZ, Xiong YC, Wen YF, Bhadauria M, Nirala SK, Attenuation of beryllium induced hepatorenal dysfunction and oxidative stress in rodents by combined effect of gallic acid and piperine, Archives of Pharmacal Research, 2007, 30(12), 1575–1583.

[51] Wacher VJ, Benet ZL Use of gallic acid esters to increase bioavailability of orally administered pharmaceutical compounds. U.S. Patent US 6180666 B1; 2001.

[52] Ogita A, Fujita K-I, Tanaka T, Enhancement of the fungicidal activity of amphotericin B by allicin: Effects on intracellular ergosterol trafficking, JPm, 2009, 75(03), 222–226.

[53] Zhang W, Tan TM, Lim LY, Impact of curcumin induced changes in P glycoprotein and CYP3A expression on the pharmacokinetics of peroral celiprolol and midazolam in rats, Drug Metabolism and Disposition, 2007, 35, 110–115.

[54] Basu NK, Human UDP – Glucuronyl transferase show a typical metabolism of mycophenolic acid and inhibition by curcumin, Drug Metabolism and Disposition, 2004, 32, 768–777.

[55] PrashithKekuda TR, Nishanth BC, Praveen Kumar SV, Kamal D, Sandeep M, Megharaj HK, Cow urine concentrate: A potent agent with antimicrobial and anthelmintic activity, Journal of Pharmacy Research, 2010, 3, 1025–1027.

[56] Cho SW, Lee SH, Choi SH, Enhanced oral bioavailability of poorly absorbed drugs. I. Screening of absorption carrier for the ceftriaxone complex, Journal of Pharmaceutical Sciences, 2004, 93, 612–620.

[57] Gowda DV, Deshpande RD, Keerthy HS, Datta V, Fabrication and evaluation of solid dispersion containing simvastatin, International Journal of Pharmaceutical Sciences Review and Research, July – August 2014, 27(2), 403–407.

[58] Kang MJ, Cho JY, Shim BH, Kim DK, Lee J, Bioavailability enhancing activities of natural compounds from medicinal plants, Journal of Medicinal Plants Research, 2019, 3(13), 1204–1211.

[59] Khan IA, Mirza ZM, Kumar VV, Qazi GN, Piperine, a phytochemical potentiator of ciprofloxacin against Staphylococcus aureus, Antimicrobial Agents and Chemotherapy, 2016, 50(2), 810–812.

[60] Patwardhan B, Mashelkar RA, Traditional medicine-inspired approaches to drug discovery and development: Can Ayurveda show a way forward?, Drug Discovery Today, 2009, 14, 804–811.

[61] Shaikh J, Ankola DD, Beniwal V, Singh D, Kumar MN, Nanoparticle encapsulation improves oral bioavailability of curcumin by at least 9 fold when compared to curcumin administered with piperine as absorption enhancer, European Journal of Pharmaceutical Sciences: Official Journal of the European Federation for Pharmaceutical Sciences, 2009, 37, 223–230.

[62] Bajad S, Bedi KL, Singla AK, Johri RK, Piperine inhibits gastric emptying and gastrointestinal transit in rats and mice, Planta Medica, 2011, 67, 176–179.

Madhur Kulkarni, Roopal Bhat, Suvarna Ingale, Abhijit Date

Chapter 5
Herbal bioenhancers in cancer drug delivery

Abstract: Effective chemotherapy with minimal adverse effects is the ultimate goal of cancer therapy. However, ever-mutating tumor cells, and difficult-to-target tumor environment impede the effective delivery of anticancer drugs to the tumor site. Additionally, suboptimal biopharmaceutical properties of anticancer drugs and drug-efflux transporters further make an increased therapeutic dose necessary to achieve the desired chemotherapeutic effect. The administration of higher doses of anticancer agents leads to several dose-dependent adverse effects, some of which could be more serious than the actual disease. Therefore, improving the therapeutic efficacy and reducing the dose of the anticancer drug is of paramount importance to improve the benefit vs. risk ratio of the cancer therapy. Bioenhancers are the agents that can improve the bioavailability and efficacy of the drugs upon concomitant administration. They can reduce the drug dose, cost of the therapy, drug resistance, and adverse effects. Plant-derived bioenhancers such as piperine, glycyrrhizin, quercetin, curcumin, allicin, cumin, and naringin have been explored for optimizing the therapeutic outcome of cancer chemotherapy. Bioenhancers have been shown to improve efficacy and reduce the dose of anticancer drugs by several mechanisms such as improvement of the pharmacokinetic profile of the cytotoxic agent via drug permeation enhancement across the physiological barriers, inhibition of the drug efflux transporters, and suppression or alteration of the drug metabolism, and elimination. Many of these agents have been shown to inherently possess some anticancer activity. Hence, several preclinical studies involving a combination of the bioenhancer with the anticancer agent have demonstrated synergism and improved pharmacodynamics. This chapter elucidates the mechanism of action and potential applications of various plant-based bioenhancers to augment outcomes of cancer chemotherapy.

Madhur Kulkarni, SCES's Indira College of Pharmacy, New Mumbai Pune Highway, Tathwade, Pune 411033, Maharashtra, India, e-mail: madhur.kulkarni@indiraicp.edu.in
Roopal Bhat, Suvarna Ingale, SCES's Indira College of Pharmacy, New Mumbai Pune Highway, Tathwade, Pune 411033, Maharashtra, India
Abhijit Date, Department of Pharmaceutics, The Daniel K. Inouye College of Pharmacy, University of Hawaii at Hilo, 200 W. Kawili Street, Hilo, HI 96720, USA

https://doi.org/10.1515/9783110746808-005

5.1 Introduction

Traditional systems of medicine have always contributed in a great way to the new drug development process through reverse pharmacology, leading to the identification of pharmacologically active phytochemicals from medicinal plants, with a concomitant reduction in the drug development cost [1].

A few of the most prominent gifts of ancient science of medicine like Ayurveda and traditional Chinese medicine is the concept of herbal bioenhancers. Although traditional medicine has been using herbs to decrease drug dose, drug toxicity, treatment cost, and to improve bioavailability of drug for centuries, the term bioenhancer was originally introduced by Indian scientists working at the Indian Institute of Integrative Medicine, Jammu. Bioenhancers are the substances that increase the bioavailability and, thereby, pharmacological effect of the drug with which it is given, without any of its own distinctive biological action at a low dose at which it is used [2]. An ideal and effective bioenhancer is nontoxic, effective at low doses, compatible with active pharmaceutical ingredients and excipients, free from toxicity, and easy to formulate [3].

Bioenhancers can be natural or synthetic, based on the source.

- **Natural bioenhancers:** These include substances obtained from herbs or plants such as piperine, quercetin, naringin, glycyrrhizin, genistein, niaziridin, curcumin, and sinomenine [4]. Animal-sourced bioenhancers include cow urine, ghee, and honey [5].
- **Synthetic bioenhancers:** These include substances obtained from synthetic origin like Eudragit E/HCl (neutralized Eudragit E using hydrochloric acid) [6], polyvinylpyrrolidone K-30, polyvinyl caprolactam–polyvinyl acetate–polyethylene glycol graft copolymer (Soluplus®) [7], Poloxamer 407 [8], polyethylene glycol 400 [9], Kolliphor TPGS® [10], etc.

Piperine is one of the first scientifically proven bioavailability enhancers in the world, discovered in 1979 [2]. Thereafter, several studies demonstrated 30–200% enhancement in bioavailability or efficacy of many drugs including antitubercular, antileprotic, antibiotics, NSAIDS, cardiovascular, and CNS (central nervous system) drugs, when used in combination with piperine [11]. The favorable outcomes of the piperine-based combination treatments increased the trust as well as the thrust for the herbal bioenhancers as adjuvants in therapeutics. Curcumin, quercetin, borneol, allicin, capsaicin, cumin, resveratrol, aloe vera, and naringin are some of the extensively studied plant-based bioenhancers.

Herbal bioenhancers can also be classified based on mechanism of actions as:

- **Para glycoprotein (P-gp) efflux pump inhibitors:** caraway, quercetin, naringin, and cumin.
- **CYP 450 enzymes and isoenzymes suppressor**: curcumin, gallic acid, and its esters.
- **Gastrointestinal tract absorption regulator:** aloe vera, glycyrrhiza, etc. [12].

Various mechanisms by which herbal bioenhancers work are depicted in Figure 5.1.

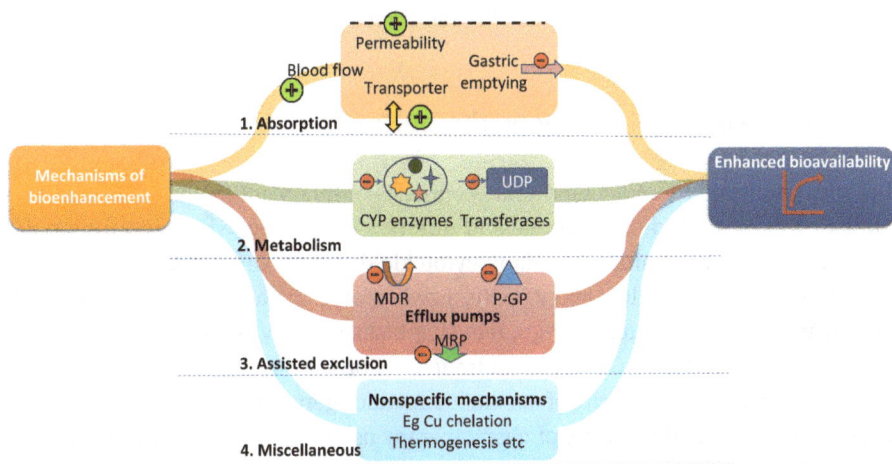

Figure 5.1: Mechanism of action of herbal bioenhancers; MDR, multidrug resistance pumps; UDP, uridine diphosphate-glucoronyl transferase; CYP, cytochrome P450 enzyme family.

5.2 Bioenhancers in cancer chemotherapy

Currently, cancer is considered a major health problem, worldwide. World Health Organization estimates cancer as the second prime cause of death globally, with about 10 million deaths per year. It is estimated that approximately one in six deaths is due to cancer, worldwide [13–15].

The current management options available for cancer include surgery, radio-therapy, chemotherapy, immunotherapy, and targeted therapy. Of these, chemo-therapy is the first-line treatment for hematological malignancies such as leukemia, lymphoma, multiple myeloma, and small cell lung cancer (SCLC). In addition, che-motherapy is used as a complementary treatment for some solid tumors to remove postsurgical residual nodules in order to prevent relapse [16, 17].

Since normal cells of the body get transformed into cancer cells, cytotoxic drugs used in cancer chemotherapy cannot discriminate between normal cells and cancer cells and, thus, exert their cytotoxic action on both, causing unpleasant side effects and, in some cases, toxicity [18]. Ideal cancer chemotherapy is expected to deliver the drug in an effective amount at the desired rate, only to tumor cells, for a duration long enough to exert cytotoxic action. With advancement in knowledge of the molecular pathogenesis of cancer, there are more than 100 anticancer drugs dis-covered and approved by the Food and Drug Administration. Some anticancer drugs act by inhibiting DNA synthesis, some by destroying the DNA; a few others inhibit cell division by interfering with various phases of the cell cycle. Still, cancer treatment is a challenging task, owing to physicochemical and biopharmaceutical issues related to bioavailability and drug resistance to cancer therapeutics [19].

5.2.1 Shortcomings of cancer chemotherapy

It is very important to understand that economics of treatment including cancer chemotherapy is related to drug dosage and duration of therapy, which in turn are dependent on drug bioavailability. Many cytotoxic drugs cannot permeate across cell membranes due to either low lipophilicity, a zwitterionic character at physiological pH, poor hydrophilicity, or efflux by P-gp, reducing bioavailability and causing therapeutic failure (Figure 5.2A) [20]. Globally, continuous efforts are taken to reduce the dose, treatment duration, and toxicity of chemotherapeutic agents by improving the bioavailability, and reducing the toxicity of chemotherapy using targeting approaches to make treatment efficacious, safe, and cost-effective, for all the sections of the society [11].

A widely used approach to enhance the therapeutic efficacy and reduce the dose is to improve the bioavailability of drugs. However, besides low bioavailability, the development of resistance to anticancer drugs via various intrinsic or extrinsic mechanisms also contributes to therapeutic failure and economic burden in cancer chemotherapy (Figure 5.2B). A logical approach to prevent the development of resistance is to use more than two anticancer agents in combination and enhance cytotoxic activity. Using drug combinations during cancer chemotherapy reduces the chances of selection of mutants and, thus inhibits the development of drug resistance.

The knowledge of physicochemical and biopharmaceutical issues related to bioavailability and drug resistance to cancer therapeutics highlights the use of bioenhancers as an important strategy to improve the therapeutic outcome of chemotherapy. The scientific data stemming from various *in vitro* cell line studies and preclinical studies in murine models, as well as clinical studies, have clearly shown the benefits of using bioenhancers as adjuvant therapy in anticancer treatment. The prominent benefits include improved efficacy, prevention of resistance, dose reduction, as well as reduced adverse reactions and toxicity associated with the cytotoxic drugs [5, 20]. In this chapter, through *in vitro*, preclinical, and/or clinical studies reported in the literature, we aim to establish the potential of the most commonly explored bioenhancers to improve the outcomes of chemotherapy.

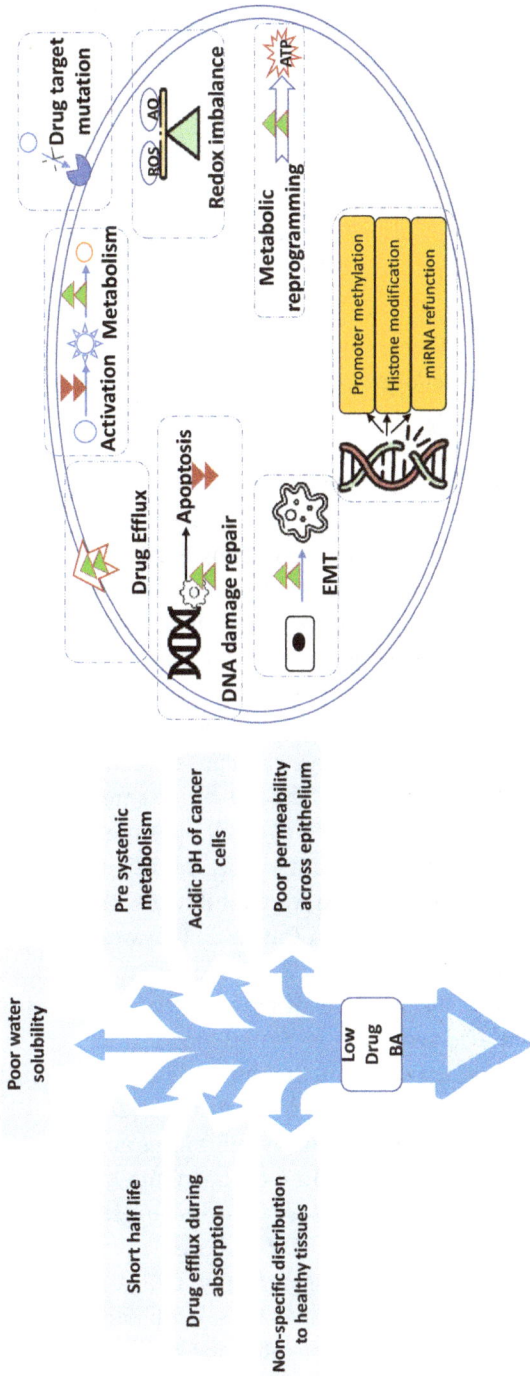

(A): Physicochemical and biopharmaceutical challenges

(B): Mechanism of development of resistance

Figure 5.2: (A) Mechanisms for development of resistance. EMT, epithelial mesenchymal transition; ROS, reactive oxygen species; AO, antioxidants; upregulation; downregulation. (B) Physicochemical and biopharmaceutical challenges resulting in low bioavailability. BA, bioavailability.

5.3 Specific examples of bioenhancers with case studies

5.3.1 Piperine

Piperine, the world's first identified natural bioenhancer [21], is one of the most potent bioenhancers. It belongs to the alkaloids class of plant secondary metabolite. It is solid in nature with a molecular weight of 285.34 Da, insoluble in water, and has a weakly basic character, tasteless at first, but leaves a burning taste after a while. Besides Piperaceae family (*P. longum* and *P. nigrum*), it has been found in numerous other plant species like *Anethum sowa*, *Rhododendron faurie*, and *Vicoa indica* [22].

5.3.1.1 Mechanisms responsible for bioenhancing effect

Piperine has been shown to act as a bioenhancer by modulating four different processes, namely, absorption, metabolism, exclusion through efflux transporters, and permeation [23]. Piperine has been proven to alter the enzyme kinetics in the intestinal wall by stimulating enzymes like glycyl–glycine dipeptidase and leucine amino peptidase. This aids in enhancement of nutritional absorption of the amino acids in the body. The non-polar/hydrophobic nature of piperine facilitates its portioning and interactions with surrounding lipids and can cause conformational changes of enzymes in the intestine. Ultra-structural studies on rats treated with piperine (10 mg/kg body wt.) showed increased length of small intestine microvilli at 2 h post absorption [24–26]. Another study has shown that solid/liquid gastric emptying in rats and gastrointestinal transit in mice were inhibited with piperine treatment in a dose and time-dependent manner, and this inhibition was found to be independent of gastric acid and pepsin secretion.

Apart from these, other nonspecific mechanisms of piperine that increase the absorption include decreased gastric acid secretion, increased blood supply of gastrointestinal tract, increased emulsification in the gut, and increase in enzymes like transpeptidase, which participate in active transport in intestinal cells [27, 28].

Piperine has also shown to alter the metabolism by inhibiting enzymes including aryl hydrocarbon hydroxylase (AHH), uridine diphosphate-glucoronyl transferase (UDP-GT), ethylmorphine-*N*-demethylase, 7-ethoxycoumarin-*O*-deethylase, 3-hydroxy- benzopyrene glucuronidation, UDP-glucose dehydrogenase (UDP-GDH), 5-lipooxygenase, cyclooxygenase-1, and, importantly, CYP3A4, which is the most prevalent metabolizing enzyme [29–31]. This enzyme inhibition prolongs the elimination and, thus, half-life of the drugs administered along with piperine [31].

Piperine is also a known inhibitor of efflux pump protein, P-gp, which is present primarily in apical epithelial cell linings of small intestine, colon, pancreas, brain, kidneys, liver, etc. [28]. This inhibition aids in further increasing the intracellular concentration of a drug. Piperine has also been demonstrated to modify the rate of glucuronidation by lowering the endogenous UDP-glucuronic acid content and also by inhibiting UDP–GT transferase activity [32]. Glucuronic acid is one of the important solubilizers in the body that prevents hydrophilic complexes from permeating across the cell membrane.

Various studies have been carried out to demonstrate the bioenhancing effect of piperine. The following sections discuss a few major *in vitro* and *in vivo* studies of piperine used in cancer therapy.

5.3.1.2 *In vitro* studies demonstrating bioenhancing effect

Piperine has shown the potential to reverse MDR *in vitro* studies by multiple mechanisms. The resistance to doxorubicin (DOX) in human alveolar basal epithelial cells A-549 and breast cancer MCF-7 cell lines was shown to reversed by piperine (50 μM) by 14.14- and 32.16-fold, respectively. The same study showed mitoxantrone re-sensitization of cells, 6.98-fold. The long-term treatment of cells with piperine inhibited transcription of the corresponding ATP-binding cassette (ABC) transporter genes [33].

Another major cause of MDR when using anticancer drugs is P-gp efflux. Many chemotherapeutic agents such as vinblastine, vincristine (VCR), docetaxel (DTX), flutamide, cyclophosphamide (CPM), paclitaxel (PTX), and ifosfamide cause P-gp overexpression, leading to MDR [34]. Piperine inhibits P-gp by competitively binding at ATP binding sites and, hence, reversing P-gp-induced MDR. In one study, two of the piperine analogs, namely, Pip1 and Pip2, were able to reverse the drug resistance *in vitro* in overexpressing P-gp cervical and colon cancer cells, when co-administered with VCR or colchicine or PTX [35].

The synergistic activity of piperine with anticancer drugs is also widely reported. In one such study, Motiwala and Rangari [36] showed that simultaneous *in vitro* exposure of PTX and piperine in MCF-7 breast cancer cells for 24 h showed better activity than sequential exposure of piperine followed by PTX (for 24 h each). The combination index (CI) of 0.58 at IC_{50} values demonstrated the synergism. This synergistic enhancement of the cytotoxic activity of PTX with piperine was shown by Pushpa Ragini et al. [37]. The group reported that piperine could reduce the IC_{50} value of PTX on MDA MB-231 cell lines (50–25 μM). The cell proliferation analysis demonstrated that PTX acts on fourth generation of cell proliferation, while PTX along with piperine acts at the third generation. This indicated that piperine acts by reducing the duration of the lag phase in PTX activity, which agrees with the results reported by Lai et al. [38] that piperine treatment led to downregulation of cyclin B1 expression and, thus, increased the percentage of cells in the G2/M phase in 4T1 cells.

Piperine when used with curcumin has also shown greater *in vitro* antimetastatic effect on breast cancer lines MDA-MB-231 and MCF-7 [39], as well as in hepatocarcinoma cells induced by diethylnitrosamine [40]. Kakarala et al [39] have shown that *in vitro* studies in normal and malignant breast cells, where piperine along with curcumin suppressed the formation of mammosphere, serial passaging, and aldehyde dehydrogenase (ALDH+) breast stem cells. The combination also completely suppressed the Wnt signaling at a concentration of 10 µM. In addition, Khamis et al., observed the synergistic anticancer activity of piperine, bee venom, and hesperidin on MCF-7 cells [41].

Furthermore, piperine combined with thymoquinone (TQ), the major active compound in the seeds of *Nigella sativa*, demonstrated a synergistic interaction, *in vitro,* on mouse epithelial breast cancer cell line EMT6/P cells. The combination therapy showed significantly reduced tumor sizes in Balb/C mice inoculated with EMT6/P cells with 60% cure percentage. However, the single treatments resulted in lower tumor percentage regression rates. This effect of the combination is mediated via the inhibition of angiogenesis, enhancing the T helper 1 response and apoptosis [42].

5.3.1.3 *In vivo* studies demonstrating bioenhancing effect

Various pharmacokinetic and pharmacodynamic effects of piperine are explored in *in vivo* and preclinical studies. It has been demonstrated that the bioavailability of anticancer drugs including, DTX, 5-fluorouracil (5-FU), etoposide, PTX, and rapamycin (RPM) can be improved by piperine both *in vitro* and *in vivo* [43–47]. In one such PK study [48], piperine was administrated via intravenous bolus at 3.5 mg/kg and via oral administration at 35 mg/kg, while DOX was intravenously administrated at 7 mg/kg to Sprague-Dawley rats. It was found that the AUC and C_0 of DOX and half-life of piperine significantly increased after their combination use, suggesting potential enhanced bioavailability of not only DOX but also piperine, which may result in enhanced pharmacological effects, overall.

To study the CYP3A4 inhibitory activity of piperine, Peter Makhov and the group studied the DTX pharmacokinetic activity in male SCID mice. Administration of piperine significantly increased and prolonged the mouse plasma levels of DTX compared with mice that received DTX alone. Most notably, the difference in DTX plasma levels was most pronounced between 1 and 2 h after DTX administration, i.e., plasma DTX levels in the piperine/DTX group was 7-fold higher than in the DTX-only group. An *in vivo* xenograft model of human CRPC was utilized to assess the antitumor effect of DTX when coadministered with piperine. Two weeks after the initiation of treatment, coadministration of piperine and DTX resulted in the most significant inhibition of tumor growth (188% vs. 461%, $P = 0.003$), as compared to the control which received only vegetable oil. Importantly, the resulting increase in the mean plasma concentrations of DTX in mice fed with piperine did not result in augmented mice toxicity during treatment. Thus, by inhibiting hepatic clearance of DTX, piperine resulted in increased DTX serum levels, without increasing DTX-mediated toxicities [45].

In the 5 L rat hepatoma model, piperine was found to directly interact with CYP1A1 enzyme and inhibit the benzo(a)pyrene metabolism and, hence, decreased the aggressiveness of cancer [49]. A significant reduction in activity of phase-I enzymes (i.e., CYP-450 family) was caused after oral administration of piperine in mice model of lung carcinogenesis. This also showed increase in glutathione peroxidase (GPX) and glutathione reductase (GR), which are glutathione-metabolizing enzymes. These enzymes are responsible for protecting cells against a variety of endogenous and exogenous toxic compounds, such as reactive oxygen species (ROS) and chemical carcinogens [50].

RPM, which can be potentially used for breast cancer treatment, is susceptible to P-gp efflux, which causes poor oral bioavailability. To overcome this limitation [47], piperine was incorporated in poly (D, L-lactide-co-glycolide) (PLGA) nanoparticles of RPM to improve the oral bioavailability and efficacy. The RPM uptake was shown to increase in presence of piperine in the ex vivo study using everted gut sac method. The *in vivo* pharmacokinetic studies in rats showed a better absorption profile of RPM (10 mg/kg for both RPM and piperine), and around 4.8-fold increase in oral bioavailability. This combinatorial enhancement was also observed in *in vitro* cell line study on human-derived breast cancer cell lines, which indicated better efficacy of nanoparticles, as compared to control free drug solution.

Etoposide, a very commonly used anticancer agent, shows a wide variability among oral bioavailability from patient to patient. In a study [51], a piperine analogue, namely, 4-ethyl 5-(3, 4-methylenedioxyphenyl)-2E, 4E-pentadienoic acid piperidide (PA-1) (20 mg/kg) was co-dosed with etoposide (20 mg/kg) in mice. The results (Table 5.1) indicated around 2.32-fold enhancement of the absolute bioavailability after coadministration. In another mechanistic investigation of *in vitro* and *in vivo* models, PA-1 significantly reduced the intestinal efflux clearance, efflux rate, and the total plasma clearance of etoposide in a single pass *in situ* perfusion experiment. However, the PAMPA assay showed no alteration in the passive diffusion pattern in presence of PA-1. PA-1 was also found to inhibit reactions like deethylation and demethylation catalyzed by NADPH and erythromycin *N*-demethylase, ethoxyresorufin-*O*-deethylase (EROD), and 7-methoxycoumarin-*O*-demethylase (MOCD). PA-1 showed no cytotoxicity towards the mucosal membrane and showed no adverse effect. Hence, this indicated that the oral bioavailability enhancement of etoposide in presence of PA-1 could be due to its ability to modify P-gp efflux and CYP 3A4-mediated drug metabolism.

Curcumin, which has been a reported anticancer agent, exhibits low oral bioavailability. The potential of piperine to enhance its bioavailability has been explored in many studies. In one such study [52], piperine (10 mg/kg) and curcumin (50 mg/kg, 100 mg/kg), alone and in combination were evaluated in Dalton lymphoma ascites (DLA)-bearing mice. Treatment with curcumin at two different concentrations and piperine alone has shown reduced tumor volumes and increased lifespan of tested animals.

Table 5.1: Pharmacokinetic parameters (etoposide): RB (relative bioavailability) = ($AUC_{etoposide}$ + PA-1/$AUC_{etoposide}$) × 100; AB (absolute bioavailability) = ($AUC_{oral}/AUC_{i.v.}$) ×($Dose_{i.v.}/Dose_{oral}$) × 100. *$p < 0.001$ versus control, **$p < 0.01$ versus control.

Parameter	Etoposide (20 mg/kg. p.o.)		Etoposide (20 mg/kg i.v.)
	Control	**In the presence of PA-1 (20 mg/kg. p.o.)**	
AUC (ng h/mL)	3,142.84 ± 196.40	7,293.55 ± 224.96*	11,562.1 ± 572.65
C_{max} (ng/mL)	639.17 ± 23.83	1,235.0 ± 56.59*	
T_{max} (h)	0.50	0.50	
$T_{1/2}$ (h)	6.61 ± 0.88	13.57 ± 1.62**	1.94 ± 0.29
Cl (mL/h/kg)	6,732.67 ± 590.82	4,073.1 ± 294.66**	1,744.33 ± 79.85
V_d (mL/kg)	63,557.94 ± 3,192.04	46,654.36 ± 2,878.36**	4,883.18 ± 294.42
RB (%)	100	232.06	
AB (%)	27.18	63.08	

Reprinted from *Simultaneous determination of etoposide and a piperine analogue (PA-1) by UPLC–qTOF-MS: Evidence that PA-1 enhances the oral bioavailability of etoposide in mice*, by B.S. Sachin et al. [51].

The therapeutic dose of piperine, according to reported studies, is in the range of 3–399 mg, without any toxicity. The bioenhancing activity of piperine has been studied for dose of approximately 15-20 mg/day, in divided doses [32]. However, it is shown that significant inhibition of the monooxygenases is apparent for lower concentration of piperine than the higher ones [53].

A pilot clinical study [54] assessed the effect of a combinatorial treatment of curcumin, taurine, and piperine on circulating levels of interleukin-10 (IL-10), and microRNAs miR-141 and miR-21 in hepatocellular carcinoma (HCC) patients. Oral dose of 4 g curcumin, 500 mg taurine, and 40 mg piperine was administered daily to 20 HCC patients for three successive 30-day trials. The results showed that combined treatment was able to produce a significant decrease in the levels of serum IL-10 and miR-21, with overall increase in median survival rate.

In another *in vivo* and pilot clinical study, effect of piperine on the pharmacokinetics of curcumin is studied in rats and healthy human volunteers [55]. Concomitant administration of piperine 20 mg/kg along with curcumin in the dose 2 g/kg to rats increased the serum concentration at 1 and 2 h (1.55 ± 0.21 and 1.50 ± 0.25 pg/mL), respectively, being significantly higher than curcumin concentration, in the absence of piperine (1.00 ± 0.26 pg/mL). The bioavailability was increased by 154%, while T_{max} was also significantly increased ($P < 0.02$), and elimination $T_{1/2}$ was significantly decreased ($P < 0.02$).

5.3.2 Curcumin

Curcumin (diferuloylmethane) is a plant-derived phenolic, colored component, obtained from the rhizomes of *Curcuma longa* Linn, family Zingiberaceae. Curcumin exhibits a range of therapeutic activities including antioxidant, analgesic, anti-inflammatory and antiseptic, hepatoprotective, and anticarcinogenic. Apart from being established as an active phytoconstituent, curcumin is also used as a bioenhancer for antimicrobial and anticancer drugs [56].

5.3.2.1 Mechanisms responsible for bioenhancing effect

The mechanisms through which curcumin enhances the bioavailability of drugs include, modulating the function and expression of drug-metabolizing enzymes (CYP3A4), alterations in P-gps, and reduction in activity of nonspecific drug-metabolizing enzymes like alcohol and aldehyde dehydrogenase. Curcumin is also a biologically active copper chelator [57]. In recent times, curcumin has got significant attention as a safer therapeutic adjuvant in the field of cancer chemotherapy. The combination of curcumin with the anticancer agents may not only improve their anticancer efficacy, but could also exert a synergistic effect, owing to curcumin's inherent anticancer activity. The mechanisms by which curcumin propagates its anticancer activity include regulation of cell cycle, induction of apoptosis and autophagy, interruption of molecular signaling, and inhibition of metastasis [56].

5.3.2.2 *In vitro* studies demonstrating bioenhancing effect

Several *in vitro* studies on the combination of curcumin and the standard chemotherapeutic agents have been conducted in various cancer cell lines. There have been encouraging study outcomes that have substantiated the role of curcumin and the underlying mechanisms as a bioenhancer via various mechanisms. Studies in HepG2 (human hepatoma cell line) cell lines showed only about 20% cell inhibition in the presence of 0.1 µg/mL of adriamycin (ADM) [58]. On the other hand, 35% cell inhibition in the presence of curcumin (20 µM) and ADM – 0.05 µg/mL was observed. A similar experiment conducted in normal hepatic LO2 cells did not lead to any cytotoxicity, thus confirming the safety of the curcumin + ADM combination. Presence of curcumin aided in a higher cytotoxic response of ADM at a 50% lower dose. A ratio of B-cell lymphoma 2 (Bcl-2) and Bcl-2-associated X protein (Bax) regulates cell apoptosis. Bcl-2 is responsible for the prevention of mitochondrial membrane depolarization, which eventually prevents apoptosis. In contrast, Bax-induced depolarization of mitochondrial membrane leads to cytochrome c release and apoptosis.

Reduction in Bcl-2/Bax ratio and promotion of caspase-3 were the main anticancer mechanisms of the ADM-curcumin combination in HepG2 cells. Additionally, the combination demonstrated significant mitochondrial fragmentation as well as the activation of the autophagic process, as confirmed by upregulation of LC3-II levels.

Tang et al., evaluated the action of curcumin on the gastric cancer cell line resistant to VCR (SGC7901/VCR) [59]. These cells showed 45 times higher resistance to VCR than the nonresistant cells. The IC_{50} value of VCR in the SGC7901/VCR cell line dropped to 11.60 μM/L, 3.28 μM/L, and 1.53 μM/L in the presence of 5, 10, and 20 μM/L curcumin, respectively. The IC_{50} value of VCR was as high as 22.27 μM/L, in the absence of curcumin. Similarly, the combination treatment of VCR with 5, 10, and 20 μM/L of curcumin led to 20.5%, 32.7%, and 43.7%, apoptosis of SGC7901/VCR cells, respectively, unlike only 12.3% in the absence of curcumin. Curcumin treatment (10 μM/L) increased the accumulation of Rh123 in SGC7901/VCR cells by 40%, owing to downregulation of P-gp expression from 42.73% to 17.69%. Curcumin-VCR co-treatment improved caspase-3 activity by 44%. Thus, curcumin-induced reduction in P-gp function and expression, as well as higher caspase-3 activation increased the sensitivity of tumor cells to VCR. Cisplatin (DDP)-induced formation of cross-links within the strands of DNA is responsible for the cytotoxic effect of the drug. However, reduced copper transporter-1 (CTR1) expression in the DDP-resistant tumor cells is responsible for lower cellular uptake of DDP and lower DNA intercalation. Curcumin, when combined with DDP, led to the enhanced drug efficacy. Incubation of A549 cells with curcumin for 12 h followed by DDP treatment resulted in a 1.4-fold increase in DDP accumulation inside the A549 cells. Increased expression of CTR1 induced by curcumin is considered responsible for enhancing the uptake of platinum ions in the tumor [60]. DOX is a widely used anticancer drug in breast cancer treatment, but the development of DOX resistance limits its chemotherapeutic potential. Due to the narrow therapeutic index, DOX therapy is associated with higher adverse effects. Wen C. et al., studied the outcome of curcumin treatment on MCF-7/DOX and MDA-MB-231/DOX cell lines [61].These human breast cancer cells lines were resistant to DOX treatment. However, the combination treatment of DOX and curcumin reversed the resistance, due to suppressed efflux of DOX by ABC subfamily B member 4 (ABCB4). Curcumin significantly reduced the IC_{50} value of DOX, especially in ABCB4 overexpressing cells, i.e., 10.9 ± 1.4 μM in MCF-7/DOX and 14.1 ± 1.6 μM in MDA-MB-231/DOX cells. Curcumin inhibited the efflux function of ABCB4 by 30% in 120 min, without altering the protein expression, but by reducing the ATPase activity of ABCB4. The intracellular concentration of DOX was significantly increased by curcumin treatment in DOX-resistant cell lines. In a research work undertaken by Zhou et al., [65] curcumin was shown to enhance the anticancer activity of mitomycin C (MMC) in breast cancer cells *in vitro* and *in vivo*. Studies in MCF-7 cells indicated a 50% dose reduction for MMC in the presence of curcumin to attain equal cytotoxic effect as in the case of a single treatment. Coadministration of curcumin and MMC showed 71.17% of cells in the G1 phase, as opposed to treatment with curcumin or MMC alone that resulted in 51.51% and 47.47% cells, in the G1 growth phase. Suppression of cyclin D1, cyclin E, cyclin A, CDK2, CDK4, and p38 MAPK (mitogen-activated protein kinase) pathway was responsible for the anti-proliferative effect of the combination therapy. In another study [62], lipid nano-particles containing a combination of DOX and curcumin showed IC_{50} of 1.16 µg/mL, as compared to 2.74 µg/mL of free DOX in HepG2 cells. On the other hand, the nanoparticles exhibited a decreased cytotoxicity in human normal liver cells L02, compared to free DOX, which could be due to the slow release of DOX (65% within 48 h) from the novel formulation. Gemcitabine treatment of pancreatic cancer, though efficacious, develops resistance over time. Gemcitabine treatment alone showed ≤5% apoptosis in pancreatic cancer cell lines BxPC-3, Panc-1, MIA PaCa-2, and MPanc-96. But, when combined with curcumin, the apoptosis levels rose to 40 ± 3.06% in BxPC-3, 55 ± 1.73% in Panc-1, 98 ± 1.15% in MIA PaCa-2, and 32 ± 2.65% in MPanc-96 cell lines [63].

5.3.2.3 *In vivo* studies demonstrating bioenhancing effect

Several *in vivo* studies have been carried out to demonstrate the bioenhancing effect of curcumin, when combined with various anticancer drugs. This section contains a summary of prominent studies that established curcumin's bioenhancing effect to improve outcomes of chemotherapy.

In the A549 xenograft mouse model, the addition of a lower dose of curcumin (50 mg/kg) to DDP (2.5 mg/kg) showed a 2-fold higher reduction in tumor volume, compared to DDP treatment alone over a period of 7 days. The combined treatment resulted in a 1.5-fold increase in DDP levels in the tumor tissue (8.2 mg Pt/g DNA versus 5.4 mg Pt/g DNA in DDP-treated mice). Curcumin co-treatment mitigated the unwanted effects of a reduction in body weight, spleen, and liver weights observed in DDP-treated animals.

The chemotherapeutic agent, DTX is a P-gp substrate, and its metabolism is primarily catalyzed by the CYP3A enzyme. Curcumin has the potential to downregulate P-gp and CYP3A protein levels. In the preclinical study conducted by Yan and coworkers, four days of oral pretreatment with 100 mg/kg of curcumin, before 30 mg/kg oral administration of DTX was highly effective in improving the pharmacokinetics of the drug [64]. The C_{max} and AUC of DTX were enhanced about 10- and 8-fold, respectively, in rats pretreated with oral curcumin. The absolute bioavailability of DTX rose to 40% in curcumin pretreated rats, which was almost 8-fold higher in comparison to the 5.5% availability in the DTX control group. However, curcumin administration just 30 min before the oral administration of DTX did not change the pharmacokinetic profile of the drug. With 100 mg/kg/day curcumin administration for 4 days before the DTX treatment, the cumulative systemic absorption of curcumin was sufficient to inhibit the manifestation and activity of P-gp and

CYP3A. As a consequence, the oral bioavailability of DTX improved remarkably due to improved oral uptake and suppression of the metabolism [64].

With the advent of nanotechnology, nanocarriers such as liposomes, solid lipid nanoparticles, polymeric nanoparticles, and nanomicelles have been explored to improve the permeation, bioavailability as well as targeting potential of the drugs. The application of nanotechnology in cancer chemotherapeutics is most common and obvious, considering the aforementioned shortcomings of anticancer drugs. The work carried out by Zhao X. et al., involved the preparation of nanoparticles of DOX [62]. The *in vivo* study in the mouse model of liver cancer included treatment with saline, DOX nanoparticles (2 mg/kg), and DOX/curcumin nanoparticles (1:1; equivalent to 2 mg/kg of DOX) for 20 weeks. Treatment with DOX nanoparticles showed almost 2-fold reduction in tumor number as well as tumor size, as compared to the saline group. In the case of the group treated with DOX/curcumin nanoparticles, the tumor growth suppression was significantly greater than the group treated with DOX nanoparticles ($P < 0.05$). This proved the synergistic effect of DOX and curcumin on liver cancer. Therefore, the simultaneous delivery of DOX and chemosensitizer using nanocarriers could provide better treatment option for liver cancer.

In the breast tumor-bearing mice, intraperitoneal (i.p.) administration of 100 mg/kg curcumin along with 1.5 mg/kg MMC for 4 weeks resulted in 60.4% and 68.6% higher reduction in tumor weight, as compared to MMC and curcumin treatment alone. Combination treatment demonstrated no significant weight reduction in the animals. Curcumin combination therapy, thus, led to synergistic anticancer effects with a reduction in dose and the side effects of MMC [65].

MDA-MB-231 breast cancer cells were treated with a combination of curcumin and PTX. Curcumin (10 µM) terminated activation of NF-κB by inhibiting PTX-driven degradation of IκBά. This resulted in a better cytotoxic effect of the drug on cancer cell growth. The effect of combination therapy with 7 mg/kg PTX (once a week, i.p. injection) and 100 mg/kg curcumin (daily oral administration) was investigated in a nude mouse xenograft model. After 5 weeks of treatment, combination therapy significantly decreased the tumor diameter to 0.32 ± 0.64 cm, unlike in the case of the mice treated with drug or curcumin alone that showed the tumor diameter of 1.27 ± 0.79 cm and 1.33 ± 0.51 cm, respectively. Matrix metalloprotease 9 (MMP 9) is associated with tumor proliferation, angiogenesis, and metastasis. A decreased drug-induced activation of MMP 9, NF-κB, and degradation of IκBά observed in the mice receiving the combination of curcumin and PTX were responsible for lower drug resistance and better anticancer effect. The combination could help in the dose reduction of PTX, which is the most desirable outcome from the toxicity point of view [66].

During *in vitro* studies, the combination of curcumin with gemcitabine was shown to enhance the effect of gemcitabine and overcome chemoresistance. *In vivo* studies of curcumin and gemcitabine co-therapy was conducted in nude mice bearing orthotopic pancreatic tumors. Curcumin-induced inhibition of proliferation, angiogenesis, and downregulation of NF-κB-related gene products resulted in better antitumor efficacy of gemcitabine. Pancreatic cancer cells (MIA PaCa-2) were implanted in the pancreatic tails of nude mice. The tumor-bearing mice were treated with once daily oral dose of curcumin (1 g/kg) and twice weekly i.p. dose of gemcitabine (25 mg/kg). At the end of five weeks of the combination treatment, significant reduction in tumor volume ($P = 0.008$ versus control; $P = 0.036$ versus gemcitabine alone) and Ki-67 proliferation index ($P = 0.030$ versus control) were observed. The combination treatment was also highly effective in the inhibition of angiogenesis, as indicated by a decrease in CD31 + microvessel density ($P = 0.018$ versus control). Suppressed expression of NF-κB-regulated gene products (cyclin D1, c-myc, Bcl-2, Bcl-xL, cellular inhibitor of apoptosis protein-1, cyclooxygenase-2, MMP, and vascular endothelial growth factor (VEGF)) was considered responsible for the improved anticancer effect of dual therapy [63].

5.3.3 Borneol

Borneol is a bicyclic monoterpene that is naturally sourced from *Blumea balsamifera*, *Cinnamonum camphora* (L.) Presl, *Dryobalanops aromatica* Gaertner, and the volatile oils extracted from various other plant sources [67]. In recent times, it is synthesized using camphor or turpentine oil as a starting material. Traditional Chinese medicine has been using borneol with the names of Bing-Pian or Long–Nao for the past 1,000 years. The key role of borneol in the traditional prescriptions is that of a messenger or carrier to deliver the other actives mainly to the CNS. Cerebral ischemia, stroke, cerebral edema, coma, unconsciousness, and dementia are some of the CNS disorders wherein there is strong evidence of borneol use [68]. Apart from this, borneol has also shown promising antimicrobial, anti-inflammatory, analgesic, antidiabetic, antihypertensive, and anticancer activities.

5.3.3.1 Mechanisms responsible for bioenhancing effect

Borneol's known brain permeation-enhancing ability has led to further investigations into its possible role as a permeation enhancer across other physiological barriers, such as gastrointestinal and nasal mucosa, skin, and even corneal epithelium. Interaction with lipophilic cell membranes, inhibition of efflux pumps like P-gp, MDR, MRP (multidrug-resistant proteins), etc., and modulation of tight junction proteins are some of the key mechanisms by which borneol enhances the drug transport across the aforementioned barriers via trans as well as paracellular routes [69]. Borneol has been explored to enhance the anticancer activity of various plant-sourced and synthetic chemotherapeutic agents. These agents, when delivered with borneol either in conventional vehicles or after encapsulation in nanocarriers, have shown promising anticancer activity in preclinical studies. There are a few *in vitro* cell line studies that indicate anticancer activity of borneol in combination with other natural or synthetic cytotoxic agents. However, there has been no reported study proving the anticancer effect of borneol alone.

5.3.3.2 *In vitro* studies demonstrating bioenhancing effect

PTX, though preferred in treating various cancers, faces the challenge of multidrug resistance mediated by overexpression of P-gp efflux transporters in cancer cells. Borneol with its known P-gp inhibitory action was, hence, incorporated along with PTX in nanoparticles comprising pegylated-polyamidoamine (PAMAM) dendrimer (PB/NPs). Appreciably higher cytotoxicity and apoptosis were observed for PB/NPs compared to PTX nanoformulation plus free borneol (P/NPs + B) in A2780/PTX cells. The IC_{50} of 4.10 and 5.72 μM was observed for PB/NPs and P/NPs + B. Similarly, studies in the rat model of ovarian cancer revealed 3.68- and 3.22-fold higher drug accumulation in the tumor, when administered as PB/NPs and P/NPs + B, respectively, as compared to free drug. I.V. administration of PB/NPs and P/NPs + B showed 1.77 and 1.62 times higher total apoptosis rate, compared to the P + B group (free PTX plus borneol group) [70].

5.3.3.3 *In vivo* studies demonstrating bioenhancing effect

Glioma is a notably malignant, aggressive, and a predominant brain tumor in humans. It is surgically as well as medically difficult to access and, hence, shows poor prognosis and lower survival rates. Standard chemotherapeutic agents rarely achieve desired therapeutic concentrations in the brain, owing to their restricted delivery through BBB. This often necessitates the administration of higher drug dose that further leads to increase in the dose-related side effects. Borneol, with its established brain permeation-enhancing activity, has shown significant promise as a bioenhancer for the treatment of brain tumors. By improving the brain distribution of the drugs, it can improve their efficacy and reduce the dose and dose-related side effects. The work done by Cao et al., strongly propagates the role of borneol as a bioenhancer as well as a chemosensitizer in improving the therapeutic outcome of DOX in the treatment of glioma [71]. Studies in U 251 human glioma cells indicated 1.5 times higher uptake of DOX in the presence of 80 μg/mL borneol and subsequent higher cell death. In the mouse model of glioma, injection of a combination of borneol (40 mg/kg) and DOX (2 mg/kg) on alternate days, during the 21 days of study duration brought about a 50% reduction in the tumor volume and 30% reduction in tumor weight. The combination therapy was twice as effective as DOX treatment alone, which could be attributed to better brain permeation of DOX in presence of borneol. Borneol-induced excess synthesis of ROS was responsible for synergizing the anticancer effect of DOX. Combination of DOX and borneol resulted in significant DNA damage and arrest of cancer cell growth in G2/M phase.

DOX was incorporated into the phospholipid micelles of 1, 2-distearoyl-sn-glycero-3-phosphoethanolamine-N-[methoxy (polyethylene glycol)-2000] (DOX-PM) [72]. The formulation was also prepared by incorporating borneol along with the drug (DOX BO-PM). All the formulations were administered intravenously at the DOX dose of 5 mg/kg on every 3rd day on five successive occasions. Higher antitumor efficacy was confirmed in DOX BO-PMs, compared to DOX-PMs and plain DOX solution in the glioblastoma mouse model. DOX BO-PMs group showed the highest apoptosis, followed by the DOX PMs group, with the least being reported by the DOX solution group. DOX BO- PM showed the least tissue necrosis, hemorrhage in the brain tissue, and tumor volume among all the treatments (Figure 5.3). There was no indication of metastasis to the liver and lung tissues among the DOX BO- PM-treated animals, whereas, in saline-, DOX solution-, and DOX- PMs-treated mice, liver metastasis was observed. The DOX BO- PM showed very low weight loss among the treatment mice, whereas, the plain solution showed a rapid decline in weight.

Borneol, due to its BBB and tumor permeation-enhancing effects, could achieve a higher distribution of DOX in the brain and the tumor. This aided in enhancing the anticancer efficacy of the drug and, at the same time, reducing the side effects in other tissues. Carmustine (CMS), an alkylating agent, is often used in the treatment of glioma. The half-life of merely 15–30 min offsets its efficacy by reducing the retention at the site of action. The drug is also associated with high systemic toxicity. Guo et al., conjugated borneol and short peptide protein (Pep-1) with distearoyl phosphoethanolamine-polyethylene glycol to create a nanomeric micellar formulation (Bor-DSPE-PEG-Pep-1) [73]. Borneol aided in swift brain permeation of the drug, whereas Pep-1, with its high affinity towards IL 13 receptors, expressed abundantly on glioma cells assisted in targeting the glioma. The formulation (Pep-1/BO/CMS-PM) containing 2 mg/kg dose of CMS was administered by i.v. route on every third day on five successive occasions in mice bearing orthotopic glioma. The animals displayed the longest survival period of 53 days, whereas all the animals in CMS-treated group showed 45 days of survival. The micellar formulation exhibited three times higher tumor uptake and, hence, higher cytotoxicity than the free CMS treatment. This could be attributed to borneol-assisted improved penetration and pep-1-induced higher drug retention and higher tumor targeting. Most importantly, the formulation did not result in body weight loss or damage to the liver and spleen, or any abnormal alterations in biochemical assays indicating the safety of CMS therapy in the form of the novel formulation. In another study conducted by Wu et al., intragastric administration of 5 mg borneol in mice, 15 min prior to the i.v. administration of DOX liposomes, increased the brain-to-plasma ratio of the drug by almost 2.5-fold and enhanced the brain bioavailability of DOX to 114.54%, respectively, in comparison with only liposomes administration [74]. Borneol, owing to its

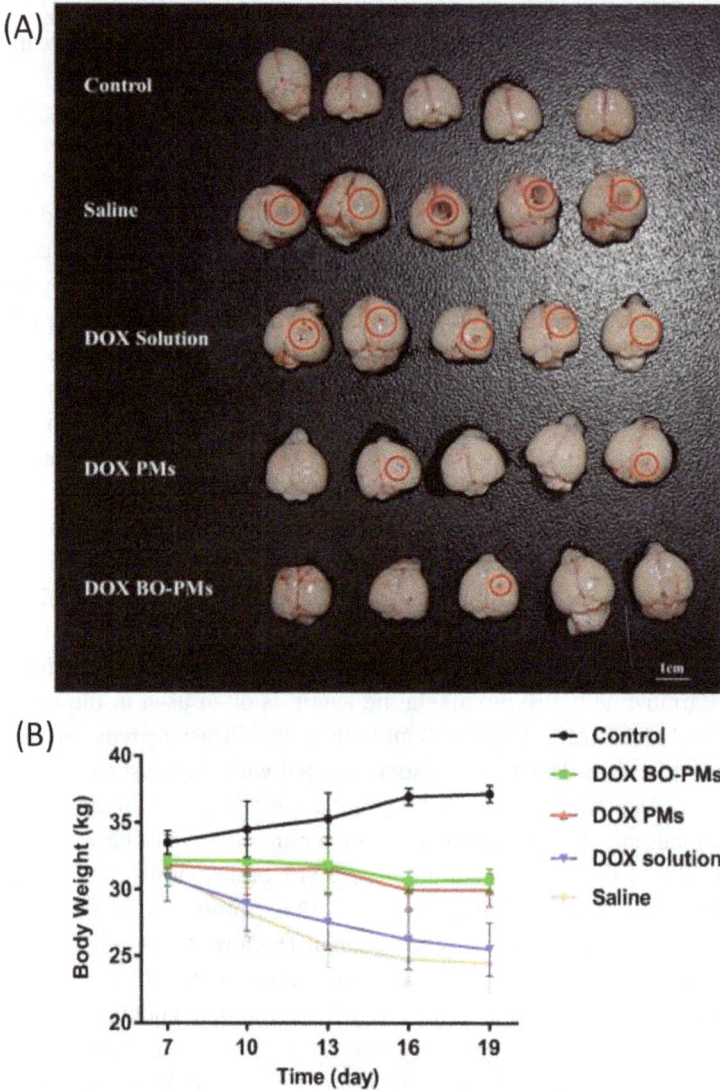

Figure 5.3: Combination of borneol and doxorubicin nanoparticles in the treatment of brain tumor in mice (A) Brain tissue of the mice treated with saline, DOX solution, DOX PMs and DOX BO-PMs compared with the control group; scale bar = 1 cm. (B) Body weight of the tumor-bearing mice in the different groups (n = 8). The control group consisted of mice without tumors treated with normal saline during surgery. Reprinted with permission from *Improving glioblastoma therapeutic outcomes via doxorubicin-loaded nanomicelles modified with borneol* by Meng, L. et al [72].

brain permeation enhancing ability, could be a great aid in achieving higher efficacy in the treatment of brain tumors.

Xu et al., [75] designed an i.v. formulation of DOX (5 mg dose) for glioma targeting. PAMAM dendrimer conjugated with borneol (BO) and folic acid (FA) was chosen as a vehicle (FA-BO-PAMAM/DOX) to incorporate the drug, apart from improving the brain permeation, helped in reducing the toxicity of the vehicle by reducing the net positive charge on the dendrimer. Folic acid conjugation led to 3 times reduction in IC_{50} of DOX compared to FA-unmodified DOX dendrimers, by improving its uptake in C6 glioma cells. FA-BO-PAMAM significantly improved the pharmacokinetics and brain distribution of DOX. The overall bioavailability of the novel formulation was 11.7-fold higher, compared to the DOX solution. DOX delivery in FA-BO-PAMAM carrier resulted in elimination half-life of 12.6 h and mean residence time of 16.58 h, which were 2.5 and 2.9 times higher, compared to free DOX. The AUC of the drug in the tumor was 127.38 ± 10.87 µg h/g as compared to negligible concentration when administered as a free drug. I.V. administration of FA-BO-PAMAM/DOX (5 mg dose) 4 times at 3 days interval in the rats bearing glioma resulted in a 57.44% inhibitory ratio of tumor growth, as compared to 17.70% for free DOX. The median survival recorded for the treated mice was 28 days in comparison with 18 days in the group treated with DOX solution. The bioenhancing effect of borneol was thus pronounced, when combined with nanotechnology-based delivery of anticancer drugs.

5.3.4 Quercetin

Quercetin, an aglycone form of flavonoid glycosides, has a ubiquitous and high quantity presence in natural sources, which makes it easy to obtain; it offers effective antioxidant action within the family of flavanols due to its singular chemical structure. Some of the top sources include apples, peppers, red wine, dark cherries, tomatoes, cruciferous and leafy green veggies, citrus fruits, whole grains, legumes, herbs, and more. It has been used as a nutritional supplement and has shown to have anticancer, anti-inflammatory, antiallergic, and cardioprotective properties [76–80]. It has also demonstrated antidiabetic, immunomodulatory, and antihypertensive activities in some reported studies [81–85].

5.3.4.1 *In vivo* studies demonstrating bioenhancing effect and its mechanism

Quercetin is an inhibitor of CYP3A4 and a modulator of P-gp and MDR transporters [12, 86, 87]. The bioavailability of PTX and its water-soluble prodrug (7-mPEG 5000-succinyloxymethyloxycarbonyl-PTX) along with quercetin after the oral administration was studied in rats [88]. PTX (40 mg/kg) and an equivalent amount of prodrug were administered orally to rats pretreated with quercetin (2, 10, 20 mg/kg). The AUC of PTX and the peak concentrations C_{max} of PTX pretreated with quercetin were significantly higher than the control with significantly longer half-life and mean residence times. The relative bioavailability of PTX after administration of the prodrug to rats pretreated with quercetin was 1.25- to 2.02-fold higher than the prodrug control. The absolute bio-availability of PTX was increased significantly by quercetin (2, 10, 20 mg/kg) from 8.0 to 10.1 and 16.2%. Hence, it seems that the development of oral PTX preparations with quercetin is feasible, which is more convenient than the i.v. dosage forms.

Table 5.2: Pharmacokinetic parameters of tamoxifen following the oral administration of tamoxifen (10 mg/kg), with or without quercetin (2.5, 7.5, or 15 mg/kg) in rats.

Parameters	Tamoxifen control	Quercetin coadministration (mg/kg)			i.v. (2 mg/kg)
		2.5	7.5	15	
AUC (ng h/mL)	1,802 ± 507	2,430 ± 634*	2,891 ± 728*	2,161 ± 572	2,402 ± 617
C_{max} (ng/mL)	73 ± 18.1	90 ± 24.1*	99 ± 28.6*	85 ± 24.0	
T_{max} (h)	2	2	2	2	
K_a	17.8 ± 5.0	24.4 ± 6.6*	29.4 ± 8.0*	21.5 ± 5.4	
$T_{1/2}$ (h)	19.9 ± 5.1	22.5 ± 7.0	24.4 ± 7.3	21.2 ± 7.1	15.9 ± 4.6
AB (%)	15.0 ± 4.1	20.2 ± 5.4*	24.1 ± 6.5*	18.0 ± 4.6	100
RB (%)	100	135	161	120	

Mean ± S.D. ($n = 6$). AUC, area under the plasma concentration–time curve from 0 h to infinity; C_{max}, peak concentration; T_{max}, time to reach peak concentration; K_a, absorption rate constant; $t_{1/2}$, terminal half-life; AB (%), absolute bioavailability; RB (%), relative bioavailability; compared $AUC_{coadmin}$ to $AUC_{control}$. *Statistically significant at $p < 0.05$ when compared with the control. Reprinted from *Enhanced bioavailability of tamoxifen after oral administration of tamoxifen with quercetin in rats*, by S. C Shin et al. [91].

Etoposide, a semisynthetic anticancer drug, is used to treat SCLC, lymphoma, and acute leukemia [89]. It is also a substrate for efflux pumps like cytochrome P450 (CYP) 3A and P-gp. The pharmacokinetics of etoposide in the presence of quercetin were investigated in rats [90]. Etoposide was given to rats i.v. (3 mg/kg) or orally (9 mg/kg), with or without quercetin (1, 5, or 15 mg/kg). The pharmacokinetic param-eters of etoposide were not significantly altered in the presence of quercetin in the i. v. group but showed significant enhancement in the oral group. This enhanced oral bioavailability could be due to suppression of CYP3A-catalyzed metabolism in the intestine and inhibition of P-gp-mediated efflux by quercetin. The presence of quercetin notably increased the AUC of orally administered etoposide from 43.0% or 53.2%. Quercetin also showed significant decrease in the total body clearance of oral etoposide. The absolute bioavailability of etoposide was significantly increased (13.6%) as compared to the control group (8.87%) in the presence of quercetin.

Orally, bioavailability of tamoxifen is limited (~15%) due to the first-pass metabolism; also, it is a substrate for MDR transporters, which leads to efflux in the intestines and liver. A study investigated the bioavailability and pharmacokinetics of tamoxifen and one of its metabolites, 4-hydroxytamoxifen, in the presence of quercetin in rats [91]. The oral coadministration of quercetin (2.5 and 7.5 mg/kg) with tamoxifen (10 mg/kg) significantly ($p < 0.05$) increased the C_{max}, absorption rate constant, and the AUC of tamoxifen (Table 5.2). The absolute bioavailability of tamoxifen was markedly high (~1.5 times higher) in the presence of quercetin than in the control group. However, no significant changes were observed in T_{max} and $t_{1/2}$ of tamoxifen, when administrated with quercetin. The AUC of 4-hydroxytamoxifen was increased significantly by coadministration of 7.5 mg/kg quercetin. Since the ratio of AUC of 4-hydroxytamoxifens to tamoxifen (metabolite ratios) was significantly lower, this indicated that quercetin inhibits both first-pass metabolism of tamoxifen and MDR transporters efflux.

The effect of oral quercetin on the bioavailability and pharmacokinetics of DOX was studied by oral (50 mg/kg) and i.v. (10 mg/kg) administration in rats [92]. Quercetin showed notable increase in the AUC, C_{max}, while $t_{1/2}$ and T_{max} of DOX showed no significant changes. Quercetin increased the absolute bioavailability of DOX, and the relative oral bioavailability was increased 1.32- to 2.36-fold. However, i.v. DOX pharmacokinetics was not affected by quercetin. These results were again attributed to quercetin inhibition of P-gp and CYP3A, which increases the oral bioavailability. Hence, this study suggested that oral DOX formulation can potentially be used as a more convenient alternative to i.v. dosage form.

Epigallocatechin gallate (EGCG), a main anticancer component in green tea, has poor bioavailability in rats and humans due to oxidation, metabolism, and efflux. Supplementation of green tea extract with quercetin (4.33 g + 9.6 mg, respectively) in rats almost doubled the plasma C_{max}. The $AUC_{0-24\ h}$ was raised by around 55%, indicating quercetin increased the bioavailability of EGCG in rats [93].

5.3.5 Capsaicin

Capsaicin, an active component found in chili peppers (*Capsicum annum*), shows analgesic properties. The "heat sensation" of capsaicin arises due to the binding of capsaicin to transient receptor potential vanilloid (TRPV) ion-channel receptors. Capsaicin has been used as a ligand to activate several types of ion-channel receptors. Exposure to high doses of capsaicin (above 100 mg capsaicin per kg body weight) for a prolonged time has been shown to cause peptic ulcers, accelerate the development of prostate, stomach, duodenal, and liver cancers, and enhance breast cancer metastasis. However, several convergent studies have indicated that low doses of capsaicin display a cancer-chemopreventive, antineoplastic activity. Apart from anticancer activity, the oral or local administration of capsaicin has been shown to reduce inflammation and pain from rheumatoid arthritis, fibromyalgia, and chemical hyperalgesia [94].

5.3.5.1 *In vitro* studies demonstrating bioenhancing effect and the underlying mechanisms

The anticancer effect of capsaicin was studied in the DDP-resistant human gastric cancer cell line SNU-668. The results indicated that the treatment of cells with DDP in combination with capsaicin showed increased cell death via apoptosis, as compared to mono treatment of either capsaicin or DDP [95]. Moreover, the combination also helped overcome Aurora-A protein-mediated DDP resistance.

SCLC, which shows a good response to chemotherapy and radiation therapy in early stages, suffers the drawback of remission with no response to the first-line treatment. Recent clinical studies have investigated the possibility of camptothecin-based combination therapy as a first-line treatment for SCLC patients. The synergistic activity of camptothecin and capsaicin is observed in both classical and variant SCLC cell lines, *in vivo* and in human SCLC tumors xenotransplanted on chicken chorioallantoic membrane models. The concentration of 10 µM capsaicin selected did not induce any cell death in SCLC, but when combined with varying concentrations of camptothecin, showed synergistically enhanced apoptosis, within a range of concentrations. The synergistic activity of capsaicin and camptothecin is mediated by the elevation of intracellular calcium and the calpain pathway [96].

The recurrence of bladder cancer after surgery, with or without chemotherapy, remains a major challenge in bladder cancer treatment. Long Zhen et al., determined the function of transient receptor potential vanilloid 1 (TRPV1), a known tumor suppressor in T24 and 5,637 (human bladder carcinoma) cell lines. Capsaicin, which is known to act on TRPV1 as agonist, was screened at various amounts (0–200 µM) for the period of 48 h [97]. The MTT (3-(4,5-dimethylthiazol-2-yl)-2,5-diphenyltetrazolium bromide) assay showed that 5,637 cells were sensitive to capsaicin treatment (IC_{50} = 149 µM), while T24 cells were insensitive (IC_{50} = 266 µM). The 5,637 cells that highly express TRPV1C showed significant growth inhibition with capsaicin, while T24 cells which express TRPV1 at low levels failed to show a significant inhibition. The cell viability assay showed that IC_{50} values dropped to 335 nM, with co-treatment of capsaicin and pirarubicin, one of the main drugs used in urinary bladder topical therapy, while the group treated with pirarubicin alone showed IC_{50} value of 566 nM. Therefore, activation of TRPV1 enhanced the cytotoxic effects of pirarubicin and sensitized bladder cancer cells to pirarubicin.

In another combination study of capsaicin along with 5-FU on gastric cancer cells, capsaicin was shown to be capable of causing multifold decrease in the IC_{50} value of 5-FU in the gastric cancer cell line, HGC-27 [98]. The effect of capsaicin on the sensitivity of cholangiocarcinoma (CCA) cells to 5-FU is analyzed *in vitro* and *in vivo* in a xenograft model [99]. The combination of capsaicin and 5-FU was found to be synergistic in QBC939 cells (CI < 1). A combination of 40 µM 5-FU with 80 µM capsaicin showed IC_{50} of 35 µM, as compared to IC_{50} value of 126 µM with just 5-FU. The combined treatment also suppressed tumor growth (~75% reduction as compared to control) in the CCA xenograft to a greater extent, than 5-FU alone. Further investigation revealed that capsaicin activates the phosphoinositide 3-kinase (PI3K)/protein kinase B (AKT)/mammalian target of RPM (mTOR) pathway in CCA cells. Hence, capsaicin can be used synergistically as co-therapy in CCA.

The effect of the combination of capsaicin with DDP has been studied in osteosarcoma (OS) cells *in vitro* and *in vivo* [100]. The combination of DDP with capsaicin showed synergy in OS cell cytotoxicity by significant effects on cell cycle arrest, apoptosis induction, and cell invasion inhibition. The co-treatment also showed to trigger pro-survival autophagy through ROS/JNK (c-Jun N-terminal kinase) and p-AKT/mTOR signaling in OS cells.

5.3.5.2 *In vivo* studies demonstrating bioenhancing effect

The *in vivo* studies on xenograft model included 20 mice randomized into the 4 groups (each group having 5 mice) and given the following regimen: oral gavage of 200 µL of PBS containing capsaicin (20 mg/kg), i.p. injection of 200 µL of 0.9% saline solution containing DDP (4 mg/kg), and a combinational group where both capsaicin and DDP were administered as mentioned above, respectively, and control group where mice were left untreated. The regimen consisted of seven treatments and groups were treated every 3 days. All treated groups showed around ~ 80% reduction in tumor volume, as compared to the control group. However, the greatest inhibition was observed with combination of DDP and capsaicin synergistically, as compared to monotreatment with either.

5.3.6 Allicin

Allicin is sourced from cloves of *Allium sativum*, commonly known as white garlic. *A. ampeloprasum* L., *A. ursinum* L., *A. vineale* L., and *A. victorialis* L., respectively, known as elephant garlic, wild garlic, field garlic, and alpine leek are also some of the major sources of allicin. It is a volatile oil that comprises sulfur-containing diallyl thiosulfinate. It attributes a characteristic pungent smell and taste to the garlic, with its mere presence of about 4–5 mg in a clove. Allicin is not present as such in the garlic but is formed *in situ* by the reaction between allin and allinase when the clove is crushed. Being hydrophobic in nature, it can easily permeate physiological barriers *in vivo* and hence shows various pharmacological activities. Strong antibacterial, antifungal, antiparasitic, and antiviral activities have been established for allicin. It has a plethora of cardiovascular activities like vasorelaxation, prevention of hyperlipidemia, cardiac hypertrophy, angiogenesis, and platelet aggregation. The potent antioxidant property of allicin helps it exhibit a substantial role in the treatment of inflammatory and neurodegenerative disorders. Besides this, preclinical studies of allicin have shown its application in reversing insulin resistance and controlling blood sugar levels, in the case of type 2 diabetes mellitus. Several preclinical studies have indicated protective activities of allicin against several drugs and chemicals. A recent study in the murine model demonstrated the potential role of allicin pretreatment in preventing DOX-associated cardiac toxicity [101].

5.3.6.1 Mechanisms responsible for bioenhancing effect

A number of *in vitro* cell lines and *in vivo* animal studies have indicated anticancer activity for allicin. The major mechanisms by which it elicits its antiproliferative and apoptotic effects include p53-mediated autophagic cell death, generation of ROS and subsequent DNA damage, modulation of extracellular signal-regulated protein kinase/MAP, mitogen-activated protein pathway, Bcl2/Bax mitochondrial pathway, and activation of p38 MAP kinase pathway. Immunomodulating effects of allicin have been exploited in improving the outcome of standard and immune-based chemotherapy. Though the bioenhancing effect of allicin is not strongly evident through the reported literature, several studies have shown the synergistic effect of combination therapy of standard drugs and allicin as an adjuvant [101].

5.3.6.2 *In vivo* studies demonstrating bioenhancing effect

Wang et al., performed a study in the rodent pancreatic cancer model. It involved individual and combined treatment of recombinant interleukin 2 (rIL2) and allicin [102]. Weekly once i.v. injection of 10 µg/kg rIL2 and 10 mg/kg allicin over 4 weeks led to a noteworthy reduction in tumor weight and volume. Though rIL2 treatment led to 63.3% tumor size reduction and 39.1% apoptosis rate of tumor cells, as compared to control, combined treatment resulted in 77.9% tumor size reduction and 60.9% apoptosis rate. Also, 60% of mice in the combination treatment group showed 55 days' survival whereas none from the rIL2 group survived on day 55. Pancreatic cancer is known to suppress the innate and adaptive immune response of the body and hence proliferates. Immune therapy plays a vital role in treating this cancer. Hence, allicin with its established immune-boosting effect, when combined with the rIL2 therapy, led to significant enhancement in the innate immunity, as confirmed by 129% higher numbers of natural killer cells compared to the tumor control group. About 87.2% higher concentration IFN-gamma, and 106.6% and 131.4% higher CD4 and CD8 cells in the peripheral blood confirmed the boosted adaptive immune response in the group of animals treated with a combination of allicin and rIL-2.

Gao et al., [103] studied allicin (10 mg/kg/day, gastric administration) as an adjuvant to CPM i.v. therapy (60 mg/kg, weekly once) in neuroblastoma-bearing mice [103]. The treatment continued for 4 weeks during which the weights of tumor, liver, spleen, and thymus tissues were monitored besides the blood levels of CD3, CD4, CD8, and VEGF. The combination therapy compared to monotherapy was more effective in reducing the tumor weight and improving the liver, spleen, and thymus tissue weights. The model group showed close to 600 pg/mL level of VEGF, whereas the CPM treatment alone and CPM + allicin therapy exhibited 200 and 100 pg/mL levels, respectively. The VEGF levels of combination therapy were comparable to that of the normal nontumor-bearing animals. CD3, CD4, and CD8 levels in the combination therapy group were notably higher compared to the monotherapy group. All these results indicated better management of neuroblastoma by the chemotherapy in the presence of allicin as an adjuvant.

Combination therapy of allicin (5 mg/kg/day, every 2 days a week, for 3 weeks) and 5-FU (20 mg/kg/day; 5 consecutive days) led to a significant reduction in HCC in the mouse model [104]. At the end of 3 weeks of the study, the average tumor volume and weight in the 5-FU + allicin group showed more than 50% reduction compared to that of the 5-FU treatment group (Figure 5.4A–C). Similarly, the combination treatment showed a 30% higher extent of apoptosis than the 5-FU monotherapy. Allicin-induced ROS generation, downregulation of the Bcl2 pathway, and induction of caspase-3 activity were the major mechanisms responsible for the synergistic effect of the allicin + 5-FU therapy. Hepatoprotective effect of allicin may further provide a greater advantage to allicin as an enhancer in the treatment of HCC [105].

Combination treatment of allicin (10 mg/kg/day) orally for 4 weeks along with 1 mg/kg tamoxifen i.p. injection on weekly basis led to synergistic antitumor activity in a mouse model of Eherlich ascites carcinoma [105]. The combination therapy resulted in 82% tumor inhibition as compared to 70% inhibition in tamoxifen-treated animals. Oral treatment of rats with 45 mg/kg/day tamoxifen for 17 consecutive days produced liver intoxication, as confirmed by elevated levels of alanine transaminase, aspartate amino transferase, alkaline phosphatase, lactate dehydrogenase, total bilirubin, and TNF-α (tumor necrosis factor α) levels, along with depleted levels of endogenous antioxidants – superoxide dismutase, GSH (glutathione), and total protein. Concomitant allicin administration (50 mg/kg/day) significantly improved the levels of antioxidants and reduced oxidative stress and inflammation, as confirmed by lowered levels of the liver enzymes as well as TNF-α.

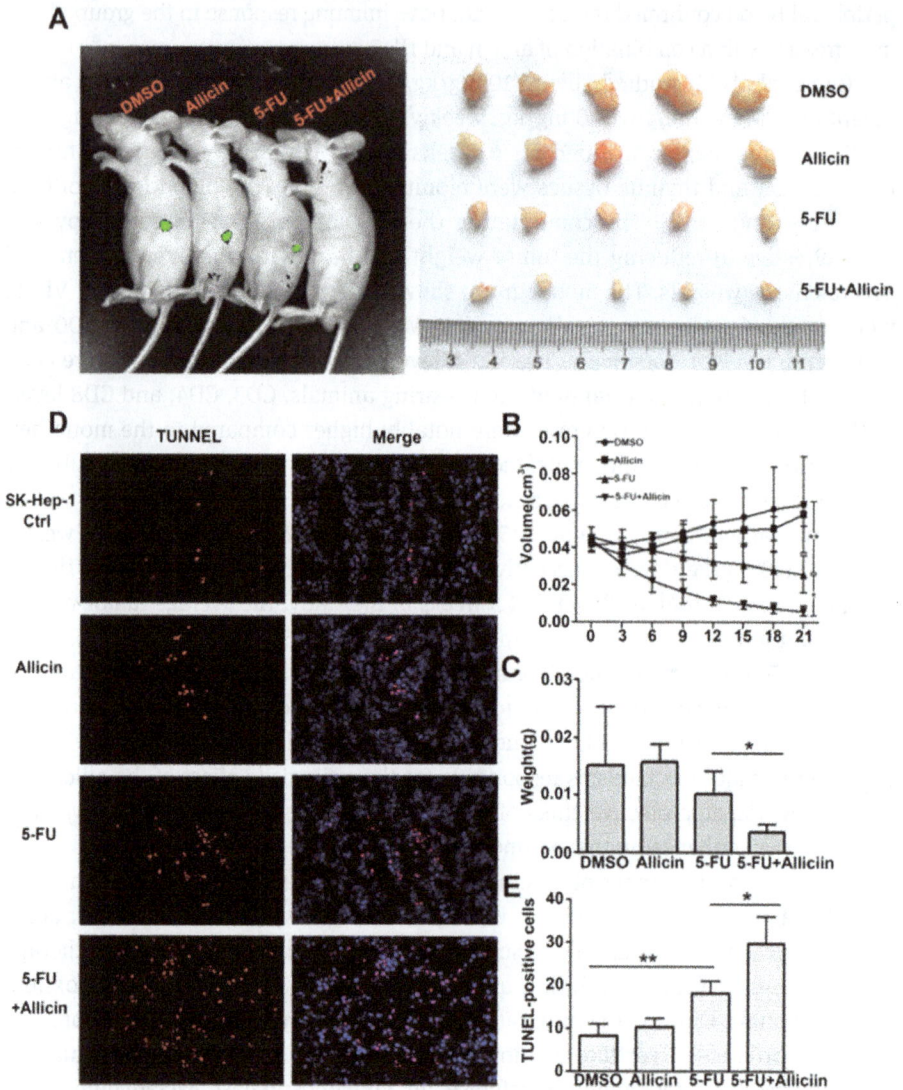

Figure 5.4: Co-treatment of allicin and 5-FU significantly inhibited HCC tumor growth *in vivo*
(A) Treatments with DMSO (as control), allicin, 5-FU, or with a combination of allicin and 5-FU, tumors isolated from each group on day 21. (B) Continuous quantification of tumor volumes of each group during the experiment. (*$P < 0.05$, **$P < 0.01$). (C) The weight of xenograft tumors were showed as the mean ± SD of five tumors excised from each group. (*$P < 0.05$). Reprinted with permission from *Allicin sensitizes hepatocellular cancer cells to anti-tumor activity of 5-fluorouracil through ROS-mediated mitochondrial pathway* by Zou, X. et al. [104].

5.3.7 Cumin

Nigella sativa (family: Ranunculaceae) is commonly known as black cumin or black seed. Cumin seeds and extracts thereof show several pharmacological properties, such as anti-inflammatory, anticancer, antifertility, antidiabetic, antimicrobial, antihistaminic, and hypotensive effects. 2-Methyl-5-isopropyl-1, 4-benzoquinone (TQ) is a monoterpene that constitutes 30–48% of cumin's bioactive composition. TQ has a promising anticancer and chemosensitizing effect. ROS generation, cell cycle arrest, immunomodulation, and promotion of apoptosis are the mechanisms that govern the anticancer activity of this phytoconstituent. It exhibits anticancer activity by regulating different targets, such as NF-κB, peroxisome proliferator-activated receptor-γ (PPARγ), and c-Myc, which lead to caspases protein activation. TQ has shown to express tumor suppressor genes (TSG), such as p53, phosphatase, and tensin homolog (PTEN). It also controls angiogenesis and cancer metastasis through activation of JNK and p38. Additionally, it boosts the immune system and reduces the side effects associated with conventional anticancer therapy [106, 107].

5.3.7.1 *In vitro* studies demonstrating bioenhancing effects and underlying mechanisms

A combination of DTX (10 nM) with TQ (60 µM) exhibited a synergistic cytotoxic effect in DU-145 (Human prostate cancer cells) [108]. The treatment of DTX alone and TQ alone showed a 43.5% and 15% reduction in cell proliferation, respectively, whereas the combination resulted in a 68.5% decrease in cell proliferation. The combination, DTX alone, and TQ alone showed 20-, 8.2-, and 3.3-fold higher DNA fragmentation, respectively, compared to the untreated controls. The apoptotic and cytotoxic effect of the combination was attributed to the blockage of the PI3K/Akt signaling pathway in DU-145 cells. The human pancreatic cancer cell lines-BxPC-3, HPAC, and COLO-357 were pretreated with 25 mmol/L of TQ for 48 h. Further treatment of cells with gemcitabine or oxaliplatin culminated in 60–80% growth inhibition [109]. Gemcitabine or oxaliplatin treatment alone showed only 15- 25% inhibition. Also, TQ aggravated drug-induced activation of NF-κB. Antineoplastic activity of TQ alone, and more importantly, in combination with DDP was evident against non-SCLC (NCI-H460) and small lung cancer cell line (NCI-H146) [110]. TQ and DDP together at 100 and 5 µM concentration, respectively, could prevent the growth of NCI-H460 cells by about 89% at 72 h. After 24 h, TQ was shown to bring apoptosis in both cell lines, viz. NCI-H460 (87.59%) and NCI-H146 (88.1%). TQ was found to be instrumental in bringing down the levels of ENA-78 and Gro-alpha, the cytokines related to neo-angiogenesis. The *in vitro* studies of TQ in different types of cancer cell lines, thus, helped in understanding the mechanistic role of cumin as a bioenhancer.

5.3.7.2 *In vivo* studies demonstrating bioenhancing effect

The chemosensitizing effect of TQ to chemotherapeutic agents, gemcitabine or oxaliplatin, has been studied *in vivo* by Banerjee et al., using an orthotopic model of pancreatic cancer [109]. *In vivo* studies involved intragastric administration of 3.0 mg/day TQ in mice bearing pancreatic tumors. Oxaliplatin (5 mg/kg) and gemcitabine (50 mg/kg) were intraperitoneally administered on once-a-week basis, up to 5 weeks. Administration of TQ, oxaliplatin, or gemcitabine alone caused 38%, 58%, and 66% reduction in tumor weight, respectively, while the combination of TQ with either gemcitabine or oxaliplatin caused close to 85% and 71% reduction compared to the model group. Additionally, both the combination treatments showed 50% lower lymph node metastases in mice, compared to the monotherapy. These results suggest that TQ treatment with gemcitabine or oxaliplatin caused greater antitumor activity and could reduce gemcitabine and oxaliplatin-induced NF-kB activation, resulting in higher chemosensitization of pancreatic tumors towards the drug therapy.

In a non-SCLC (NCI-H460) xenograft study, female SCID mice treated with once weekly i.p. dose of DDP (2.5 mg/kg), in conjunction with subcutaneous doses of TQ, showed a significant reduction in tumor volume, after 26 days. TQ monotherapy was ineffective and mildly effective in reducing the tumor volume at 5 and 20 mg/kg doses, respectively. Treatment with DDP at 2.5 mg/kg dose could reduce the tumor volume appreciably ($p < 0.001$). The combination of TQ and DDP in 5 and 2.5 mg/kg doses suppressed the tumor volume by 59%, compared to control. Higher dose of TQ (20 mg/kg), when coupled with 2.5 mg/kg DDP further reduced the tumor volume by 79%, without additional toxicity to the mice. Also, mice treated with this combination showed a notable reduction in the ratio of phosphor-Ser529 NF-kB/NF-kB. TQ-induced inhibition of NF-kB expression aided in overcoming DDP resistance and improving its efficacy in treating lung cancer.

Azoxymethane-induced colorectal cancer model in rats was selected for the evaluation of the therapeutic efficacy of TQ, 5-FU, and their combination therapy [111]. TQ (35 mg/kg/day, 3 day/week) was administered during the 7th and 15th weeks post-azoxymethane injection, while 5-FU was given during the 9th and 10th weeks (12 mg/kg/day for 4 days; then four doses of 6 mg/kg every other day). The untreated rats showed a significant number of tumors (29.16 ± 2.92) in the colon, whereas the rats treated with 5-FU or TQ alone showed a significantly decreased number of the grown tumors (16.66 ± 4.4 and 17.5 ± 2.1). The combination of 5-FU/TQ resulted in the least number of tumors (13.8 ± 2.8). Similarly, 5-FU/TQ combination therapy showed 16.4 ± 3.1 number of aberrant crypts foci (ACF), which were notably lower than 23.2 ± 3.7 and 25.0 ± 4.0 foci seen after 5-FU or TQ monotherapy. Similarly, the highest attenuating effect on the development of preneoplastic large ACF and mucin-depleted foci, as well as tubular adenomas was observed with 5-FU/TQ dual therapy. Both 5-FU and TQ worked in tandem to lower the expression of β-catenin, Wnt (Wingless-related integration site), COX-2, inducible nitric oxide synthase (iNOS), NF-kB, thiobarbituric acid reactive substance (TBARS), and VEGF. The duo enhanced the levels of antitumorigenic constituents like dickkopf-like protein 1 (DKK-1), cyclin-dependent kinase inhibitor 1A (CDNK-1A), transforming growth factor β1 and β II, Smad4, and glutathione peroxidase (GPx).

5.3.8 Glycyrrhizin

Glycyrrhizin is a triterpenoid saponin glycoside obtained from dried, peeled or unpeeled, root and stolon of *Glycyrrhiza glabra*. It comprises two glucuronic acid molecules and one molecule of glycyrrhetinic acid. It is used often as a sweetener and a flavoring agent in the food and pharma industry. Glycyrrhizin is known to have diverse pharmacological actions such as expectorant, diuretic, laxative, immunity-boosting, antioxidant, and anti-inflammatory [20]. Scientific literature also indicates its role in hepatoprotection and the treatment of hepatic steatosis, viral hepatitis, and hepatoma [112]. Several *in vitro* and preclinical studies suggest an inhibitory effect of glycyrrhizin on lung cancer, colon cancer, leukemia, and glioma [113]. Glycyrrhizin demonstrates bioenhancing activity when combined with various antimicrobial and anticancer agents.

5.3.8.1 Mechanisms responsible for bioenhancing effect

Glycyrrhizin primarily improves the oral absorption of the drugs by inhibiting the P-gp efflux transporters present in the intestinal lining. Absorption-enhancing activity of glycyrrhizin depends upon its conversion to glycyrrhetic acid by the action of β-glucuronidase, an intestinal bacterial enzyme. A number of *in vitro* and preclinical studies have shown the improved efficacy of anticancer therapy in the presence of glycyrrhizin.

5.3.8.2 *In vitro* studies demonstrating bioenhancing effect

Treatment of triple-negative breast cancer cells (MDA-MB-231) with 20 µM/L glycyrrhetinic acid (GA) 6 h prior to the treatment with etoposide helped in lowering the IC_{50} of etoposide from 2.5 to 1.32 ± 0.29 µM/L. GA-induced GSH depletion and modulation of MAPK and AKT pathways accounted for enhancement of topoisomerase IIα expression [114]. Enhanced topoisomerase improved the chemo sensitization of the cancer cells to etoposide, resulting in a better antineoplastic effect. Glycyrrhizin, however, did not show this effect.

The co-treatment with glycyrrhizin inhibited the DDP efflux from the HCC cells and altered the DDP resistance. DDP-resistant Huh7 HCC cell line exhibited 14.1-

fold higher resistance to DDP due to 6.29-, 3.2-, 11.3-, and 3.39-fold higher expression of MRP2, MRP3, MRP4, and MRP5 proteins [115]. Treatment with DDP (5 µg/mL) + glycyrrhizin (100 µg/mL) decreased the Huh7 HCC cell viability to 76.8%, compared to DDP treatment alone that resulted in 100% viability. Glycyrrhizin, by acting as the competitive substrate for MRP2 and MRP3, reversed the effects of DDP resistance by inhibiting DDP efflux from the resistant HCC cells.

5.3.8.3 *In vivo* studies demonstrating bioenhancing effect

Mouse model of lung adenocarcinoma was studied for the effect of co-treatment of glycyrrhizin (15, 45, and 135 mg/kg/day) and DDP (2.5 mg/kg/day) over a period of 30 days. The combination of 135 mg/kg glycyrrhizin and DDP reversed the tumor-induced cachexia and brought the treated animals' body weight to normal, as compared to 13.6% lower average weight in case of untreated animals. The glycyrrhizin co-treatment even restored the conditions of the renal and liver toxicities, as indicated by the normal serum levels of creatinine, urea, aspartate aminotransferase, and alanine aminotransferase. Glycyrrhizin treatment was found to inhibit the thromboxane synthase (TxAS)-induced production of thromboxane A2 (TxA2). TxA2 is responsible for tumor growth, metastasis, as well as the multiple tissue toxicity, and resistance to the chemotherapy. Proliferating cell nuclear antigen (PCNA) is also responsible for tumor progression. Mice bearing lung cancer showed overexpression of both TxA2 and PCNA. Though DDP could effectively suppress PCNA, it was unable to regulate the TxA2 pathway. Increased concentration of TxA2 in the tumor environment, thus, posed resistance to the DDP chemotherapy and also led to cachexia, liver and lung toxicity. On the other hand, glycyrrhizin treatment reduced the TxAS serum levels by 30.6%, 50.2%, and 72.45% in a dose-dependent manner, as compared to model control. The lung tissue of tumor-bearing mice treated with a blend of 2.5 mg/kg/day DDP and 15 mg/kg/day glycyrrhizin demonstrated 11% lower TxAS levels, compared to the treatment with 15 mg/kg/day glycyrrhizin alone ($p < 0.05$). The combination of glycyrrhizin and DDP with a potential reduction in both TxAS and PCNA levels remarkably improved sensitivity of tumor cells toward DDP that resulted in better anticancer effect. Lower levels of these two, especially the TxAS, helped in lower tissue and organ toxicity [116].

PTX-loaded glycyrrhizic acid (GZA) nanomeric micelles prepared in a 1:10 weight ratio enhanced the oral bioavailability of PTX using GZA as a carrier. The oral absorption of PTX from the PTX-loaded GZA micelles and Taxol® were investigated in rats, after administering a dose of 20 mg/kg. The C_{max} and $AUC_{0 \to 24\,h}$ of PTX-loaded GZA micelles were 0.460 ± 0.10 µg/mL and 3.42 ± 1.02 µg · h/mL, appreciably greater than that of Taxol®. (0.095 ± 0.01 µg/mL, 0.573 ± 0.12 µg h/mL). A notable improvement in oral bioavailability of PTX from PTX-loaded GA micelles could be attributed to the influence of GZA, which effectively inhibited the drug efflux caused by the P-gp in the intestine. Since both GZA and PTX are terpenoids, GZA micelles enhanced the solubility of the drug based on the "like dissolves like" theory. The clearance of the drug from the micellar formulation was almost six times lower than that from Taxol. Micellar formulation, thus, protected the drug from rapid clearance and thus improved its systemic exposure [117].

5.3.9 Genistein

Genistein, belonging to the class of isoflavones, is found majorly in soy-based plants like kudzu (*Pueraria lobata*) and soybean (*Glycine max*). With a molecular weight of 270.24 Da, it is a pale yellow dendritic needle-like powder that is almost insoluble in water. It is a known phytoestrogen, which has been studied for several potential health effects including anti-inflammatory and anticancer benefits [118, 119]. A few studies have also shown its bioenhancer potential with cancer chemotherapeutic agents.

5.3.9.1 *In vivo* studies demonstrating bioenhancing effect

Genistein is also known to be an inhibitor of ABC transporters and metabolizing enzymes like CYP 3A4 and 2C8. PTX, as, like most anticancer drugs, is also a substrate for ABC transporters such as P-gp and is majorly metabolized by CYP 3A4 and 2C8. A study investigated the effect of orally administered genistein on the pharmacokinetics of PTX administered through oral and i.v. route in rats [120]. The presence of 10 mg/kg genistein significantly ($p < 0.05$) increased AUC (54.7% greater) of orally administered PTX, which was due to the significant ($p < 0.05$) decrease in total plasma clearance of PTX (35.2% lower). Genistein at 3.3 mg/kg increased the C_{max} of PTX (66.8% higher), whereas at 10 mg/kg it led to 91.8% higher C_{max}. Consequently, the absolute and relative bioavailabilities of PTX were increased in the presence of genistein, on oral administration (Table 5.3). Genistein also significantly enhanced the AUC (40.5% greater) and lowered the total clearance (30% lower) of PTX on i.v. administration. Hence, the presence of genistein improved the PTX activity systemically.

5.4 Clinical studies involving bioenhancers

Encouraging *in vitro* and preclinical results of various herbal bioenhancers in cancer chemotherapy tempt us to speculate favorable clinical outcomes of these combination treatments. However, very few of the aforementioned bioenhancers have been subjected to clinical studies to evaluate their impact on the outcomes of the

Table 5.3: Mean (±S.D.) pharmacokinetic parameters of paclitaxel after oral administration of paclitaxel (30 mg/kg) to rat in the presence or absence of genistein.

Parameters	Control	Paclitaxel + genistein	
		3.3 mg/kg	10 mg/kg
AUC (ng h/mL)	702 ± 184	885 ± 213	1,086 ± 249*
CL/F (mL/min kg)	712 ± 186	560 ± 143	461 ± 115*
C_{max}(ng/mL)	36.8 ± 9.5	61.4 ± 15.5*	70.6 ± 18.0**
T_{max} (h)	1.0	0.5	0.5
K_{el} (h^{-1})	0.047 ± 0.012	0.045 ± 0.011	0.043 ± 0.109
$T_{1/2}$ (h)	14.7 ± 3.71	15.4 ± 3.83	16.2 ± 4.01
F	0.016	0.020	0.025
Fr	1	1.26	1.55

Mean ± S.D. ($n = 6$), *$p < 0.05$, **$p < 0.01$ compared to control. AUC, area under the plasma concentration–time curve from 0 h to infinity; CL/F, total plasma clearance; C_{max}, peak concentration; T_{max}, time to reach peak concentration; K_{el}, elimination rate constant; $t_{1/2}$: terminal half-life; F, absolute bioavailability; Fr, relative bioavailability. Reprinted from *Effect of genistein on the pharmacokinetics of paclitaxel administered orally or intravenously in rats*, by Xiuguo Li et al [120].

cancer therapeutics. MB-6, a combination of green tea, soybean, grape seed, spirulina, *Antrodia camphorata* mycelia, and curcumin was given to 72 patients with metastatic colorectal cancer. The patients received leucovorin, 5-FU, and oxaliplatin treatment for 16 weeks [121]. The disease progression rate and incidence of adverse events were significantly lower in the group of patients who received MB-6. In another study involving 199 prostate cancer patients, the group of patients who received a dietary supplement containing green tea, pomegranate, broccoli, and curcumin for 6 months showed lower prostate-specific antigen levels compared to patients who received the placebo supplement [122]. These studies highlight the benefits of coadministration of herbal-based enhancers like green tea, soybean (containing genistein), broccoli (source of quercetin), and curcumin. Curcumin is the most clinically researched bioenhancer (Table 5.4). Several phase 1 and 2 studies of curcumin in patients with various cancers like pancreatic or biliary tract cancer [123], prostate cancer [124], colorectal liver metastatic cancer [125], breast cancer [126]), or pancreatic cancer [127] have been reported in the literature. Poor bioavailability of curcumin has compelled researchers to use high doses. However, at oral doses as high as 8 g/day, curcumin has demonstrated excellent tolerability, lack of toxic effects or adverse events, and, most importantly, patient acceptance and compliance [124, 127]. Nanotechnology-based formulation approaches such as nanosuspension [123], solid lipid nanoparticles [128], micellar preparations [129], and phytosomes

Table 5.4: Clinical trials involving curcumin and other drugs in cancer therapy.

S. no.	Study title and phase	Trial status	Condition	Drug in combination	References
1	Curcumin in combination with 5-FU for colon cancer, early phase 1	Active, not recruiting	Metastatic colon cancer	5-FU	[132]
2	Avastin/FOLFIRI in combination with curcumin in colorectal cancer patients with unresectable metastasis, phase 2	Completed	Colorectal cancer	Avastin/FOLFIRI	[133]
3	Effect of curcumin on dose-limiting toxicity and pharmacokinetics of irinotecan in patients with solid tumors, phase 1	Completed	Advanced colorectal cancer	Irinotecan	[134]
4	Curcumin in combination with chemotherapy in advanced breast cancer, phase 2	Completed	Advanced, metastatic breast cancer	PTX	[135]
5	Combining curcumin with FOLFOX chemotherapy in patients with inoperable colorectal cancer, phase 2	Completed	Colonic cancer metastasis	Chemotherapy (FOLFOX)	[136]
6	Trial of gemcitabine, curcumin and celebrex in patients with metastatic colon cancer, phase 3	Unknown	Colon neoplasm	Celecoxib	[137]
7	Gemcitabine with curcumin for pancreatic cancer, phase 2	Completed	Pancreatic cancer	Gemcitabine	[138]
8	Radiation therapy and capecitabine with or without curcumin before surgery in treating patients with rectal cancer, phase 2	Active, not recruiting	Rectal mucinous adenocarcinoma, rectal signet ring cell adenocarcinoma, recurrent rectal carcinoma	Capecitabine	[139]
9	Trial to modulate intermediate endpoint biomarkers in former smokers, phase 2	Recruiting	Lung diseases, lung cancer, protection against	Lovaza®	[140]

(continued)

Table 5.4 (continued)

S. no.	Study title and phase	Trial status	Condition	Drug in combination	References
10	Study of pembrolizumab, radiation and immune modulatory cocktail in cervical/uterine cancer, phase 2	Recruiting	Cervical cancer, endometrial cancer, uterine cancer	Pembrolizumab + vitamin D + aspirin + lansoprazole + CPM	[141]
11	Gemcitabine hydrochloride, paclitaxel albumin-stabilized nanoparticle formulation, metformin hydrochloride, and a standardized dietary supplement in treating patients with pancreatic cancer that cannot be removed by surgery, phase 1	Active, not recruiting	Pancreatic adenocarcinoma, unresectable pancreatic carcinoma, stage III and stage IV pancreatic cancer	Gemcitabine hydrochloride + PTX + metformin hydrochloride	[142]
12	Curcumin and piperine supplementation in early-stage prostate cancer patients undergoing active surveillance or patients on observation for MGUS/low-risk smoldering myeloma, phase 2	Active, not recruiting	Monoclonal Gammopathy of unknown significance, low-risk smoldering multiple myeloma or early-stage prostate cancer	Piperine	[143]
13	Pharmacokinetics of curcumin in healthy volunteers, interventional clinical trial	Completed	Healthy volunteers	Piperine and silybin	[144]

[130] have helped in a significant reduction of the curcumin doses from grams to even up to a few hundred mg, while notably improving the pharmacokinetic parameters. Coadministration of such novel curcumin formulations with standard chemotherapeutics has shown variable improvement in response rate, overall survival, quality of life, and reduction in toxic effects in patient population. A prospective phase 2 study involving 53 patients suffering from pancreatic cancer were treated with gemcitabine (1,000 mg/m^2) on days 1, 8, and 15 days of a 28 days cycle. During the entire cycle, patients were simultaneously subjected to 2 g/day oral phytosomal curcumin formulation. The study showed a 27.3% of response rate, absence of disease progression among 34% of patients, and improvement in overall progression-free survival from 8.4 M (gemcitabine monotherapy) to 10.2 M [130]. A clinical study initiated by Mahammedi and group involved 30 patients suffering from castration-resistant prostate cancer [124]. The subjects received DTX (75 mg/m^2) infusion on days 1 and 21, for 6 cycles, along with oral 5 mg prednisone twice daily, throughout the study duration. The subjects also received 6 g curcumin/day in divided doses for 7 consecutive days (4 days prior to and 3 days post the chemotherapy) in each cycle. Within the first three cycles, 65% of the subjects showed > 80% reduction in PSA, and 36% of them reached normal levels. Beneficial effects of curcumin therapy were evident in the first three cycles of chemotherapy. Table 5.4 gives an account of various clinical studies that have included curcumin as a complementary therapy. Most of the studies have proven the nontoxic and innocuous nature of curcumin, at various doses. The therapeutic benefit of curcumin, however, is either inconclusive or needs to be confirmed by incorporating larger number of subjects.

The clinical studies of piperine as a bioenhancer in anticancer activity include two main studies. The pilot studies conducted by Shoba et al., showed the effect of piperine on the pharmacokinetics of curcumin in healthy human volunteers [55]. These studies were followed by *in vivo* studies in rats. In humans, concomitant administration of piperine, 20 mg, produced much higher serum concentrations from 0.25 to 1 h post-drug, and the increase in bioavailability was as great as 2,000%. Another ongoing phase 2 clinical study is investigating whether the supplementation of curcumin plus piperine can prevent or delay the progression of prostate cancer, monoclonal gammopathy of unknown significance, or low-risk smoldering myeloma into a more aggressive cancer that requires treatment. The investigators evaluated a marker in the patient's blood called MIC-1 to determine if it could be a useful predictor of whether the disease is improving or progressing. The study would be complete by May 2023.

Piperine coadministered with a formulation containing pyrazinamide, rifampicin, and isoniazid has been tested in human volunteers, which showed that the bioavailabilities were higher in the presence of piperine. The patent claims the use of piperine in the range of 0.4-0.9% by weight of the antituberculosis or antileprosy drugs. The administration of 20 mg of piperine alongside antitubercular regimen of rifampicin (450 mg), isoniazid (300 mg), pyrazinamide (1,500 mg) resulted in increased plasma concentration of each drug by around 2-, 5-, and 2-fold respectively. This can potentially result in the development of efficient antileprosy and antituberculosis formulations, which can be prohibitively costly [131].

Bioperine®, a formulation by Sabinsa Corporation, has also been reported for its bioenhancing effects on nutrients. It is a standardized extract of fruits of *P. longum* Linn, which contains 95% of piperine. This claims to enhance the bioavailability of many nutrients like coenzyme Q10, beta-carotene, selenium, vitamin C, vitamin B6, amino acids, and herbal extract of curcumin [32].

Apart from the bioenhancers described so far, phytoconstituents like lysergol [145, 146], stevia [147], naringin [148], aloe vera [149], ginger [150], niaziridin [151], and caraway [152] have also been explored as bioenhancers. They have shown encouraging effects of antitubercular and antibiotic drugs, during *in vitro* and preclinical studies. Of note, is that these herbal constituents are not much explored in cancer chemotherapy. Thus, they hold the potential to be researched in the pursuit of improving the bioavailability of cancer therapeutics, in the future.

5.5 Summary and future perspectives

The hitherto reported *in vitro* and *in vivo* preclinical studies indicate that naturally occurring bioenhancers have the potential to improve outcomes of chemotherapy and can be explored as adjuncts. However, a few studies have demonstrated the anticancer activity of the bioenhancers at concentrations that are otherwise deemed to be biologically inactive. This discrepancy certainly poses hurdles to isolate between the inherent bioactivity and bioenhancing activity of the bioenhancer. Additionally, a few *in vivo* studies have shown that coadministration of bioenhancers with anticancer agents significantly diminished the chemotherapeutic outcomes in these preclinical studies [153, 154]. These preclinical outliers, while insignificant in terms of number, need to be accounted for, while considering the clinical translation of bioenhancers. Furthermore, it should be noted that most of the bioenhancers discussed in this chapter are rapidly metabolized on oral administration in humans. While the bioenhancers that undergo rapid metabolism on oral administration can act as a sacrificial agent to improve the pharmacokinetics of anticancer drugs, it may be difficult to achieve other bioenhancing advantages, such as P-gp efflux in the solid tumor tissues in the case of these bioenhancers. To date, no clinical trials have demonstrated the ability of natural bioenhancers to modulate P-gp efflux and reverse the resistance to anticancer drugs in humans. Poor aqueous solubility is another limiting factor for the clinical translation of bioenhancers. It should be noted that most of the preclinical studies have not reported the solubilization strategies used to deliver the desired dose of the bioenhancer, either orally or systemically. This lack of information certainly impacts the clinical translation of bioenhancers. Given these limitations, it is not so surprising to see that a clinical trial involving a curcumin dose as high as 8 g/day has not been very successful in the patients. Thus far, clinical studies have demonstrated the ability of piperine to improve the oral bioavailability of curcumin, and a trial is ongoing to evaluate the impact of piperine on the chemotherapeutic activity of curcumin in cancer patients. Furthermore, clinical trials involving bioenhancers encapsulated into suitable delivery systems have shown some promise in improving chemotherapeutic outcomes, although extensive studies are still required. While many clinical trials can be found in the clinical trials registry, the results of these studies largely remain unpublished. To summarize, naturally occurring bioenhancers have shown promise to augment chemotherapy in preclinical studies and a few clinical trials. However, extensive and systematically designed clinical trials are still required to fully unravel the potential of natural bioenhancers in chemotherapy.

Conflict of interests

All the authors declare no conflict of interests.

List of abbreviations

5-FU	5-Fluorouracil
ABC	ATP-binding cassette
ADM	Adriamycin
Bax	Bcl-2-associated X protein
Bcl-2	B-cell lymphoma 2
CI	Combination Index
CNS	Central nervous system
CPM	Cyclophosphamide
CYP 450	Cytochrome P450
CYP3A4	Cytochrome P450 3A4
DDP	Cisplatin
DNA	Deoxyribonucleic acid
DOX	Doxorubicin
DTX	Docetaxel
GSH	Glutathione
HepG2	Human hepatoma cell line
IL	Interleukin
JNK	c-Jun N-terminal kinase
MAPK	Mitogen-activated protein kinase
MDR	Multidrug resistant
MMC	Mytomycin C
MMP9	Matrix metalloproteinase 9

MRP	Multidrug-resistant proteins
NF-κB	Nuclear factor-kappa light chain enhancer of activated B cell
Nrf2	Nuclear factor-erythroid 2 p45-related factor 2
NSAIDs	Nonsteroidal anti-inflammatory drugs
P38-COX 2- PGE2	p38 mitogen-activated protein – cyclooxygenase 2- prostaglandin E2
PARP	Poly (ADP-ribose) polymerase
PEG 400	Polyethylene glycol 400
P-gp	P-glycoprotein
PTX	Paclitaxel
ROS	Reactive oxygen species
TNF-α	Tumor necrosis factor α
UDP-GDH	UDP-glucose dehydrogenase
UDP-GT	Uridine diphosphate-glucoronyl transferase
VCR	Vincristine

References

[1] Patwardhan B, Mashelkar RA, Traditional medicine-inspired approaches to drug discovery: Can Ayurveda show the way forward?, Drug Discovery Today, 2009, 14(15–16), 804–811.
[2] Atal C, A breakthrough in drug bioavailability-a clue from age old wisdom of Ayurveda, IDMA Bulletin, 1979, 10, 483–484.
[3] Grace XF, et al., Bioenhancer a new approach in modern medicine, Indo American Journal of Pharmaceutical Research, 2013, 3(12), 1576–1580.
[4] Ratndeep S, et al., Indian herbal bioenhancers: A review, Pharmacognosy Reviews, 2009, 3 (5), 80–82.
[5] Majumdar S, Kulkarni A, Kumbhar S, "Yogvahi (Bioenhancer)": An Ayurvedic Concept Used in Modern Medicines, 2018.
[6] Ha E-S, et al., Enhanced Oral Bioavailability of Resveratrol by Using Neutralized Eudragit E Solid Dispersion Prepared via Spray Drying, Antioxidants, 2021, 10(1), 90.
[7] Xu Y, et al., Preparation of Deoxycholate-Modified Docetaxel-Cimetidine Complex Chitosan Nanoparticles to Improve Oral Bioavailability, AAPS Pharmscitech, 2019, 20(7), 1–10.
[8] Meng Q, et al., Intranasal delivery of Huperzine A to the brain using lactoferrin-conjugated N-trimethylated chitosan surface-modified PLGA nanoparticles for treatment of Alzheimer's disease, International Journal of Nanomedicine, 2018, 13, 705.
[9] Mai Y, et al., Excipient-mediated alteration in drug bioavailability in the rat depends on the sex of the animal, European Journal of Pharmaceutical Sciences, 2017, 107, 249–255.
[10] Rehman S, et al., Role of P-glycoprotein inhibitors in the bioavailability enhancement of solid dispersion of darunavir, BioMed Research International, 2017, 2017, 1–17.
[11] Atal N, Bedi K, Bioenhancers: Revolutionary concept to market, Journal of Ayurveda and Integrative Medicine, 2010, 1(2), 96.
[12] Tatiraju DV, et al., Natural bioenhancers: An overview, Ournal of Pharmacognosy and Phytochemistry, 2013, 2(3), 55–60.
[13] Ferlay J, et al., Global cancer observatory: Cancer today, Lyon: International Agency for Research on Cancer, 2018, 2020.

[14] Exchange GHD, GBD results tool, Seattle (WA), Institute for Health Metrics and Evaluation, University of Washington, 2020, April 6, 2021 Available from http://ghdx.healthdata.org/gbd-results-tool.

[15] WHO. Cancer: Key facts. WHO, 2021. 2021 Accessed on April 6, 2021]; Available from: https://www.who.int/news-room/fact-sheets/detail/cancer.

[16] Holohan C, et al., Cancer drug resistance: An evolving paradigm, Nature Reviews. Cancer, 2013, 13(10), 714–726.

[17] Ankathil R, The mechanisms and challenges of cancer chemotherapy resistance: A current overview, European Journal of Molecular & Clinical Medicine, 2019, 6(1), 26–34.

[18] DiPiro JT, Talbert RL, Yee GC, Matzke GR, Wells BG, LPosey LM, Pharmacotherapy: A pathophysiologic approach, 10th ed., 2017, New York, NY, USA, McGraw-Hill.

[19] Kakde D, et al., Cancer therapeutics-opportunities, challenges and advances in drug delivery, Journal of Applied Pharmaceutical Science, 2011, 1(9), 1–10.

[20] Alexander A, et al., Role of herbal bioactives as a potential bioavailability enhancer for active pharmaceutical ingredients, Fitoterapia, 2014, 97, 1–14.

[21] Kesarwani K, Gupta R, Bioavailability enhancers of herbal origin: An overview, Asian Pacific Journal of Tropical Biomedicine, 2013, 3(4), 253–266.

[22] Chopra B, et al., Piperine and its various physicochemical and biological aspects: A review, Open Chemistry Journal, 2016, 3, 1.

[23] Mills R, Piperine multiplies the strength of many supplement and drugs, The Delano Report, Article Published by Delano Company, San Fransico, 2003, 32–5.

[24] Khajuria A, Zutshi U, Bedi K, Permeability characteristics of piperine on oral absorption – An active alkaloid from peppers and a bioavailability enhancer, Indian Journal of Experimental Biology, 1998, 36(1), 46–50.

[25] Khajuria A, Thusu N, Zutshi U, Piperine modulates permeability characteristics of intestine by inducing alterations in membrane dynamics: Influence on brush border membrane fluidity, ultrastructure and enzyme kinetics, Phytomedicine, 2002, 9(3), 224–231.

[26] Jensen-Jarolim E, et al., Hot spices influence permeability of human intestinal epithelial monolayers, The Journal of Nutrition, 1998, 128(3), 577–581.

[27] Annamalai A, Manavalan R, Effect of "Trikatu" and its individual components and piperine on gastrointestinal tracts, Indian Drugs, 1990, 27(12), 595–604.

[28] Johri R, Zutshi U, An Ayurvedic formulation 'Trikatu'and its constituents, Journal of Ethnopharmacology, 1992, 37(2), 85–91.

[29] Atal C, Dubey RK, Singh J, Biochemical basis of enhanced drug bioavailability by piperine: Evidence that piperine is a potent inhibitor of drug metabolism, Journal of Pharmacology and Experimental Therapeutics, 1985, 232(1), 258–262.

[30] Singh J, Dubey R, Atal C, Piperine-mediated inhibition of glucuronidation activity in isolated epithelial cells of the guinea-pig small intestine: Evidence that piperine lowers the endogenous UDP-glucuronic acid content, Journal of Pharmacology and Experimental Therapeutics, 1986, 236(2), 488–493.

[31] Bhardwaj RK, et al., Piperine, a major constituent of black pepper, inhibits human P-glycoprotein and CYP3A4, Journal of Pharmacology and Experimental Therapeutics, 2002, 302(2), 645–650.

[32] Majeed M, Badmaev V, Rajendran R, Use of piperine as a bioavailability enhancer. 1999, Google Patents.

[33] Li S, et al., Piperine, a piperidine alkaloid from Piper nigrum re-sensitizes P-gp, MRP1 and BCRP dependent multidrug resistant cancer cells, Phytomedicine, 2011, 19(1), 83–87.

[34] Callaghan R, Luk F, Bebawy M, Inhibition of the multidrug resistance P-glycoprotein: Time for a change of strategy?, Drug Metabolism and Disposition, 2014, 42(4), 623–631.

[35] Syed SB, et al., Targeting P-glycoprotein: Investigation of piperine analogs for overcoming drug resistance in cancer, Scientific Reports, 2017, 7(1), 1–18.
[36] Motiwala M, Rangari V, Combined effect of paclitaxel and piperine on a MCF-7 breast cancer cell line *in vitro* Evidence of a synergistic interaction, Synergy, 2015, 2(1), 1–6.
[37] Pushpa Ragini S, Naga Divya A, Anusha Ch KY, Enhancement of paclitaxel and doxorubicin cytotoxicity in breast cancer cell lines in combination with piperine treatment and analysis of expression of autophagy and apoptosis genes, Journal of Medical Science and Clinical Research, 2014, 2(2), 62–67.
[38] Lai L-H, et al., Piperine suppresses tumor growth and metastasis *in vitro* and *in vivo* in a 4T1 murine breast cancer model, Acta Pharmacologica Sinica, 2012, 33(4), 523–530.
[39] Kakarala M, et al., Targeting breast stem cells with the cancer preventive compounds curcumin and piperine, Breast Cancer Research and Treatment, 2010, 122(3), 777–785.
[40] Patial V, et al., Synergistic effect of curcumin and piperine in suppression of DENA-induced hepatocellular carcinoma in rats, Environmental Toxicology and Pharmacology, 2015, 40(2), 445–452.
[41] Khamis AA, et al., Hesperidin, piperine and bee venom synergistically potentiate the anticancer effect of tamoxifen against breast cancer cells, Biomedicine & Pharmacotherapy, 2018, 105, 1335–1343.
[42] Talib WH, Regressions of breast carcinoma syngraft following treatment with piperine in combination with thymoquinone, Scientia Pharmaceutica, 2017, 85(3), 27.
[43] Li C, et al., Enhanced anti-tumor efficacy and mechanisms associated with docetaxel-piperine combination-*in vitro* and *in vivo* investigation using a taxane-resistant prostate cancer model, Oncotarget, 2018, 9(3), 3338.
[44] Do MT, et al., Antitumor efficacy of piperine in the treatment of human HER2-overexpressing breast cancer cells, Food Chemistry, 2013, 141(3), 2591–2599.
[45] Makhov P, et al., Co-administration of piperine and docetaxel results in improved anti-tumor efficacy via inhibition of CYP3A4 activity, The Prostate, 2012, 72(6), 661–667.
[46] Najar I, et al., Involvement of P-glycoprotein and CYP 3A4 in the enhancement of etoposide bioavailability by a piperine analogue, Chemico-biological Interactions, 2011, 190(2–3), 84–90.
[47] Katiyar SS, et al., Co-delivery of rapamycin-and piperine-loaded polymeric nanoparticles for breast cancer treatment, Drug Delivery, 2016, 23(7), 2608–2616.
[48] Li C, et al., Non-linear pharmacokinetics of piperine and its herb-drug interactions with docetaxel in Sprague-Dawley rats, Journal of Pharmaceutical and Biomedical Analysis, 2016, 128, 286–293.
[49] Reen RK, et al., Piperine impairs cytochrome P4501A1 activity by direct interaction with the enzyme and not by down regulation of CYP1A1 gene expression in the rat hepatoma 5L cell line, Biochemical and Biophysical Research Communications, 1996, 218(2), 562–569.
[50] Selvendiran K, et al., Chemopreventive effect of piperine on mitochondrial TCA cycle and phase-I and glutathione-metabolizing enzymes in benzo (a) pyrene induced lung carcinogenesis in Swiss albino mice, Molecular and Cellular Biochemistry, 2005, 271(1), 101–106.
[51] Sachin B, et al., Simultaneous determination of etoposide and a piperine analogue (PA-1) by UPLC–qTOF-MS: Evidence that PA-1 enhances the oral bioavailability of etoposide in mice, Journal of Chromatography B, 2010, 878(9–10), 823–830.
[52] Danduga RCSR, Kola PK, Matli B, Anticancer activity of curcumin alone and in combination with piperine in Dalton lymphoma ascites bearing mice, Indian Journal of Experimental Biology, 2020, 58, 181–189

[53] Reen R, Singh J, *In vitro* and *in vivo* inhibition of pulmonary cytochrome P450 activities by piperine, a major ingredient of piper species, Indian Journal of Experimental Biology, 1991, 29(6), 568–573.

[54] Hatab HM, et al., A combined treatment of curcumin, piperine, and taurine alters the circulating levels of IL-10 and miR-21 in hepatocellular carcinoma patients: A pilot study, Journal of Gastrointestinal Oncology, 2019, 10(4), 766.

[55] Shoba G, et al., Influence of piperine on the pharmacokinetics of curcumin in animals and human volunteers, Planta Medica, 1998, 64(04), 353–356.

[56] Klinger NV, Mittal S, Therapeutic potential of curcumin for the treatment of brain tumors, Oxidative Medicine and Cellular Longevity, 2016, 2016, 1–14.

[57] Rainey NE, Moustapha A, Petit PX, Curcumin, a multifaceted hormetic agent, mediates an intricate crosstalk between mitochondrial turnover, autophagy, and apoptosis, Oxidative Medicine and Cellular Longevity, 2020, 2020, 3656419.

[58] Qian H, Yang Y, Wang X, Curcumin enhanced adriamycin-induced human liver-derived Hepatoma G2 cell death through activation of mitochondria-mediated apoptosis and autophagy, European Journal of Pharmaceutical Sciences, 2011, 43(3), 125–131.

[59] Tang X-Q, et al., Effect of curcumin on multidrug resistance in resistant human gastric carcinoma cell line SGC7901/VCR, Acta Pharmacologica Sinica, 2005, 26(8), 1009–1016.

[60] Zhang W, et al., Curcumin enhances cisplatin sensitivity of human NSCLC cell lines through influencing Cu-Sp1-CTR1 regulatory loop, Phytomedicine, 2018, 48, 51–61.

[61] Wen C, et al., Curcumin reverses doxorubicin resistance via inhibition the efflux function of ABCB4 in doxorubicin-resistant breast cancer cells, Molecular Medicine Reports, 2019, 19(6), 5162–5168.

[62] Zhao X, et al., Doxorubicin and curcumin co-delivery by lipid nanoparticles for enhanced treatment of diethylnitrosamine-induced hepatocellular carcinoma in mice, European Journal of Pharmaceutics and Biopharmaceutics, 2015, 93, 27–36.

[63] Kunnumakkara AB, et al., Curcumin potentiates antitumor activity of gemcitabine in an orthotopic model of pancreatic cancer through suppression of proliferation, angiogenesis, and inhibition of nuclear factor-κB–regulated gene products, Cancer Research, 2007, 67(8), 3853–3861.

[64] Yan Y-D, et al., Enhanced oral bioavailability of docetaxel in rats by four consecutive days of pre-treatment with curcumin, International Journal of Pharmaceutics, 2010, 399(1–2), 116–120.

[65] Zhou Q-M, et al., Curcumin enhanced antiproliferative effect of mitomycin C in human breast cancer MCF-7 cells *in vitro* and *in vivo*, Acta Pharmacologica Sinica, 2011, 32(11), 1402–1410.

[66] Kang HJ, et al., Curcumin suppresses the paclitaxel-induced nuclear factor-κB in breast cancer cells and potentiates the growth inhibitory effect of paclitaxel in a breast cancer nude mice model, The Breast Journal, 2009, 15(3), 223–229.

[67] Zhang Q-L, Fu BM, Zhang Z-J, Borneol, a novel agent that improves central nervous system drug delivery by enhancing blood–brain barrier permeability, Drug Delivery, 2017, 24(1), 1037–1044.

[68] Zheng Q, et al., Borneol, a messenger agent, improves central nervous system drug delivery through enhancing blood–brain barrier permeability: A preclinical systematic review and meta-analysis, Drug Delivery, 2018, 25(1), 1617–1633.

[69] Kulkarni M, et al., Borneol: A Promising Monoterpenoid in Enhancing Drug Delivery Across Various Physiological Barriers, AAPS Pharmscitech, 2021, 22(4), 1–17.

[70] Zou L, et al., Drug resistance reversal in ovarian cancer cells of paclitaxel and borneol combination therapy mediated by PEG-PAMAM nanoparticles, Oncotarget, 2017, 8(36), 60453.

[71] Cao W-Q, et al. Enhanced anticancer efficiency of doxorubicin against human glioma by natural borneol through triggering ROS-mediated signal, Biomedicine & Pharmacotherapy, 2019, 118, 109261.

[72] Meng L, et al., Improving glioblastoma therapeutic outcomes via doxorubicin-loaded nanomicelles modified with borneol, International Journal of Pharmaceutics, 2019, 567, 118485.

[73] Guo X, et al., Pep-1&borneol–bifunctionalized carmustine-loaded micelles enhance anti-glioma efficacy through tumor-targeting and BBB-penetrating, Journal of Pharmaceutical Sciences, 2019, 108(5), 1726–1735.

[74] Wu Y, et al., Effect of borneol as a penetration enhancer on brain targeting of nanoliposomes: Facilitate direct delivery to neurons, Nanomedicine, 2018, 13(21), 2709–2727.

[75] Xu X, et al., A novel doxorubicin loaded folic acid conjugated PAMAM modified with borneol, a nature dual-functional product of reducing PAMAM toxicity and boosting BBB penetration, European Journal of Pharmaceutical Sciences, 2016, 88, 178–190.

[76] Xiao X, et al., Quercetin suppresses cyclooxygenase-2 expression and angiogenesis through inactivation of P300 signaling, PloS One, 2011, 6(8), e22934.

[77] Eid HM, Haddad PS, The antidiabetic potential of quercetin: underlying mechanisms. Current Medicinal Chemistry, 2017, 24(4), 355–364.

[78] Vessal M, Hemmati M, Vasei M, Antidiabetic effects of quercetin in streptozocin-induced diabetic rats, Comparative Biochemistry and Physiology Part C: Toxicology & Pharmacology, 2003, 135(3), 357–364.

[79] Yang DK, Kang H-S, Anti-diabetic effect of cotreatment with quercetin and resveratrol in streptozotocin-induced diabetic rats, Biomolecules & Therapeutics, 2018, 26(2), 130.

[80] Bule M, et al., Antidiabetic effect of quercetin: A systematic review and meta-analysis of animal studies, Food and Chemical Toxicology, 2019, 125, 494–502.

[81] Kotob SE, et al., Quercetin and ellagic acid in gastric ulcer prevention: An integrated scheme of the potential mechanisms of action from in vivo study, Asian Journal of Pharmaceutical and Clinical Research, 2018, 11(1), 381–389.

[82] Perez-Vizcaino F, et al., Antihypertensive effects of the flavonoid quercetin, Pharmacological Reports, 2009, 61(1), 67–75.

[83] Larson AJ, Symons JD, Jalili T, Therapeutic potential of quercetin to decrease blood pressure: Review of efficacy and mechanisms, Advances in Nutrition, 2012, 3(1), 39–46.

[84] Lakhanpal P, Rai DK, Quercetin: A versatile flavonoid, Internet Journal of Medical Update, 2007, 2(2), 22–37.

[85] Bhimanwar R, Kothapalli L, Khawshi A, Quercetin as Natural Bioavailability Modulator: An Overview, Research Journal of Pharmacy and Technology, 2020, 13(4), 2045–2052.

[86] Dudhatra GB, et al., A comprehensive review on pharmacotherapeutics of herbal bioenhancers, The Scientific World Journal, 2012, 2012, 637953.

[87] Shanmugam S, Natural Bioenhancers: Current Outlook, Clinical Pharmacology & Biopharmaceutics, 2015, 4(2), 1–3.

[88] Choi J-S, Jo B-W, Kim Y-C, Enhanced paclitaxel bioavailability after oral administration of paclitaxel or prodrug to rats pretreated with quercetin, European Journal of Pharmaceutics and Biopharmaceutics, 2004, 57(2), 313–318.

[89] Clark PI, Slevin ML, The clinical pharmacology of etoposide and teniposide, Clinical Pharmacokinetics, 1987, 12(4), 223–252.

[90] Li X, Choi J-S, Effects of quercetin on the pharmacokinetics of etoposide after oral or intravenous administration of etoposide in rats, Anticancer Research, 2009, 29(4), 1411–1415.

[91] Shin S-C, Choi J-S, Li X, Enhanced bioavailability of tamoxifen after oral administration of tamoxifen with quercetin in rats, International Journal of Pharmaceutics, 2006, 313(1–2), 144–149.

[92] Choi J-S, Piao Y-J, Kang KW, Effects of quercetin on the bioavailability of doxorubicin in rats: Role of CYP3A4 and P-gp inhibition by quercetin, Archives of Pharmacal Research, 2011, 34(4), 607–613.

[93] Kale A, et al., Studies on the effects of oral administration of nutrient mixture, quercetin and red onions on the bioavailability of epigallocatechin gallate from green tea extract, Phytotherapy Research, 2010, 24(S1), S48–S55.

[94] Rollyson WD, et al., Bioavailability of capsaicin and its implications for drug delivery, Journal of Controlled Release, 2014, 196, 96–105.

[95] Huh H-C, et al., Capsaicin induces apoptosis of cisplatin-resistant stomach cancer cells by causing degradation of cisplatin-inducible Aurora-A protein, Nutrition and Cancer, 2011, 63 (7), 1095–1103.

[96] Friedman JR, et al., Capsaicin synergizes with camptothecin to induce increased apoptosis in human small cell lung cancers via the calpain pathway, Biochemical Pharmacology, 2017, 129, 54–66.

[97] Zheng L, et al., Capsaicin enhances anti-proliferation efficacy of pirarubicin via activating TRPV1 and inhibiting PCNA nuclear translocation in 5637 cells, Molecular Medicine Reports, 2016, 13(1), 881–887.

[98] Meral O, et al., Capsaicin inhibits cell proliferation by cytochrome c release in gastric cancer cells, Tumor Biology, 2014, 35(7), 6485–6492.

[99] Hong Z-F, et al., Capsaicin enhances the drug sensitivity of cholangiocarcinoma through the inhibition of chemotherapeutic-induced autophagy, PloS One, 2015, 10(5), e0121538.

[100] Wang Y, et al., Synergistic inhibitory effects of capsaicin combined with cisplatin on human osteosarcoma in culture and in xenografts, Journal of Experimental & Clinical Cancer Research, 2018, 37(1), 1–17.

[101] Salehi B, et al., Allicin and health: A comprehensive review, Trends in Food Science & Technology, 2019, 86, 502–516.

[102] Wang C-J, et al., Effect of combined treatment with recombinant interleukin-2 and allicin on pancreatic cancer, Molecular Biology Reports, 2013, 40(12), 6579–6585.

[103] Gao X-Y, et al., Effect of combined treatment with cyclophosphamidum and allicin on neuroblastoma–bearing mice, Asian Pacific Journal of Tropical Medicine, 2015, 8(2), 137–141.

[104] Zou X, et al., Allicin sensitizes hepatocellular cancer cells to anti-tumor activity of 5-fluorouracil through ROS-mediated mitochondrial pathway, Journal of Pharmacological Sciences, 2016, 131(4), 233–240.

[105] Suddek GM, Allicin enhances chemotherapeutic response and ameliorates tamoxifen-induced liver injury in experimental animals, Pharmaceutical Biology, 2014, 52(8), 1009–1014.

[106] Almajali B, et al., Thymoquinone, as a Novel Therapeutic Candidate of Cancers, Pharmaceuticals, 2021, 14(4), 369.

[107] Peterson B, et al., Drug bioavailability enhancing agents of natural origin (bioenhancers) that modulate drug membrane permeation and pre-systemic metabolism, Pharmaceutics, 2019, 11(1), 33.

[108] Dirican A, et al., Novel combination of docetaxel and thymoquinone induces synergistic cytotoxicity and apoptosis in DU-145 human prostate cancer cells by modulating PI3K–AKT pathway, Clinical and Translational Oncology, 2015, 17(2), 145–151.

[109] Banerjee S, et al., Antitumor activity of gemcitabine and oxaliplatin is augmented by thymoquinone in pancreatic cancer, Cancer Research, 2009, 69(13), 5575–5583.

[110] Jafri SH, et al., Thymoquinone and cisplatin as a therapeutic combination in lung cancer: *In vitro* and *in vivo*, Journal of Experimental & Clinical Cancer Research, 2010, 29(1), 1–11.

[111] Kensara OA, et al., Thymoquinone subdues tumor growth and potentiates the chemopreventive effect of 5-fluorouracil on the early stages of colorectal carcinogenesis in rats, Drug Design, Development and Therapy, 2016, 10, 2239.

[112] Li J-Y, et al., Glycyrrhizic acid in the treatment of liver diseases: Literature review, BioMed Research International, 2014, 2014, 872139.

[113] Su X, et al., Glycyrrhizic acid: A promising carrier material for anticancer therapy, Biomedicine & Pharmacotherapy, 2017, 95, 670–678.

[114] Cai Y, et al., The selective effect of glycyrrhizin and glycyrrhetinic acid on topoisomerase IIα and apoptosis in combination with etoposide on triple negative breast cancer MDA-MB-231 cells, European Journal of Pharmacology, 2017, 809, 87–97.

[115] Wakamatsu T, et al., The combination of glycyrrhizin and lamivudine can reverse the cisplatin resistance in hepatocellular carcinoma cells through inhibition of multidrug resistance-associated proteins, International Journal of Oncology, 2007, 31(6), 1465–1472.

[116] Deng Q-P, et al., Effects of glycyrrhizin in a mouse model of lung adenocarcinoma, Cellular Physiology and Biochemistry, 2017, 41(4), 1383–1392.

[117] Yang F-H, et al., Bioavailability enhancement of paclitaxel via a novel oral drug delivery system: Paclitaxel-loaded glycyrrhizic acid micelles, Molecules, 2015, 20(3), 4337–4356.

[118] Lambert JD, et al., Inhibition of carcinogenesis by polyphenols: Evidence from laboratory investigations, The American Journal of Clinical Nutrition, 2005, 81(1), 284S–291S.

[119] Kurzer MS, Xu X, Dietary phytoestrogens, Annual Review of Nutrition, 1997, 17(1), 353–381.

[120] Li X, Choi J-S, Effect of genistein on the pharmacokinetics of paclitaxel administered orally or intravenously in rats, International Journal of Pharmaceutics, 2007, 337(1–2), 188–193.

[121] Chen WT-L, et al., Effectiveness of a novel herbal agent MB-6 as a potential adjunct to 5-fluoracil–based chemotherapy in colorectal cancer, Nutrition Research, 2014, 34(7), 585–594.

[122] Thomas R, et al., A double-blind, placebo-controlled randomised trial evaluating the effect of a polyphenol-rich whole food supplement on PSA progression in men with prostate cancer – The UK NCRN Pomi-T study, Prostate Cancer and Prostatic Diseases, 2014, 17(2), 180–186.

[123] Kanai M, et al., A phase I/II study of gemcitabine-based chemotherapy plus curcumin for patients with gemcitabine-resistant pancreatic cancer, Cancer Chemotherapy and Pharmacology, 2011, 68(1), 157–164.

[124] Mahammedi H, et al., The new combination docetaxel, prednisone and curcumin in patients with castration-resistant prostate cancer: A pilot phase II study, Oncology, 2016, 90(2), 69–78.

[125] James MI, et al., Curcumin inhibits cancer stem cell phenotypes in ex vivo models of colorectal liver metastases, and is clinically safe and tolerable in combination with FOLFOX chemotherapy, Cancer Letters, 2015, 364(2), 135–141.

[126] Bayet-Robert M, et al., Phase I dose escalation trial of docetaxel plus curcumin in patients with advanced and metastatic breast cancer, Cancer Biology & Therapy, 2010, 9(1), 8–14.

[127] Dhillon N, et al., Phase II trial of curcumin in patients with advanced pancreatic cancer, Clinical Cancer Research, 2008, 14(14), 4491–4499.

[128] Gota VS, et al., Safety and pharmacokinetics of a solid lipid curcumin particle formulation in osteosarcoma patients and healthy volunteers, Journal of Agricultural and Food Chemistry, 2010, 58(4), 2095–2099.

[129] Dützmann S, et al., Intratumoral concentrations and effects of orally administered micellar curcuminoids in glioblastoma patients, Nutrition and Cancer, 2016, 68(6), 943–948.

[130] Pastorelli D, et al., Phytosome complex of curcumin as complementary therapy of advanced pancreatic cancer improves safety and efficacy of gemcitabine: Results of a prospective phase II trial, Pharmacological Research, 2018, 132, 72–79.

[131] Randhir SK, et al., Process for preparation of pharmaceutical composition with enhanced activity for treatment of tuberculosis and leprosy. 1993, Council of Scientific and Industrial Research CSIR.

[132] Center, B.C.A.S.C., A Pilot, Feasibility Study of Curcumin in Combination With 5FU for Patients With 5FU-Resistant Metastatic Colon Cancer. March 2016 to June 2020, https://ClinicalTrials.gov/show/NCT02724202: Dallas, Texas, United States.

[133] Jeong-Heum Baek M, Gachon University Gil Medical Center, Avastin/FOLFIRI in Combination With Curcumin in Colorectal Cancer Patients With Unresectable Metastasis. May 2015 to February 2020, https://ClinicalTrials.gov/show/NCT02439385.

[134] IU Simon Cancer Center, U.o.N.C.a.C.H.L.C.c.C., Effect of Curcumin on Dose Limiting Toxicity and Pharmacokinetics of Irinotecan in Patients With Solid Tumors. May 22, 2013 to June 22, 2020, https://ClinicalTrials.gov/show/NCT01859858: Indianapolis, Indiana, United States and Chapel Hill, North Carolina, United States.

[135] Oncology, N.C.o., Curcumin in Combination With Chemotherapy in Advanced Breast Cancer. March 20, 2017 to June 30, 2019, https://ClinicalTrials.gov/show/NCT03072992: Yerevan, Armenia.

[136] Dept Oncology, L.R.I., Combining Curcumin With FOLFOX Chemotherapy in Patients With Inoperable Colorectal Cancer. February 2012 to May 31, 2017, https://ClinicalTrials.gov/show/NCT01490996: Leicester, United Kingdom.

[137] Center, T.-A.S.M., Phase III Trial of Gemcitabine, Curcumin and Celebrex in Patients With Metastatic Colon Cancer. March 2006 to March 2007, https://ClinicalTrials.gov/show/NCT00295035.

[138] Ron Epelbaum M, Rambam Health Care Campus, Gemcitabine With Curcumin for Pancreatic Cancer. July 2004 to September 2010, https://ClinicalTrials.gov/show/NCT00192842.

[139] Center, M.D.A.C., Radiation Therapy and Capecitabine With or Without Curcumin Before Surgery in Treating Patients With Rectal Cancer. August 11, 2008 to March 31, 2023, https://ClinicalTrials.gov/show/NCT00745134: Houston, Texas, United States.

[140] H. Lee Moffitt Cancer Center and Research Institute, F., United States, Phase II Trial to Modulate Intermediate Endpoint Biomarkers in Former and Current Smokers. June 5, 2019 to October 2023, https://ClinicalTrials.gov/show/NCT03598309: Tampa, Florida, United States.

[141] University Hospital Antwerp, I.J.B., University Hospital Gent and CMSE Namur, Study of Pembrolizumab, Radiation and Immune Modulatory Cocktail in Cervical/Uterine Cancer. July 1, 2017 to June 2022, https://ClinicalTrials.gov/show/NCT03192059: Antwerp, Brussels, Gent and Namur, Belgium.

[142] (NCI), C.o.H.M.C.a.N.C.I., Gemcitabine Hydrochloride, Paclitaxel Albumin-Stabilized Nanoparticle Formulation, Metformin Hydrochloride, and a Standardized Dietary Supplement in Treating Patients With Pancreatic Cancer That Cannot be Removed by Surgery. January 14, 2016 to December 31, 2021, https://ClinicalTrials.gov/show/NCT02336087: City of Hope Medical Center Duarte, California, United States, City of Hope Rancho Cucamonga, Rancho Cucamonga, California, United States, City of Hope South Pasadena, South Pasadena, California, United States, City of Hope West Covina, West Covina, California, United States.

[143] Peter Van Veldhuizen UOR, Curcumin and Piperine in Patients on Surveillance for Monoclonal Gammopathy, Smoldering Myeloma or Prostate Cancer. September 1, 2021 to May 31, 2023, https://ClinicalTrials.gov/show/NCT04731844: University of Rochester, Rochester, New York, United States.

[144] Laura Sullivan MGH, Pharmacokinetics of Curcumin in Healthy Volunteers. August 2005 to October 17, 2007, https://ClinicalTrials.gov/show/NCT00181662: Massachusetts General Hospital Boston, Massachusetts, United States.

[145] Patil S, et al., Simultaneous quantification of berberine and lysergol by HPLC-UV: Evidence that lysergol enhances the oral bioavailability of berberine in rats, Biomedical Chromatography, 2012, 26(10), 1170–1175.

[146] Khanuja S, et al., Antibiotic pharmaceutical composition with lysergol as bio-enhancer and method of treatment. 2007, Google Patents.

[147] Gokaraju GR, et al., Bio-availability/bio-efficacy enhancing activity of Stevia rebaudiana and extracts and fractions and compounds thereof. 2010, Google Patents.

[148] Choi J-S, Shin S-C, Enhanced paclitaxel bioavailability after oral coadministration of paclitaxel prodrug with naringin to rats, International Journal of Pharmaceutics, 2005, 292(1–2), 149–156.

[149] Haasbroek A, et al., Intestinal drug absorption enhancement by Aloe vera gel and whole leaf extract: *In vitro* investigations into the mechanisms of action, Pharmaceutics, 2019, 11(1), 36.

[150] Qazi G, et al., Bioavailability enhancing activity of Zingiber officinale Linn and its extracts/ fractions thereof. 2003, Google Patents.

[151] Khanuja SPS, et al., Nitrile glycoside useful as a bioenhancer of drugs and nutrients, process of its isolation from Moringa oleifera. 2005, Google Patents.

[152] Choudhary N, et al., Effect of Carum carvi, a herbal bioenhancer on pharmacokinetics of antitubercular drugs: A study in healthy human volunteers, Perspectives in Clinical Research, 2014, 5(2), 80.

[153] Karan R, Bhargava V, Garg S, Effect of Trikatu (Piperine) on the pharmacokinetic profile of isoniazid in rabbits, Indian Journal of Pharmacology, 1998, 30(4), 254.

[154] Hussaarts KG, et al., Impact of curcumin (with or without piperine) on the pharmacokinetics of tamoxifen, Cancers, 2019, 11(3), 403.

Rupali A. Patil, Krutika H. Pardeshi, Harshal P. Chavan,
Sunil V. Amrutkar

Chapter 6
Pharmacotherapeutics and pharmacokinetics of herbal bioenhancers

Abstract: Novel chemical substances with high therapeutic efficacy are revealed, but, due to low solubility and/or membrane permeability, numerous substances of these have unfavorable pharmacokinetic properties. The lipid-like barrier created by the epithelial mucosal layers limits membrane permeability, which must be traversed by therapeutic molecules. Another reason for low bioavailability is the presystemic metabolism of drugs, principally by cytochrome P450 enzymes situated in the intestinal enterocytes and liver hepatocytes. Herbal bioenhancers are the agents that have no intrinsic pharmacological activity at their therapeutic doses used but when coadministered with various classes of drugs such as anticancer, antibiotics, antifungal, antituberculosis, antiviral, antihypertensives, and antihyperlipidemic, are capable of enhancing their bioavailability and bioefficacy. This chapter accentuates the numerous natural compounds that can be exploited as an effective bioenhancer, such as sinomenine, lysergol, quercetin, and glycyrrhizin that are capable of enhancing pharmacokinetic parameters of various drugs, mainly by improving the absorption and reducing the metabolism. They have also been shown to help with nutraceutical absorption, such as amino acids, minerals, vitamins, and few herbal compounds. Often, poor bioavailable drugs are administered by parenteral route. Bioenhancers can be used to improve the bioavailability of these poor bioavailable drugs and their administration by an alternative route such as the oral, nasal, buccal, pulmonary route can be explored. This results in a reduction in dosage of small drug molecules and treatment costs. There is an extensive variety of untouched plants that can be tested for their ability to enhance drug bioavailability. These herbal bioenhancers' toxicity characteristics must not be overlooked. Further research is needed to address these concerns and give a safe and effective therapeutic dose to produce the intended pharmacological response.

Rupali A. Patil, GES's Sir Dr. M. S. Gosavi College of Pharmaceutical Education and Research, Nashik, Maharashtra 425005, India, e-mail: ruupalipatil@gmail.com
Krutika H. Pardeshi, Harshal P. Chavan, Sunil V. Amrutkar, GES's Sir Dr. M. S. Gosavi College of Pharmaceutical Education and Research, Nashik, Maharashtra 425005, India

https://doi.org/10.1515/9783110746808-006

6.1 Introduction

6.1.1 Introduction to bioenhancer

All aspects of pharmaceutical science are focused on discovering new dosage forms with different modes of action. New drug development strategies have emerged as a result to reduce the cost and the treatment linked to drug dosage. As a result, treatments are becoming more accessible to a wider range of people, including those who are financially disadvantaged. Increased drug bioavailability is one way to reduce drug dosage and, as a result, drug toxicity and expense [1].

Bioavailability is defined as the rate and extent to which any substance reaches the systemic circulation and triggers the desired action. Various classes of drugs have diverse potent therapeutic activity, but their use has always been a critical feature of drug development programmers due to their low bioavailability. Physiochemical properties of the drug such as poor solubility and membrane permeability, first-pass metabolism, stability issues of the drug in the gastrointestinal tract [GIT], biological barriers such as hepatic and metabolizing enzymes present in the intestine, and drug transporters are major factors responsible for poor bioavailability of a drug. In addition to this, P-glycoprotein(P-gp) is also responsible for reduced bioavailability of the therapeutically active drugs, especially anticancer drugs. The drugs administered through oral route undergo various steps such as incomplete drug absorption and first-pass metabolism, which ultimately leads to low bioavailability [2].

Herbal medicine has been practiced for hundreds of years. There is an increasing interest in combining potent drugs with natural bioenhancers to increase their bioavailability [1, 3]. A bioenhancer does not have any pharmacological activity, but, when combined, it is able to improve the bioavailability and effectiveness of a medication. In 1929, Bose reported the potential of *Adhatoda vasaka* as antihistaminic when he combined it with long pepper [3].

The "bioavailability enhancer" concept, which uses the component of natural origin was derived from Ayurveda. One example is "Trikatu" which is well known to everyone. Trikatu is an ancient Ayurvedic preparation. A drug's ability to cross biological membranes is hampered by many factors such as zwitterionic characteristics observed at physiological pH, weak aqueous solubility, or the presence of P-gp on the outer surface. As a result, natural bioenhancers such as genistein, sinomenine, curcumin, and piperine are increasingly being used to boost the pharmacokinetic and pharmacodynamic profile of medications, and hence the bioavailability of different potent drugs [4].

6.1.2 Classification of bioenhancers [5, 6]

Table 6.1: Classification of natural bioenhancer.

Based on plant origin	Based on animal origin	Based on mechanism of action	
Piperine [black pepper]	Cow urine distillate [Kamdhenu ark]	P-glycoprotein (P-GP) efflux pump and other pump inhibitors	Caraway, quercetin, naringin, *Cuminum cyminum* (black cumin), genistein and sinomenine
Gingerol (ginger)		Inhibitors of enzyme including CYP-450 enzyme and its isoenzymes	Sinomenine, naringin, quercetin, gallic acid
Glycyrrhizin (liquorice)		Regulates function of gastrointestinal tract to increase absorption	Niaziridin (drumstick pods), glycyrrhizin (liquorice), *Aloe vera* (aloe), *Zingiber officinale* (ginger).
Caraway (cumin)			
Black cumin			
Quercetin			
Niaziridin			
Capsaicin			
Stevia			
Allicin (garlic)			
Curcumin (turmeric)			
Aloe vera			
Lysergol (morning glory plant)			
Simomenine			
Genistein			
Naringin			
Ocimum sanctum (tulsi)			
Ammannia multiflora (jerry-jerry)			
Tribulus terrestris (gokhru)			

6.1.3 Ideal properties of a bioenhancer

A bioenhancer should have the following properties, ideally:
1. It should not be toxic when administered to a human or an animal.
2. It should show effect at very low concentrations.
3. It must be simple to formulate.
4. The most significant property of any bioenhancer is to increase the absorption of the drug and, thereby, its pharmacological activity. [7]

6.1.4 Bioavailability enhancing effect of a bioenhancer

In order to enhance bioavailability, pharmaceutical industry focuses on factors such as poor absorption, long-term administration, toxicity, and cost of drug. Increase in bioavailability is directly related to the therapeutic efficacy of drug. The cost of the drug and its toxicity, by minimizing the concentration of drug, can be reduced by improving the bioavailability. Drugs with poor bioavailability remain sub-therapeutic as the required dose is not able to reach the systemic circulation, whereas a large dose of drug can lead to serious side effects. Poor water solubility, poor permeability, drug degradation in the gastric or intestinal fluids, and drug metabolism are problems associated with poor oral bioavailability. Using the bioavailability approach, lowering a dose and the dose frequency can be achieved for a specific drug.

Bioenhancers improve the pharmacological activity of various drugs and nutraceuticals by improving their bioavailability/bioefficacy.

Various approaches of bioenhancers for improvement of bioavailability are as follows:
1. They promote drug absorption by the gastrointestinal tract.
2. They inhibit metabolism of the drug by the liver or the intestinal enzyme.
3. They modify the immune system to substantially reduce the dose of the drug.
4. They promote penetration of drug into pathogen even when pathogens become persistors within the macrophages, as in the case of *Mycobacterium tuberculosis* and other pathogens.
5. They inhibit pathogen capability of the tissue in order to reject the drug, such as antimicrobial and anticancer agents.
6. They modify the signaling process between the pathogen and the host, which improves the route of the drug to reach to the pathogen.
7. They also increase the binding property between the drug and the target sites, such as RNA, receptors, and DNA, and increase the bioavailability of the drug.

Bioenhancers may also assist in the movement of nutrients and medications to the brain, crossing the BBB, which is helpful in the treatment of CNS disorders, including epilepsy and cerebral infections. Bioavailability enhancement has gained widespread acceptance since the olden days, by combining primary and secondary medicinal agents. Hence, employing herbal bioenhancers is a modern approach based on Ayurveda literature, and it involves enhancing the bioavailability of drugs [5, 8].

6.1.5 Mechanism of bioavailability enhancement by a bioenhancer

Several mechanisms are involved in improving the bioavailability of drugs by natural bioenhancers. Herbal enhancers can have similar or dissimilar modes of action (Figure 6.1).

1. Bioenhancers act on the gastrointestinal tract, which improves drug absorption, whereas antimicrobial bioenhancers act by altering the metabolic process of the drug.
2. Bioenhancers inhibit the transit, emptying, and motility of GIT. They also enhance the uptake of the amino acid by increasing the activity of gamma-glutamyl transpeptidase (GGT).

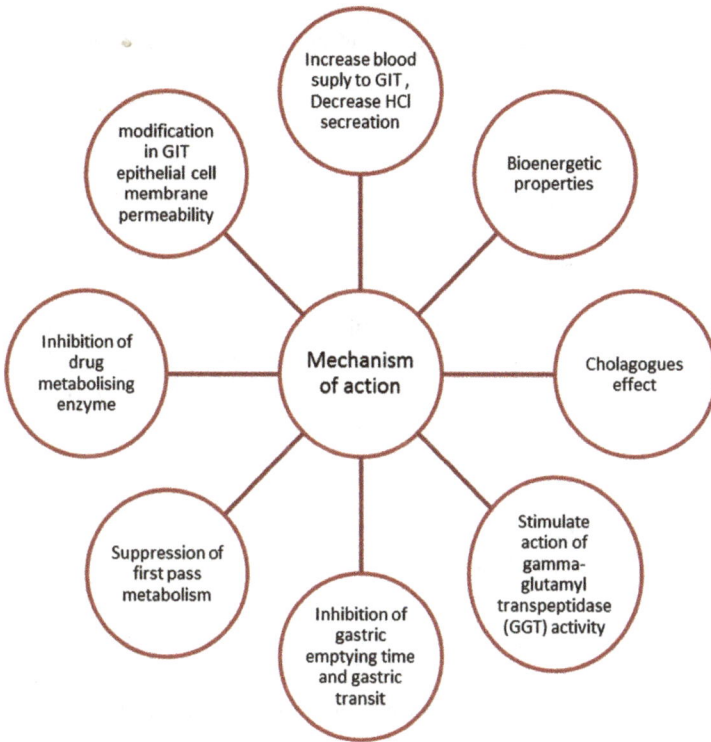

Figure 6.1: Bioavailability improvement approaches.

3. Bioenhancers increase the blood circulation and, thereby, improve the oral bioavailability of drugs.
4. Bioenhancers also act by inhibiting efflux pump P-gp to move out the drug so
   as to prevent the drug from reaching the targeted site [8–10].

6.2 Piperine as model bioenhancer

Biological source [1]	Piperine is a perennial shrub (family: Piperaceae) found in black pepper and long pepper.
IUPAC name [11]	1-[5-[1,3-Benzodioxol-5-yl]-1-oxo-2,4-pentadienyl] piperidine
Chemical constituent [11]	Piperine, sarmentosine, piperamide, piperamine, trichostar, sarmentine, chavicine
Biological/pharmacological activity [11]	Antihistaminic, antioxidant, antitumor, antipyretic, antifungal, antihypertensive, antidiabetic, antidiarrheal, antiobesity, hepatoprotective, larvicidal, antiasthmatic, and bioavailability enhancer
Mechanism of action as bioenhancer [12]	Inhibition of the efflux pump of the drug and receptor binding of DNA

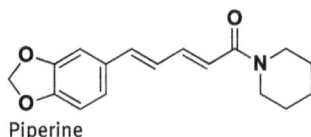

Piperine

Piperine (*Piper nigrum*) is a common household spice. Its major constituent is 1-pe-
peroyl piperidine, has a pungent taste, and has a majority of pharmacological actions. It was discovered by Hans Christian in the eighteenth century. It is known as
Kali mirch in Hindi, Marich in Nepali language, and Pippali in Sanskrit. Piperine is
used in the traditional Unani and Ayurveda medicine for a long time [11].

6.2.1 Pharmacokinetics of piperine

Piperine has been used in many studies to explore its pharmacokinetic profile in
both humans and laboratory animals.
1. In the first investigation to find the pharmacokinetic parameters of Piperine
   was coadministered with Benjakul formulation, a Thai traditional medicine
   made up of herbs that can be used as a complementary medicine alongside
   other herbal medicines to treat cancer. Benjakul formulation composed of five

plats including *Piper chaba* Hunter is a fruit, *Piper sarmentosum* Roxb and *Plumbago indica* Linn. roots, *Piper interruptum* Opiz steam, and rhizome of *Zingiber officinale* Roscoe.

Healthy male and female Thai subjects ($n = 20$) distributed equally, with age group 20–38 years, weighing between 42 and 84 kg, and with body mass index between 17.9 and 25.9 kg/m² were selected in the study. Two dosage regimens, 100 mg (6 mg piperine) and 200 mg (12 mg piperine) of Benjakul formulation were selected.

After an oral dose of Benjakul formulation, piperine exhibited good absorption through the gastrointestinal tract. Following a 200 mg Benjakul dose, the first C_{max} of piperine was 1,078 ng/mL, which was significantly higher than the 100 mg (467 ng/mL). In T_{max}, the first maximum concentration was found to be approximately 1 h. Following a 200 mg dose, the area under the curve 0–48 h was 10,216 ng h/mL, which was substantially higher than the 100 mg dose (4,288 ng h/mL) [13].

2. The pharmacokinetic study of piperine carried out in rats using single oral dose of *Piper longum* 12 mg/200 mg weight showed that piperine was well absorbed, rapidly distributed, and slowly eliminated. Mean C_{max} was 4,292 ng/mL, t_{max} was 2.45 h, $AUC_{0-\alpha}$ was 23.10 µg/mL, and $t_{1/2}$ was found to be 4.10 h. Piperine is well absorbed in humans compared to rats [14].

3. Piperine metabolites, including piperonyl alcohol, vanillic acid, piperonylic acid, and piperonal were found in the unbounded form in urine within 96 h, but piperic acid was found within 6 h in the bile, when administered orally at a dose 170 mg/kg. Piperine was mainly excreted from urine and bile by the glucuronidation–sulfation pathways [15].

4. Piperine absorption was around 96% when administered at a dose of 170 mg/kg body weight in rats, according to tissue distribution and elimination studies. Piperine is spread across different tissues after absorption, including the kidney, spleen, liver, small intestine, and the stomach [16].

6.2.2 Inhibition of drug metabolizing enzymes by piperine

Study of piperine on the biotransforming action due to the effect of different enzymes in the hepatic tissue has been carried out outside and inside the body.

Piperine inhibits aryl-hydrocarbon hydroxylation, ethylmorphine *N*-demethylation, *O*-de-ethylation of 7-ethoxycoumarin, and 3-hydroxybenzo[a]pyrene glucuronidation in rats. Hepatic microsomal aryl hydrocarbon hydroxylase (AHH) is inhibited by piperine in treated and nontreated rats. Similarly, ethylmorphine-*N*-demethylase noncompetitive inhibition was observed from the microsomal enzyme of rat liver with an apparent $K_i = 35$ mM and $K_m = 0.8$. Piperine is a nonspecific metabolism regulator drug. Hepatic AHH and uridine diphosphate-glucuronyl transferase activity were inhibited by piperine in rats when given orally. Maximum inhibition of AHH was observed within an hour, which returned to normal within 6 h. The possibility of this is due to the inhibition of transferase enzyme. Hence, it was found that piperine inhibits the metabolism of the drug.

Piperine inhibition activity is because of the presence of conjugated double bonds when compared with SAR of its analogues; the presence of a side chain in the molecule serves a major role in piperine inhibition. Piperine, on the other hand, was less efficient in reducing UDP-GA levels in isolated rat hepatocytes than in guinea pig small intestinal enterocytes. Piperine is a significant inhibitor of UDPGDH because of the presence of the double bond in the structure, and it also has a stronger effect on intestinal glucuronidation in rats. Piperine also inhibits P-gp and CYP: 450 causing first-pass elimination of many drugs. Examples of enzymes metabolized by piperine include CYP 1-A1, CYP 1-B1/B2, CYP 3- A4, and CYP 2-E1 [21].

6.3 Impact of piperine as bioenhancer on pharmacokinetics of other drugs and nutraceuticals

6.3.1 Piperine and pentobarbitone

In comparison to the controls, the effect of piperine on pentobarbitone-treated hypnotic rats considerably increased their sleeping time. No significant change was observed in sleeping time induced by barbital sodium. Piperine may block microsomal enzyme present in the liver and increase the sleeping time when pentobarbitone is taken [22].

6.3.2 Piperine and β-carotene

A crossover study design was carried out to study the ability of Piperine to improve the serum response of β-carotene. Subjects were administered a daily dose of 15 mg of β-carotene along with 5 mg of piperine. Intersubject variability was minimized by selecting healthy adult male volunteers with β-carotene serum value < 0.0001. Pharmacokinetic values of β-carotene administered along with piperine were found to be 49.8 pg/dL compared to β-carotene alone. Coadministration of β-carotene along with piperine showed a 60% greater increase in AUC as compared to β-carotene alone [23].

6.3.3 Piperine and curcumin

Curcumin is rapidly metabolized by the liver and the intestinal wall, and hence it has low oral bioavailability. Piperine was combined with curcumin to increase its bioavailability and was orally given to healthy human volunteers and also in rats. At a dose of 2 g/kg, curcumin was administered to rats, resulting in moderate serum curcumin concentration for 4 h. When piperine 20 mg/kg was administered along with curcumin, the serum curcumin concentration increased. The time to reach C_{max} significantly increased and $t_{1/2}$ significantly decreased with a 154% increase in the bioavailability. Hence, Piperine can be administered in rats as well as humans without any adverse effects in order to increase bioavailability, serum concentration, and absorption of curcumin [24].

6.3.4 Piperine with rifampicin, isoniazid, and pyrazinamide

A combination of drugs, including 100–300 mg of isoniazid, 100–300 mg of rifampicin and 100–300 mg of pyrazinamide, at a dose of 5–29 mg was studied for the effect of piperine on its pharmacokinetic parameter in human volunteers. $AUC_{0-\infty}$ ($\mu g \cdot h/mL$) of rifampicin, isoniazid, and pyrazinamide, increased from 104.6 ± 4.38, 41.0 ± 1.72, 318.6 ± 8.05 to 143.95 ± 5.85 94.95 ± 3.98 435.41 ± 9.7, respectively. C_{max} ($\mu g/mL$) of rifampicin 8.20 ± 0.46, isoniazid 2.56 ± 0.17, and pyrazinamide 27.0 ± 2.0 were increased to 17.6 ± 1.09, 10.07 ± 0.90, 40.0 ± 1.64, respectively [25].

6.3.5 Piperine with amoxicillin trihydrate and cefotaxime

Beta-lactam antibiotics, including amoxicillin trihydrate and cefotaxime sodium bioavailability, were found to be greatly increased in rats when administered along with piperine. The enhanced bioavailability of these antibiotics is reflected in their pharmacokinetic parameters, such as t_{max}, C_{max}, $t_{1/2}$, and AUC. The effect of piperine on microsomal metabolizing enzymes or the enzyme system may explain the increased bioavailability [26].

6.3.6 Piperine with EGCG (epigallocatechin gallate)

Piperine enhances tea polyphenol EGCG (cardioprotective) bioavailability in mice. Intra gastric administration of EGCG at a dose of 163.8 μmol/kg and of piperine at a dose of 70.2 μmol/kg in mice increased the C_{max} and AUC as compared to administering alone. The mechanism of the action of piperine to increase bioavailability is due to the inhibition of glucuronidation. When EGCG was administered alone, the resulting C_{max} was found to be 37.50 nmol/g at 60 min, which then decreased to 5.14 nmol/g at 90 min. When piperine and EGCG are coadministered, C_{max} increased to 31.60 nmol/g at 90 min, and maintained up to 180 min [27].

6.3.7 Piperine and oxytetracycline

To study pharmacokinetics of oxytetracycline (10 mg/kg), White Leghorn birds were used, after 7 days of piperine oral treatment (15 mg/kg). *Bacillus cereus* var. *mycoides* was used to study plasma oxytetracycline concentrations using the microbial assay technique. The plasma levels of oxytetracycline showed that animals administered with piperine have significantly higher AUC, MRT, and AUMC. The clearance decreased by 21%. Treatment with piperine also reduced the loading dose by 33.3% and maintenance dose by 39% [28].

6.3.8 Piperine and ciprofloxacin

Rabbits were used to observe the change in the pharmacokinetic parameter of ciprofloxacin, when combined with piperine. Piperine has a statistically important effect on the pharmacokinetics of ciprofloxacin, as shown by changes in C_{max}, t_{max}, AUC, and K_{el}. The study showed that ciprofloxacin absorption and ciprofloxacin metabolism inhibition are two effects of piperine administration. Piperine increased the bioavailability of ciprofloxacin. Piperine's potentiating effect, in combination with ciprofloxacin, using *Staphylococcus aureus* was investigated, as well as its possible function as an efflux pump inhibitor. Piperine decreased the mutation prevention concentration and MIC, i.e., minimum inhibitory concentration of ciprofloxacin for *S. aureus*, when used in conjunction with ciprofloxacin [29].

6.3.9 Piperine and nevirapine

Nevirapine [potent non-nucleoside inhibitor of HIV-1 reverse transcriptase] was also studied for the effect of piperine administration on eight healthy male subjects using randomized crossover-controlled design. Nevirapine was rapidly absorbed in both the groups that are treated with nevirapine 200 mg + placebo (group I) and nevirapine 200 mg + piperine 20 mg (group II). t_{max} for group I was found to be 4.13 h and for group II, it was 6.0 h. Nevirapine C_{max} increased by 120%, AUC_t by 167%, AUC_∞ and C_{last} values increased by 170% and 146%, respectively in group II subjects. Other pharmacokinetic parameters of nevirapine such as t_{max}, $t_{1/2}$ were not changed when coadministered with piperine [20].

6.3.10 Piperine with diclofenac sodium and pentazocine

The effect of piperine on the analgesic activity of pentazocine and diclofenac sodium was evaluated using healthy albino mice (25–30 kg) divided into four groups ($n = 8$). Acetic acid-induced writhing model was used to evaluate the peripheral analgesic activity, using 5 mg/kg of diclofenac sodium along with piperine extract from *P. nigrum* at a dose of 10 mg/kg and a combination at a dose of (5 + 10 mg/kg), both orally. The tail flick method was used to check the central analgesic activity at a dose of 5 mg/kg for pentazocine. Administration of *P. nigrum* extract per se. No significant analgesic activity was observed. When the extract was used in combination with diclofenac sodium, a significant decrease (78.43%) in acetic acid-induced writhing model was found, in comparison with the drug alone. The results indicate that the extract of *P. nigrum* can be used to improve pentazocine and diclofenac sodium's analgesic activity [30].

6.3.11 Piperine with ampicillin and norfloxacin

To study the effect of piperine on ampicillin and norfloxacin bioavailability, piperine at a concentration of 20 mg/kg was given orally along with the drugs. The bioavailability increased after the oral administration of ampicillin and norfloxacin in the animal model. An increase in the C_{max} of ampicillin was observed, from 44.6 ± 0.27 to 251.2 ± 0.28 µg/mL, when coadministered with piperine. Similarly, an increase in C_{max} of norfloxacin was observed, from 11 ± 0.26 to 16.1 ± 0.27 µg/mL, when co-administered with piperine. The AUC for ampicillin increased from 103.7 ± 0.52 to 350.49 ± 0.47 µg · h/mL. For norfloxacin, there was a significant increase in AUC, from 63.98 ± 0.51 to 111.69 ± 0.54 µg · h/mL [31].

6.3.12 Piperine and fexofenadine

On a rat model, the pharmacokinetic characteristics of fexofenadine at a dose of 10 mg/kg along with piperine at 10–20 mg/kg were investigated. The combination of fexofenadine and piperine enhanced fexofenadine's AUC by 180% to 190%, while C_{max} and $t_{1/2}$ remained unchanged. With the coadministration of piperine with fexofenadine, the bioavailability increased by around twofold. In the presence of piperine, t_{max} appears to be higher, which could be owing to the delayed stomach emptying. Therefore, piperine significantly increased fexofenadine oral exposure in rats. P-gp cellular efflux inhibition during absorption through the intestine suggests that using piperine/piperine-containing food with fexofenadine is necessarily required in order to avoid possible drug-diet interactions [32].

6.3.13 Piperine and metronidazole

Male rabbits weighing between 2 and 5 kg were used to study the effect of bioavailability of metronidazole using piperine. The rabbits were divided into three groups; one group was treated with an oral administration of distilled water, and the other two groups were administered with metronidazole alone and in combination with piperine, respectively. C_{max} ng/mL of metronidazole increased from 3,805.89 ± 233.8 to 6,007.07 ± 348.8 (combined with piperine) and AUC from 45,073.75 ± 713.7 to 84,980.98 ± 345.6 ng · h/mL, when given in combination. Hence, piperine can be used to improve the bioavailability of metronidazole [33].

6.3.14 Piperine and ibuprofen

Coadministration of piperine at a dose of 10 mg/kg with ibuprofen enhanced the antinociceptive activity in the Writhing and Formalin test. The amount of piperine in plasma increased, when combined with ibuprofen, because of the synergistic anti-nociception effect of the two drugs. This study suggests the combination of piperine as a bioenhancer with ibuprofen [34].

6.3.15 Piperine and oxyresveratrol

The bioenhancing property of piperine on oxyresveratrol (stilbenoid) was studied using 8–12 weeks aged Wistar rats (male), administered with 100 mg/kg oxyresveratrol and oxyresveratrol 100 mg/kg plus piperine 10 mg/kg via the oral route. Within 1–2 h after oral administration, the coadministration with piperine showed a significantly higher C_{max} (roughly 1,500 g/L), and increased the oxyresveratrol bioavailability 2-fold. After the administration of oxyresveratrol per se or with piperine, oxyresveratrol glucuronide showed in urinary excretion – the primary route of excretion. After the IV treatment of oxyresveratrol 10 mg/kg plus piperine 1 mg/kg, the generation of glucuronide metabolites from oxyresveratrol appeared to be reduced [18].

6.3.16 Piperine with propranolol and theophylline

Pharmacokinetics and bioavailability of propranolol (beta-blocker) and theophylline (phosphodiesterase inhibitor) are checked along with its administration with piperine using crossover study design. Twelve healthy nonsmoking subjects (18–45 years; 45–66 kg) were selected for the study. Six subjects were placed in each group and they received a single oral dose of propranolol (140 mg) or theophylline (150 mg) alone or in combination with piperine 20 mg daily for 7 days. An earlier t_{max}, and a higher C_{max} and AUC were observed in the subjects who received piperine and propranolol. C_{max} of propranolol increased from 45 to 95 ng/mL whereas AUC increased from 561 to 1,140 µg/mL. No significant change in was observed $t_{1/2}$. It produced a higher C_{max}, i.e., from 4.55 to 7.36 ng/mL, a longer elimination half-life, i.e., from 6.55 to 10.78 ng/mL h, and a higher AUC, from 43 to 85 µg/mL with theophylline due to reversible inhibition of drug metabolism enzyme [17].

6.3.17 Piperine and metformin

Adult Swiss albino mice (20–30 kg) were selected to study the improved pharmacokinetic parameters of metformin (antidiabetic) using piperine as a bioenhancer. Mice were administered with alloxan monohydrate 150 mg/kg body weight to induce diabetes. A combination of piperine 10 mg/kg with a standard dose of metformin at a dose of 250 mg/kg showed better results in lowering blood glucose levels at mid-study interval and end of the study; day 14 and 28 levels were compared for metformin alone, which indicates that piperine significantly enhanced the activity of metformin. When a reduced dose, i.e., 125 mg/kg of metformin was used along with piperine, low concentration of blood glucose level was observed as compared to the control. This combination showed greater result in decreasing blood glucose level by 17% as compared to 12.5% by metformin (250 mg/kg) alone [19]. Bioenhancement of drugs by piperine and animals or humans on which the study has been conducted are listed in Table 6.2. Nutraceuticals bioenhanced by piperine are listed in Table 6.3.

Table 6.2: Drugs bioenhanced by piperine [21].

Drug	Animals/humans on which the study has been carried out
Vacisine	Rat model
Phenytoin, theophylline, sulfadiazine, tetracycline	Human volunteer
Pentobarbitone	Human volunteer
Curcumin	Human volunteer and rat model
Nimesulide	In vivo study
Indomethacin	Rat
Oxyphenylbutazone	Mice model
Phenytoin	Rabbit model
Rifampicin	Rat model
Amoxycillin trihydrate and cefotaxime	Rat model

Table 6.2 (continued)

Drug	Animals/humans on which the study has been carried out
EGCG [[−]-epigallocatechin-3-gallate]	Mice model
Oxy-tetracycline	WLH hens
Ciprofloxacin	Rabbit model
Nevirapine	In vitro study
Diclofenac sodium and pentazocine	Human model
Pefloxacin	Albino mice
Carbamazepine	Gaddi goats
Fexofenadine	Rabbit
Metronidazole	In vitro study
Ampicillin trihydrate	Rat
Resveratrol	Rabbit
Gatifloxacin	Human model
Atenolol	Rat
Ibuprofen	In vitro study
Losartan potassium	Rat model

Table 6.3: Nutraceutical bioenhanced by piperine [35, 36].

Class of nutraceuticals	Example
Fat/water vitamins	Vitamin A, vitamin B1, vitamin B2, vitamin B12, vitamin C, vitamin D, vitamin E, vitamin K, provitamin
Amino acid	Lysine, isoleucine, leucine, phenylalanine, threonine, valine, tryptophan, methionine
Minerals	Zn, I, Ca, Cu, Mg, K, iron, selenium, and manganese
Herbal products	Boswellic acid: *Boswellia serrata* Pycnogenol: *Pinus pinaster* Withanaloids: *Withania somnifera* Curcuminoides: *Curcuma longa* Ginsenosides: *Gingko biloba*

6.4 Impact of other herbal bioenhancers on pharmacokinetics of some drugs and nutraceuticals

6.4.1 *Allium sativum* (garlic): allicin

Biological source	Organosulfur allicin is compound obtained from garlic, a species in the Alliaceae family.
IUPAC name	*S*-[2-Propenyl]-2-propene-1-sulfinothioate
Chemical constituents	Alliin, allicin, *E*-ajoene, *Z*-ajoene, 2-vinyl-4*H*-1,3-dithiin, diallyl sulfide, diallyl disulfide, diallyl trisulfide, allyl methyl sulfide
Biological/pharmacological activity [38]	Antibacterial, antiviral, antioxidant, anti-inflammatory, anticancer, immunomodulatory
Mechanism of action as bioenhancer [21, 37]	1. Allicin inhibits transport of ergosterol between the cytoplasm and the plasma membrane. 2. Allicin inhibits ergosterol trafficking from the plasma vacuole membrane but enhances membrane damage by amphotericin B.

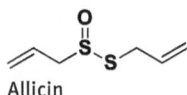

Allicin

6.4.1.1 Pharmacokinetics of allicin

Alliinase is a garlic amino acid conversion enzyme that transforms allicin to allyl sulfenic acid. Allicin is a chemical that decomposes rapidly into other oil- and water-soluble organosulfur compounds when 2 molecules of allyl sulfenic acid spontaneously condense to produce one molecule. At pH 7.0 and 35 °C, the pure alliinase enzyme is most active, while it becomes inactive at pH values below 3.5 or when heated. A lot of factors affect allicin absorption, including its steric structure, solubility, pK_a, molecular size, and hydrophilicity. Because of its reactivity with thiol groups in proteins, such as those found in enzymes, allicin has a significant antibacterial impact.

Volunteers' breath contained allyl methyl sulfide, diallyl sulfide, diallyl disulfide, diallyl trisulfide, and dimethyl sulfide within 2–3 h after consuming 38 g raw garlic. The [AUC] of the primary breath Allicin metabolite [Allyl methyl sulfide] was checked in healthy people after 32 h of ingestion [6 female and 7 males]. BA/BE values of allicin were found to be (36–104%) in enteric tablets, nonenteric tablets (80–111%), and garlic powder capsules (26–109%) [39].

6.4.1.2 Pharmacokinetics of allicin administered with another drug

1. Allicin and omeprazole

Eighteen healthy Chinese male volunteers were administered with 180 mg of allicin for 14 days to see how it affected omeprazole [a proton pump inhibitor], and CYP2C19 and CYP3A4 activities. The study used HPLC and a two-phase randomized crossover trial. Treatment with allicin increased omeprazole's peak plasma concentration [C_{max}] by 54.29% and AUC 0-∞ by 48.19%. The C_{max} and AUC $_{0-\infty}$ of omeprazole showed no significant change. This study concluded that allicin is helpful to reduce the metabolism of omeprazole by inhibiting CYP2C19 activity [40].

2. Allicin and Cu^{2+}

Cu^{2+} has fungicidal activity for *S. cerevisiae* cells, and its action increased with allicin. Cu^{2+} fungicidal action was unaffected by substances such as l-cysteine, N-acetyl-cysteine, or dithiothreitol. Cu^{2+} produced deadly ROS, i.e., reactive oxygen species in *Saccharomyces cerevisiae* cells; however, intracellular oxidative stress during the time of cell death was seen when Cu^{2+} and allicin were combined [41].

3. Amphotericin B with allicin

The effect of allicin on the fungicidal action of amphotericin B against the yeast *Saccharomyces cerevisiae* shows considerably enhanced activity. Amphotericin B and allicin concentrations were observed to be increased, i.e., 0.5 + 120 g/mL In addition to modifying the permeability of the plasma membrane, amphotericin B induced disruption to the vacuole membrane, resulting in the organelles to be recognized as little separate particles. Allicin improved Amphotericin B-induced damage to the structure of the vacuole membrane. Allicin can improve antifungal activity of amphotericin B against *C. albicans* and *Aspergillus fumigatus*, two dangerous fungi [42].

6.4.2 *Aloe barb adensis* (Aloe vera)

IUPAC name	S-[2-Propenyl]-2-propene-1-sulfinothioate
Chemical constituent [43]	30% aloin is present in aloe (mixture of barbaloin, isobarbaloin, and β-barbaioin)
Biological/pharmacological activity	Moisturization, anti-acne, skin rejuvenation, skin hydration, anti-inflammatory, wound healing, hail inflammation, athletes' foot, skin repair, antiaging, protection from radiation
Mechanism of action as bioenhancer [44]	Facilitator of absorption and regulator of GIT function

(continued)

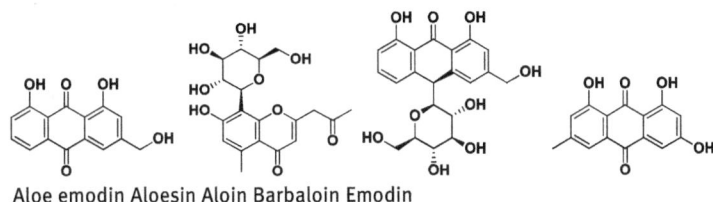

Aloe emodin Aloesin Aloin Barbaloin Emodin

6.4.2.1 Aloe vera and vitamin C, E

Aloe vera gel consists of vitamin C and vitamin E. It increases the plasma concentrations and enhances absorption. The Aloe vera gel extract was particularly successful in increasing ascorbate absorption. Its plasma concentration was substantially prolonged, even after an overnight fast of 24 h. Vitamin E absorption and plasma concentration were improved by both the gel and whole-leaf extracts, particularly after 8 h. AUC of vitamin C (339 ± 124 µMh) increased up to (272 ± 144 µMh) and ($1{,}031 \pm 513$ µMh) when administered along with aloe leaf extract and aloe gel, respectively. AUC of vitamin E (19.3 ± 23.2 µMh) increased (38.3 ± 17.0 µMh and 71.3 ± 22.5 µMh) when administered along with aloe leaf extract and aloe gel, respectively [45].

6.4.2.2 Aloe vera and protein peptide drug

The leaf extract of *Aloe vera* improved the transport of FITC-dextran with a molecular weight of 4 kDa [FD-4], with not more than 10 kDa molecular weight, demonstrating lowest drug absorption in terms of molecular size. Confocal laser scanning microscopy imaging revealed an accumulation of FD 4 in Caco-2 cells after treatment with the *Aloe vera* gel and whole-leaf extract, implying that FD-4 was transported paracellularly after the interaction of *Aloe vera* gel and whole-leaf extract with epithelial cell monolayers. Fluorescence labeling revealed changes in F-actin distribution in the cytoskeleton of Caco-2 cell monolayers, showing that the *Aloe vera* gel and whole-leaf extract's absorption-increasing impact is due to close junction remodeling [46].

6.4.3 *Ammannia multiflora* (jerry-jerry)

Biological source [47]	Leaves of the plant of *Ammannia* genus belonging to the Lythraceae family
Chemical constituent [47]	Glycoside, flavonoids, carbohydrates, steroids, phenol, and sterols

(continued)

Biological/pharmacological activity [47]	Antipyretic, antidiuretic, antimicrobial, antirheumatic, anticancer, rubefacient, antimalarial
Mechanism of action as bioenhancer	Under investigation

Ammannia multiflora extract (methanolic) shows bioenhancing activity when administered with nalidixic acid. Using methanolic extract, researchers were able to isolate a novel chemical called 2,5-bis-[3,3′-hydroxyaryl] tetrahydrofuran, also known as ammaniol, as well as nine other recognized chemicals. In addition, 4-hydroxy-tetralon was transformed into five semisynthetic acyl derivatives – 1A, 1B, 1 C, 1D, and 1E – which were investigated for bioenhancing action in conjunction with nalidixic acid against two *Escherichia coli* strains, CA8000 and DH5. The methanolic extract of *A. multiflora* as well as compounds 1 and 9 had considerable bio enhancing effect and lowered the dose of nalidixic acid 4-fold, whereas compounds 5, 6, 10, and semisynthetic derivatives 1A-1E lowered the dose of Nalidixic acid 2-fold. Compound 5 was further evaluated for anti-mycobacterial activity against *Mycobacterium H37Rv*, which showed moderate action (MIC 25 g/mL) [48].

6.4.4 Capmul

Glyceryl caprate, i.e., capmul is made from edible oils and fats. It is a common ingredient in lip balms. Capmul MCM C10 increased ceftriaxone bioavailability by 55–79% in rats [49].

6.4.5 *Capsicum annum* (chili peppers): capsaicin [50]

Biological source	Capsaicin is the active phenolic compound found in *Capsicum annum*
IUPAC name	8-Methyl-*N*-vanillyl-6-nonenamide
	Causes irritation to humans, mammals and produces a feeling of tissue burning when it comes in contact with tissues

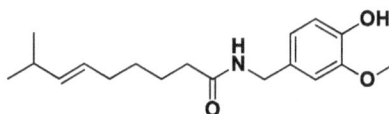

Capsaicin

6.4.5.1 Capsaicin and theophylline

Sustained-release gelatin capsule of theophylline was used to investigate the absorption and bioavailability in rabbits ($n = 10$, male) with oral administration of theophylline (20 mg/kg), along with capsaicin and the drug alone. Coadministration of capsicum increased AUC to 138.32 mg × h/L, C_{max} to 8.78 mg/L, mean residence time (MRT) to 20.98 h. Each rabbit showed increase in theophylline plasma levels after receiving the next dose of the capsicum suspension 11 h after the first dose. This meant that theophylline bioavailability increased with capsaicin [51].

6.4.5.2 Capsaicin and aspirin

According to the study carried out by Cruz et al., capsaicin reduces the oral bioavailability of aspirin in rats [52].

6.4.5.3 Capsaicin and ciprofloxacin

A study involving the effect of capsaicin on ciprofloxacin concluded that there is no or very little effect of capsaicin on the bioavailability of ciprofloxacin [53].

6.4.6 Cow urine distillate

The urine/gomutra of a cow (*Bos indicus*) has been extensively explained in Ayurveda and classified as a beneficial medicinal substance/secretion of animal origin, with multiple therapeutic capabilities, in the "Sushruta Samhita," "Ashtanga Sangraha," and other Ayurvedic literatures. In Ayurveda, Gomutra is referred to as "Sanjivani" and "Amrita." Gomutra is a nontoxic waste product made up of 95% water, 2.5% urea, and the remaining 2.5% includes salts, hormones, minerals, and enzymes.

Cow urine has two US patents for its pharmaceutical properties, especially as a bioenhancer and as an antibiotic, antifungal, and anticancer agent (nos. 6,896,907 and 6,410,059].

Mechanism of action as bioenhancer: Increases drug absorption across cell membrane [54].

6.4.6.1 Cow urine distillate and cadmium toxicity, bioenhancer of zinc

Antimicrobial, antifungal, and anticancer medications work better using cow urine distillate as a bioenhancer. Cow urine exhibits antitoxic effect on the toxicity of cadmium chloride and is used as zinc bioenhancer. Cadmium chloride has no effect on the fertility of mature male mice. Cadmium chloride, zinc sulfate, and cow urine-treated animals had a 90% reproductive rate, 100% viability, and 100% lactation indices. The fertility index in the cadmium chloride, zinc sulfate, and cow urine-treated animals was estimated to be 88%. Cow urine distillate was found to be effective against cadmium chloride toxicity and it also reduced the amount of cadmium chloride in the body [54].

6.4.6.2 Cow urine and gonadotropin-releasing hormone

Cow urine distillate has great effect in increasing the activity of gonadotropin-releasing hormone on the motility and count of sperm, gonado-somatic indices, and morphology of sperm in male mice. Cow urine distillate has immunomodulatory characteristics; hence it boosted this effect [55].

6.4.6.3 Cow urine and rifampicin

The absorption of antibiotics, including rifampicin, tetracycline, and ampicillin through the gut wall improved 2–7 times when cow urine distillate was used. Rifampicin bioavailability increased 80 times when cow urine was used (0.05 µg/mL), ampicillin bioavailability increased 11.6 times at 0.05 µg/mL, and cotrimoxazole bioavailability increased 5-fold at 0.88 µg/mL Cow urine improves bioavailability by promoting drug absorption through the cell membranes [56].

6.4.7 *Curcuma longa* (turmeric): curcumin

Biological source [57]	Turmeric consists of (*Curcuma longa* L. syn *C. domestica* Val.) and rhizome belongs to the Zingiberaceae family
IUPAC name [60]	Curcumin [diferuloylmethane; 1,7-bis[4-hydroxy-3-methoxyphenyl]-1,6-heptadiene-3,5-dione]
Chemical constituent [57]	Curcumin, dimethoxy curcumin and bisdemethoxycurcumin, collectively known as curcuminoids (3–6%)
Biological/pharmacological activity [57]	Anti-inflammatory, antiparasitic, antispasmodic, anticarcinogenic, anticancer, antitumor, antioxidant, antiprotozoal, and wound healing property.

(continued)

Biological source [57]	Turmeric consists of (*Curcuma longa* L. syn *C. domestica* Val.) and rhizome belongs to the Zingiberaceae family
Mechanism of action as bioenhancer [58, 59]	1. Curcumin has the ability to suppress liver CYP3A4. 2. Curcumin induces changes in the drug transporter P-glycoprotein. 3. Curcumin suppresses nonspecific drug metabolizing enzyme

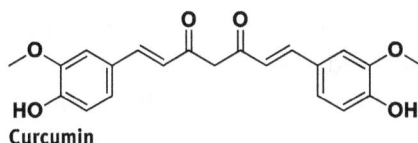

Curcumin

Turmeric (*Curcuma longa*) is one of the most widely used spices and condiments in Indian cuisine. It is a primary orange hydrophobic polyphenol pigment that is used in pharmaceutical ointments and creams to impart color. Curcumin is used in antimicrobial and anticancer medicines as a bioenhancer [3].

Demethoxy-curcumin, bisdemethoxycurcumin, and volatile oils have also been classified as curcumin constituents. Commercially, curcumin contains 75% curcumin, 15% dimethoxy-curcumin, and 5% bisdemethoxycurcumin [60].

6.4.7.1 Pharmacokinetics of curcumin

Curcumin oral supplement has been used in many studies to explore its pharmacokinetic profile in both humans and laboratory animals.

1. A single dose of curcumin-standardized powder extract was given to 24 healthy people at a dose of 0.5–12 g. Curcumin was not identified in the serum of participants who were given 8 g dose of Curcumin. Increase in dose to 10 g in 2 out of 24 subjects showed curcumin level of about 30 ng/mL (1 h), 40 ng/mL (2 h), and 50 ng/mL (4 h). After a dose of 12 grams, the level of curcumin was found to be about 30 ng/mL (1 h), 60 ng/mL (2 h), and 50 ng/mL (4 h) ng/mL [61].
2. Curcuma extract (36–180 mg) of pure curcumin was given to 15 patients with chemotherapy-resistant colorectal cancer for up to 4 months. Curcumin was found in feces only after 29 days, with concentrations ranging between 64 and 1,054 nmol/g. Curcumin or its metabolites were not detectable in blood and urine [62].
3. A study of pharmacokinetics of curcumin was carried out using four human participants for pharmacokinetics parameters when foods such as sandwich, soup, and oat bar were consumed. Each subject consumed 3 g (100 mg of curcumin). Curcumin was studied in three different ways. Dimethoxy curcumin glucuronide, curcumin glucuronide, and curcumin sulfate were observed in all

four volunteers with a C_{max} of 47.6, 1.9, and 2.1 nM, respectively within 30 min post-food [62].

4. Colorectal cancer patients having hepatic metastases were treated with 0.45–3.6 g of curcumin before surgery. Curcumin and its conjugates, curcumin glucuronide and curcumin sulfate, were present in trace amounts (<0.01 µM) in the portal circulation and live. [63].

5. To study the pharmacokinetics of curcumin after an oral dose of 10–12 g, healthy volunteers were selected and a study was performed. Only one subject showed free curcumin in the plasma after 30 min of administration. The amounts of curcumin glucuronide and curcumin sulfate at T_{max} at 10 g were 2.04 µg/mL and 1.06 µg/mL, respectively, and at 12 g, were 1.40 µg/mL and 0.87 µg/mL, respectively [64].

6. Curcumin is able to cross BBB because of its lipophilicity. As a consequence, it can enter the brain in biologically effective amounts, resulting in neuroprotection. Curcumin concentration in the brain was discovered in a study on murine models. At 30, 60, and 120 min after receiving an oral dose of curcumin of 50 mg/kg, mice had a brain concentration below the limit of detection. In comparison, when 100 mg/kg of curcumin was injected intraperitoneally, the concentration varied from 4–5 µg/g tissue in 20–40 min [65].

6.4.7.2 Effect of curcumin on pharmacokinetics of coadministered drugs

6.4.7.2.1 Curcumin with tamoxifen and endoxifen in patients with breast cancer

A study of pharmacokinetics of tamoxifen and endoxifen(antiestrogen), alone and administered along with curcumin alone and curcumin-piperine, on 17 patients with breast cancer was conducted in three cycles. Patients were given tamoxifen (20–30 mg) monotherapy in cycle 1, curcumin concomitantly with tamoxifen (1,200 mg, 3 times/day) in cycle 2, and tamoxifen concomitantly with curcumin and piperine in cycle 3. $AUC_{0-24\ h}$ and C_{max} of tamoxifen decreased to 8.4% and 7.1% in patients treated with tamoxifen plus curcumin, compared to tamoxifen per se. Furthermore, endoxifens AUC_{0-24h} and C_{max} decreased to 7.7% and 7.1%, respectively, with concomitant curcumin treatment. Along with curcumin and piperine, tamoxifen AUC_{0-24h} and C_{max} decreased 12.8% and 11.1%. The endoxifen $AUC_{0-24\ h}$ and C_{max} decreased to 12.4% and 9.8%, respectively. Patients should be informed to discontinue consuming turmeric-rich food during tamoxifen treatment, or the treatment effectiveness of tamoxifen should be closely checked, according to the findings [66].

6.4.7.2.2 Curcumin and marbofloxacin in chicken

A study of curcumin on chickens divided into four groups ($n = 7$) was carried out. Groups I and II were administered marbofloxacin (category: broad-spectrum bactericidal agent) (5 mg/kg) iv and orally, while animals in groups III and IV received a similar dose of Marbofloxacin IV and orally, after oral pretreatment with curcumin (100 mg/kg/day for 10 days). Marbofloxacin MRT were 10.11 and 8.8 h and the elimination half-lives ($t_{1/2}$) were 7.26 and 6.43 h after IV administration, in curcumin pretreated and normal chickens, respectively. The maximal serum concentrations [C_{max}] of marbofloxacin after oral administrations were 1.92 and 1.32 µg/mL, with time-to-peak concentration [T_{max}] values of 0.39 and 1.11 h and absolute bioavailability were less than 100% and 71.10% in curcumin pretreated and normal chickens, respectively. Delay in the excretion of marbofloxacin in chickens may be due to suppression of CYP3A4 drug metabolizing enzymes. Increased absorption may be due to the ability of curcumin to influence drug transporter protein [P-gp] in the intestine [67].

6.4.7.2.3 Curcumin and loratadine

Effects of coadministration of curcumin on the pharmacokinetics of loratadine (category: antihistaminic) were studied in 8 weeks old male Sprague Dawley rats, divided into seven groups ($n = 6$). The oral group was administered without or with 2–8 mg of curcumin and 4 mg/kg of loratadine dissolved in water; IV group also received with or without the same amount of curcumin and 1 mg/kg of loratadine dissolved in 0.9% NaCl solution. Curcumin coadministration improved absorption and other pharmacokinetic parameters of loratadine as compared to loratadine treatment alone. Curcumin increases $AUC_{0-\infty}$ of loratadine by 39.4–66.7%. C_{max} was enhanced by 34.2–61.5%. Loratadine's absolute and relative bioavailability values were significantly increasing to 40.0–66.1% and 1.39–1.67%, respectively. In the presence of curcumin, however, there were no major improvements in T_{max} and $t_{1/2}$ of loratadine. Increase in the oral bioavailability of loratadine can be observed by the inhibition of metabolism characterized by CYP3 A4 enzyme and P-gp efflux pump inhibition in the small intestine by curcumin [68].

6.4.7.2.4 Curcumin and talinolol in healthy Chinese volunteers

A study was carried out involving eighteen healthy subjects according to their A-B-C-B-1 multidrug resistance genotype. They were grouped as follows: wild-type homozygotes, C-34-35-C [CC, $n = 6$]; variant heterozygotes, C-34-35-T [CT, $n = 6$]; and variant homozygotes, T-34-35-T [TT, $n = 6$]. $AUC_{0-48\ h}$ and $AUC_{0-\infty}$ of talinolol (category: β1-selective adrenoceptor blocking agent) increased by 67.0% and 80.8%, respectively, when administered with curcumin. C_{max} of talinolol was found to be increased with curcumin administration, whereas clearance was decreased by 25.9% without any change in t_{max} and $t_{1/2}$. Talinolol $AUC_{0-48\ h}$ and $AUC_{0-\infty}$ C_{max} were significantly increased and $CL_{oral/F}$ decreased in TT subjects [69].

6.4.8 *Zingiber officinale* (ginger): gingerols

Biological source [70]	*Ginger* is a tuberous or nontuberous rhizome belonging to the family Zingiberaceae.
IUPAC name [70]	2-Methyl-5-[6-methylhept-5-en-2-yl]cyclohexa-1,3-diene
Chemical constituents [70]	Zingiberene, camphene, zingiberole, citral, cineole, bisabolene, phellandrene, borneol, limonene and camphene, citronellol, geraniol, and linalool are important constituents of ginger. Other active constituents are vitamin B6, vitamin C, mucilages, proteins, calcium magnesium, phosphorus, potassium, sulfur, and linoleic acid.
Biological/pharmacological activity [70]	Antiulcer, antimutagenic, antimicrobial antibiotic, antifungal, anticarcinogenic, antidiabetic, antiemetic thrombotic, and fertility enhancer
Mechanism of action as bioenhancer [70]	Regulates intestinal function and facilitates absorption

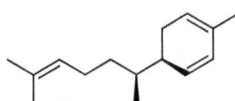 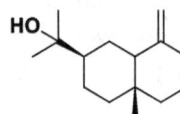

Zingiberene Zingiberole

1. *Z. officinale* extract enhanced the bioavailability of pharmaceutical compounds. Piperine increases bioavailability by 10–85%, when combined with *Z. officinale*, whereas *Z. officinale* alone has a bioenhancing impact of 30–75%. The extracts of *Z. officinale* maintain their bioenhancing activity, whether piperine is present or not. Ginger's bioactive fraction increased the bioavailability of the anticancer drugs, such as methotrexate, by 87% to 110%. It was also discovered that it improved the Acyclovir [Antiviral drug] concentration in blood by 82%. Herbal preparations and nutraceuticals' bioavailability is also improved by the bioactive fraction. Vitamin A bioavailability was enhanced by 30% when combined with ginger extract. As a result, ginger can be used per se or in conjunction with piperine to increase the bioavailability of a variety of essential therapeutic groups of drugs, such as antibiotics and nutraceuticals [71].

2. The percentage enhancement in bioavailability of certain drugs in combination with extracts of *Z. officinale* alone and in combination with piperine [21, 72] is shown in the next table (Table 6.4).

Table 6.4: Percentage enhancement in the bioavailability of certain drugs in combination with the extracts of *Z. officinale* alone and with piperine [21].

Compound	Percent enhancement in bioavailability	
	Z. officinale	*Z. officinale* + piperine
Erythromycin	68	65
Amoxycillin	80	90
Kanamycin	65	92
Ciprofloxacin	68	70
Fluconazole	120	110
Alprazolam	76	70
5-Fluorouracil	110	93
Doxorubicin	72	75
Propranolol	76	104
Nimesulide	144	165
Rifampicin	65	98
Salbutamol	78	92
Bromhexine	46	75
Ethionamide	56	48
Diclofenac	90	140
Dexamethasone	76	80
Cyclosporin A	116	120
Ranitidine	147	165
Retinol	30	40
Vitamin E	27	25
Folic acid	34	35
β-Carotene	36	55
Curcumin	43	70
Tinospora cordifolia	67	112
Asparagus racemosus	44	60

Table 6.4 (continued)

Compound	Percent enhancement in bioavailability	
	Z. officinale	*Z. officinale* + piperine
Lysine	15	40
Potassium	21	19
Zinc	19	20

6.4.9 *Glycyrrhiza glabra* (Liquorice): glycyrrhizin

Biological source	It is a glycoside. It is obtained from stolon and roots of liquorice, belonging to family *Glycyrrhiza glabra*.
Chemical constituents	Glycyrrhizin, a triterpenoid saponin, is the primary component of roots. Liquorice contains proteins, simple sugar, amino acids, and polysaccharides, as well as mineral salts such as Ca, K, Mg, I, Na, silicon, selenium, manganese, Zn, Cu, as well as resins, sterols, pectins, gums, and starches.
Pharmacological activity [73]	It has an expectorant effect and can help with bronchitis, gastritis, sore throat, allergies, asthma, peptic ulcers, and rheumatism. It aids in the detoxification of drugs by the liver and is used to treat liver disease. It is a diuretic and laxative, as well as strengthens the immune system and stimulates the adrenal gland. Glycyrrhizin's main applications include the treatment of peptic ulcers and stomach disorders as well as the treatment of respiratory and intestinal passages. Antihepatotoxic, anti-inflammatory, anticancer, and antiviral actions are some of the properties of glycyrrhizin.
IUPAC name [74]	[3β,20β]-20-Carboxy-11-oxo-30-norolean-12-en-3-yl2-*O*-β-D-glucopyranuronosyl-α-D-glucopyranosiduronic acid
Mechanism of action as bioenhancer	Glycyrrhizin is converted by β-glucuronidase present in intestine to glycyrrhetic acid and, thereby, enhances absorption.

(continued)

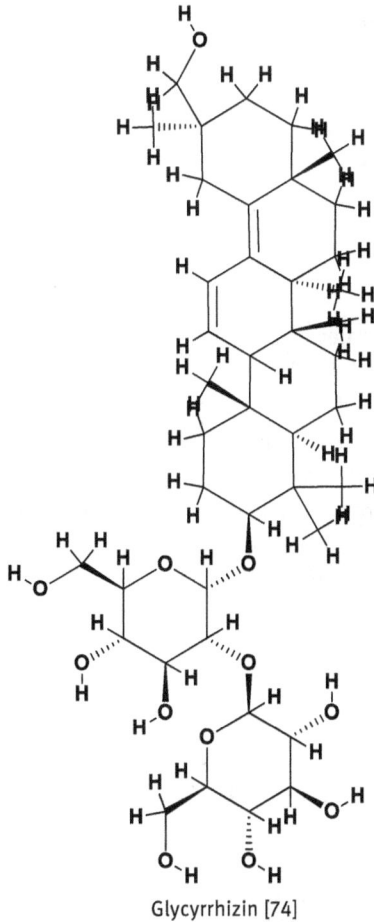

Glycyrrhizin [74]

6.4.9.1 Glycyrrhizin and aconitine

The effect on the pharmacokinetics of aconitine in rats through three different routes was assessed using diammonium glycyrrhizinate. Oral administration of glycyrrhizinate with aconitine, the C_{max}, AUC and bioavailability increased 1.64-fold, 1.63-fold, and 1.85-fold, respectively, while the half-life, and clearance (Cl) did not alter significantly. No changes in the pharmacokinetic parameters were observed during the Tail vein and Hepatic vein administration of glycyrrhizinate. This increase in absorption is due to the suppression of the P-gp efflux transporter in the intestine [74].

6.4.9.2 Glycyrrhizin and paclitaxel

An experiment involving MCF-7 cell line, enhancement in activity because of glycyr-rhizin, was observed in "taxol" (paclitaxel). Taxol's anticancer activity increased 5-fold in terms of inhibiting MCF-7 cancer cell growth and multiplication. Taxol at a dose of 0.01 g/mL in the presence of glycyrrhizin at a dose of 1 g/mL inhibits cancerous cell growth more effectively than Taxol (0.05 g/mL) treatment alone [75, 76].

6.4.9.3 Glycyrrhizin and antibiotic

Khanuja et al. also revealed the effectiveness of the glycyrrhizin. It increased the efficacy and bioavailability of a few anticancer drugs, antibiotics, and anti-infectives. Glycyrrhizin facilitates the absorption of antibiotic and other agents through biological membranes, increasing their levels in plasma and, thereby, their bioavailability. It increases the effectiveness of antibiotics such as rifampicin 3.8-fold, ampicillin 12.6-fold, tetracycline 1.9-fold, and nalidixic acid against *Escherichia coli*. It also enhances the bioavailability of rifampicin 6.5-fold, tetracycline 1.5-fold and Nalidixic acid 6.8-fold against Gram-positive bacteria such as *B. subtilis* and *M. smegmatis*. Glycyrrhizin also makes antifungal drugs such as clotrimazole more effective against *Candida albicans* [77].

6.4.10 *Ipomoea* spp.: *Rivea corymbosa, ipomoea violacea*, and *Ipomoea muricata* [morning glory plant): lysergol

Biological source	Morning glory plant (*Ipomoea* spp.)
Biological/pharmacological activity	It enhances the antibiotic activity and is a promising herbal bioenhancer.
IUPAC	9,10-Dihydro-6-methylergoline8-methanol
Mechanism of action as bioenhancer [78]	Still under investigation

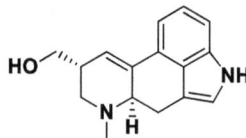

Lysergol [80]

A dose of 1–10 g/mL of lysergol is effective as a bioenhancer is; however, 10 g/mL is preferable. Rifampicin, tetracycline, and ampicillin are all bioenhancers. Lysergol boosts an antibiotic's antibacterial properties. It works against *E. coli, B. subtilis*, and *M. smegmatis*, as well as Gram-positive and Gram-negative microorganisms. It enhances the antibacterial activity of rifampicin [10 g/mL] 6–12 times against *E. coli* and 3–4.6 times against *B. subtilis*. Rifampicin increases the activity against *M. smegmatis* 4.5–6 times when used at a concentration of 0.4 g/mL [78]. Oral administration of lysergol to SD rats improved the bioavailability of berberine [79–81].

6.4.11 Peppermint oil

Biological source [82, 83]	Peppermint oil is extracted from perennial herb leaves, *M. arvensis* var. piperascens and *Mentha piperita* L. (family: Labiatae).
Chemical constituent [82, 83]	Cineole 3.5–14.0%, menthone 14.0–32.0%, limonene 1.0–5.0%, isomenthone 1.5–10.0%, menthofuran 1.0–9.0%, menthyl isopulegol max. 0.2%, acetate 2.8–10.0%, menthol 30.0–55.0%, pulegone max. 4.0%, and carvone max. 1.0%.
Biological/pharmacological activity [83]	Hot flushes in women, irritable bowel syndrome, antimicrobial, antiplasmid, larvicidal, and mosquito-repellent action, and treatment of nervous disorders and indigestion
Mechanism of action as bioenhancer	Under investigation

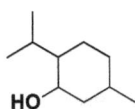

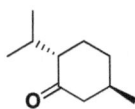

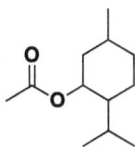

Menthol Menthone Menthyl acetate 1,8 cineole

The effect of peppermint oil on D-tocopheryl polysuccinate [TPGS] and ketoconazole on the oral bioavailability of cyclosporine was studied in rats. Peppermint oil, 100 mg/kg, increased C_{max} and the area under the concentration versus time curve of cyclosporine from 0.60 to 1.6 g/L and from 8.3 to 24.3 g h/mL, respectively. T_{max} increased from 2 to 6 h. Coadministration of D-tocopheryl polysuccinate at a dose of 50 mg/kg with cyclosporine in a saline vehicle-enhanced AUC0 from 28.5 to 59.7 g h/L and increased the C_{max} of cyclosporine from 1.3 to 2.9 g/mL. The t_{max} remained the same [3 h]. The terminal half-life increased by 44% [15.4 h compared 10.7 h] and the MRT increased by 24%. Use of Ketoconazole in range of 10–20 mg/kg had no effect on the absorption of cyclosporine. Poor cyclosporine metabolism in rat intestinal tissue indicates that cytochrome P450-3A inhibition is not the sole way peppermint oil improves cyclosporine oral bioavailability [82].

6.4.12 *Sinomenium acutum (Cocculus divers)*: sinomenine

Biological source [84]	Sinomenine (alkaloid) found in climbing plant *Sinomenium acutum*, located near Japan and China.
Biological/pharmacological activity [85]	Sinomenine is used in Japan and China for the treatment of rheumatic and arthritic diseases.
IUPAC [85]	7,8-Didehydro-4-hydroxy-3,7-dimethoxy17-methylmorphinan-6-one
Mechanism of action as bioenhancer [85]	Sinomenine is P-glycoprotein Inhibitor

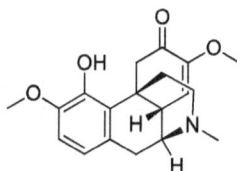

Sinomenine [84]

The activity of sinomenine on the augmentation of paeoniflorin absorption in the gut was investigated using co-incubate tissue culture experiments. Coadministration of paeoniflorin and Sinomenine increased paeoniflorin absorption against *Candida albicans*. Administration of a 150 mg of paeoniflorin per kg body weight alone and with 90 mg/kg sinomenine hydrochloride to SD rats found that, sinomenine administration increases the C_{max} of paeoniflorin and AUC0 $-\infty$, delays t_{max}, and decreases CL and Vd. sinomenine has been postulated as a strategy for improving paeoniflorin bioavailability by decreasing paeoniflorin efflux transport via P-gp [84].

6.4.13 *Stevia rebaudiana (Stevia)*

Biological source [86]	*Stevia rebaudiana* sweetener (*Stevia*) obtained from honey leaf available in South America.
Biological/pharmacological activity [86]	This is widely used in China and Japan for rheumatic arthritis treatment and also in other arthritic diseases.
Chemical constituents [86]	It consists of a glycoside known as steviol, which is sweeter than sucrose (200 times). Rebaudioside, austroinulin, steviol, and dulcoside are also present.

(continued)

Mechanism of action as bioenhancer [86]	Under investigation

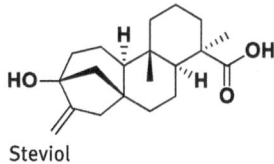

Steviol

When taken alone or in combination with piperine, stevia extracts improve the bio-efficacy of pharmaceutical compounds. Stevia in 1–80% amount is used as a bioenhancer. Bioenhancer obtained from stevia plant is around 30 mg/kg, while piperine is in the range of 8 mg/kg. Stevia increases the bioavailability of antibiotics, cardiovascular agents, anti-inflammatory, antitumor, antiulcer, and antidiabetic drugs [87].

6.4.14 Pod of drumstick (*Moringa oleifera*) niaziridins

Biological source [90]	Niaziridins is novel glycoside, which is obtained from drumstick pods (*Moringa oleifera*) after fractional distillation.
Biological/pharmacological activity [88]	Antifertility, spasmolytic, antimicrobial, diuretic, antioxidant and hepatoprotective, anticancer, anti-inflammatory, antifungal, and antiulcer.

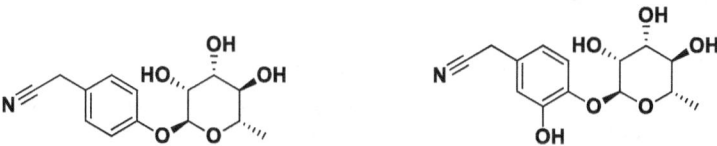

Niazirin [89] Niaziridin [89]

Niaziridin enhances the bioavailability of antibiotics such as tetracycline, rifampicin, Nalidixic acid, and ampicillin, when used against *Mycobacterium smegmatis* and *Bacillus subtilis*. It improves the antibiotic facilitation against *Escherichia coli*. It also enhances 5-fold the effectiveness of antifungal drugs from the category of azole such as clotrimazole against *Candida albicans*. It also enhances and facilitates the absorption of cyanocobalamin vitamin. As a result, niaziridin, in conjunction with antifungal and antitubercular medications, can be used to improve pharmacological effects. Hepatitis, jaundice, peripheral neuritis, gastric inflammation, and other serious side effects are caused due to antitubercular drugs such as rifampicin and isoniazid at higher doses. As a bioenhancer, niaziridin lowers the overall dose of such medications, allowing it to be used to minimize drug-related side effects [88, 89].

6.4.15 *Cuminum cyminum* (jeera): oil

Biological source [91]	Caraway (cumin) obtained from dried ripe fruits of *Carum carvi* (Umbelliferae)
IUPAC [91]	Carvone [5*R*]-2-methyl-5-prop-1-en-2-ylcyclohex-2-en-1-one and limonene [4*R*]-1-methyl-4-prop-1-en-2-ylcyclohexene
Biological/pharmacological activity [91]	Antioxidant, antidiabetic, anti-inflammatory, anticonvulsant, diuretic, antimicrobial, and carminative activity. Various antitubercular, antibiotic, antifungal, anticancer, antiviral, antihistaminic, and anti-inflammatory medicines have been shown to increase bioavailability. It contains essential oil [3–7%], protein [20%], fatty acids [10–18%], carbohydrate [15%], phenolic acids, and flavonoids. Its aqueous extract contains tannins, alkaloids, and terpenoids.

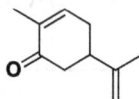

Carvone

The capacity of the caraway to enhance permeability through mucosa and its effect on the P-gp transporter could explain its bioenhancing properties [93]. Caraway was studied as a bioenhancer to improve the bioavailability of fixed-dose combination of antitubercular drugs in humans. A capsule containing the caraway extract [100 mg] was coadministered with a combination of a drug containing 450 mg of rifampicin, 300 mg of isoniazid, and 100 mg of pyrazinamide, orally, and their various pharmacokinetic parameters were studied. Bioavailability indices such as C_{max}, and the AUC for rifampicin, isoniazid and pyrazinamide increased by 32.22% and 32.16%, 36.01% and 29.06%, and 33.22% and 27.92%, respectively, while the half-life remained unchanged [92, 93].

6.4.16 Quercetin

Biological source [94]	Quercetin is a derivative of flavonoid glycosides present in citrus fruits, apples, vegetables, peppers, leaves, red wine, tomatoes, dark cherries, grains, etc.
IUPAC name [94]	Quercetin [2-[3,4- dihydroxyphenyl]-3,5,7-trihydroxy-4 H-chromen-4-one]
Biological/pharmacological activity [94]	It has been used as a dietary supplement and exhibits antioxidant, anticancer, cardiovascular protection, antiviral, anti-inflammatory, antidiabetic, antiallergic, antiulcer, antihypertensive, anti-infective, gastroprotective effects, and immunomodulatory activities

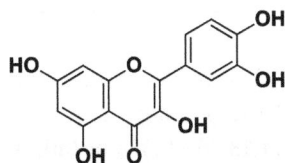

Quercetin

Quercetin not only exhibits poor and extremely variable bioavailability [0–50%] but also gets rapidly eliminated [half-life: 1–2 h] after ingestion. The pharmacological response predicted from in vitro studies are unlikely to be mirrored when the study is done on actual animal models because quercetin is quickly and thoroughly metabolized. In rats, there is no substantial phase I metabolism of quercetin. Quercetin produces metabolites that are more polar than the parent molecule and are eliminated faster due to the conjugation reaction.

It acts by inhibiting the cytochrome P3A4 and modulating the P-gp that results in the enhancement of bioavailability of numerous API [94].

Various pharmacokinetic properties of tamoxifen at a dose of 10 mg/kg were investigated in plasma, with and without quercetin, after oral administration. The dose of quercetin administered was 2.5, 7.5, and 15 mg/kg of body weight. When quercetin was coadministered with tamoxifen, K_a, C_{max}, AUC increased to 21.5 ± 5.4 per h, 85 ± 24 ng/mL, 2161 ± 572 ng/mL h, respectively. Absolute bioavailability of Tamoxifen was in the range of 18–24% when compared with the control group [15%]. Similarly, the relative bioavailability of tamoxifen along with quercetin improved by 1.20–1.61-fold when compared to tamoxifen alone. Quercetin increased the bioavailability of tamoxifen by enhancing its absorption through the intestine and reducing its first-pass metabolism [95].

6.4.17 *Ocimum sanctum* (tulsi)

Biological source [96]	Eugenol is obtained from fresh and dried leaves of *Ocimum sanctum* Linn. (family: Labiatae), commonly known as Tulsi.
Biological/pharmacological activity [96]	Antidiabetic, anticoagulant, antioxidant, antimicrobial, antifertility, antioxidant, anti-inflammatory, immunomodulatory, antistress and anthelmintic.

Eugenol

Administration of metformin at a dose of 100 mg/kg, alone and along with a combination of *O. sanctum* at a dose of 100–150 mg/kg, orally, increased C_{max} from 15.96 ± 1.066 ng/mL to 19.8 ± 0.825 and 18.76 ± 0.328 ng/mL AUC increased from 156.09 [ng/mL h] to 219.075 and 194.66 [ng/mL h], and T_{max} decreased from 4.0 to 3.027 and 3.51, respectively, as compared to the control group [97, 98].

Future perspectives

The idea of bioenhancers, originated from the use of "Trikatu" a formulation composed of *Piper nigrum*, *Piper longum*, and *Zingiber officinale* in 1:1:1 ratio in Ayurveda, has promoted the enhancement of bioavailability in a variety of medicines [99]. Many scientists and pharmaceutical companies are now focusing on improving the bioavailability of potent drugs having poor bioavailability. Naturally available bioenhancers offer a revolutionary concept for lowering the medication doses and making care more affordable and accessible to a wider group of people. Natural bioenhancers also help to prevent bacteria from developing drug resistance, which is a big concern for humans. A potent bioenhancer, quercetin, has been used for several classes of drugs but is still not able to show its action in the analgesic category of drugs and drugs acting on CNS. Naringin acts as a bioenhancer of CVS, anticancer drugs and steroidal medications, although, research on its efficacy against antiviral, anti-inflammatory, and CNS medications is yet to be conducted. Similarly, some other bioenhancers such as genistein, Niaziridin, and sinomenine, have not been thoroughly investigated in a variety of drugs. These bioenhancing agents can be employed for improving the pharmacokinetics of a wide range of pharmaceuticals, including respiratory, gastrointestinal, and steroid medications. To demonstrate the influence of natural bioenhancers, the bulk of recent studies have centered on the oral route. The only bioenhancer studied extensively with wide classes of drugs is

Piperine. There are a number of other herbal bioenhancers that are yet to be fully utilized in a number of key areas. Many disorders such as Alzheimer's disease and Parkinson's disease have grown increasingly and there is a significant requirement for drugs that can directly affect the central nervous system (CNS). In order to work on the CNS, medicines must be able to pass the blood–brain barrier. Experiments with these bioenhancers to boost the bioavailability via various modes of drug administration are required. The use of herbal bioenhancers in novel drug delivery systems is in high demand due to the advances in technology [100].

This chapter focused on various pharmacokinetic parameters of herbal bioenhancers along with several examples of drugs whose bioavailability is enhanced using herbal bioenhancers. More experimental studies need to conducted in order to isolate and study the active constituent from a wide range of unknown plants, which can act as bioenhancer for several classes of drugs and nutraceuticals. The exact mode of action of several bioenhancers is still under investigation. They are also assessed to find out the true mode of action of bioenhancers. Several bioenhancers have been linked to toxicity when used in high doses or over long periods of time. Research is needed to overcome these natural bioenhancers' toxicity characteristics and administer a safe dose [101].

List of abbreviations

μg	Microgram
μM	Micromolar
AHH	Aryl hydrocarbon hydroxylase
AUC	Area under curve
AUC_∞	AUC (from zero to infinity)
AUC_t	Area under the curve over a time interval
AUMC	Area under the first moment curve
BA	Bioavailability
BE	Bioequivalence
$CL_{oral/F}$	Oral clearance
C_{max}	Maximum concentration
CNS	Central nervous system
CVS	Cardiovascular system
CYP-450	Cytochrome P450
DNA	Deoxyribonucleic acid
EGCG	Epigallocatechin Gallate
GDH	Glucose 6-dehydrogenase
GGT	Gamma-glutamyl transpeptidase
GIT	Gastrointestinal tract
HPLC	High-performance liquid chromatography
h	Hour

IUPAC	The International Union of Pure and Applied Chemistry
K_{el}	Elimination rate constant
kg	Kilogram
m^2	Square meter
MCF	Michigan Cancer Foundation
mg	Milligram
MIC	Minimum inhibitory concentration
mL	Milliliter
MRT	Mean residence time
ng	Nanogram
nM	Nanomolar
pg/dL	Picogram/deciliter
P-gp	P-glycoprotein
RNA	Ribonucleic acid
SAR	Structure–activity relationship
SD rats	Sprague Dawley rats
$t_{1/2}$	Half-life
t_{max}	Time to maximum plasma concentration
TPGS	Tocopheryl polyethylene glycol succinate
UDP-GA	Uridine diphosphate glucuronic acid
μmol/kg	Micromole per kilogram

References

[1]   Atal N, Bedi KL, Bioenhancers, revolutionary concept to market, Journal of Ayurveda and Integrative Medicine, 2010, 1(2), 96.

[2]   Shanmugam S, Natural bioenhancers, current outlook, Clinical Pharmacology and Biopharmaceutics, 2015, 4(2), 1–3.

[3]   Raeissi SD, Li J, Hidalgo IJ, The role of an α-Amino group on H+-dependent transepithelial transport of cephalosporins in Caco-2 cells, Journal of Pharmacy and Pharmacology, 1999, 51(1), 35–40.

[4]   Alexander A, Qureshi A, Kumari L, Vaishnav P, Sharma M, Saraf S, Saraf S, Role of herbal bioactives as a potential bioavailability enhancer for active pharmaceutical ingredients, Fitoterapia, 2014, 97, 1–4.

[5]   Tatiraju DV, Bagade VB, Karambelkar PJ, Jadhav VM, Kadam V, Natural bioenhancers, an overview, Journal of Pharmacognosy and Phytochemistry, 2013, 2(3), 55–60.

[6]   Prasad R, Singh A, Gupta N, Tarke C, Role of bioenhancers in tuberculosis, International Journal of Health Sciences and Research, 2016, 3076.

[7]   Gopal V, Prakash Yoganandam G, Velvizhi Thilagam T, Bio-enhancer, a pharmacognostic perspective, European Journal of Biochemistry, 2016, 3(1), 33–38.

[8]   Dudhatra GB, Mody SK, Awale MM, Patel HB, Modi CM, Kumar A, Kamani DR, Chauhan BN, A comprehensive review on pharmacotherapeutics of herbal bioenhancers, Scientific World Journal, 2012, 17(2012).

[9]   Singh R, Devi S, Patel J, Patel U, Bhavsar S, Thaker A, Indian herbal bioenhancers, a review, Pharmacognosy Reviews, 2009, 3(5), 90.

[10] Khajuria A, Thusu N, Zutshi U, Piperine modulates permeability characteristics of intestine by inducing alterations in membrane dynamics, influence on brush border membrane fluidity, ultrastructure and enzyme kinetics, Phytomedicine, 2002, 9(3), 224–231.
[11] Joshi DR, Shrestha AC, Adhikari N, A review on diversified use of the king of spices, *Piper nigrum* [Black Pepper], International journal of pharmaceutical sciences research, 2018, 9(10), 4089–4101.
[12] Bajad S, Bedi KL, Singla AK, Johri RK, Piperine inhibits gastric emptying and gastrointestinal transit in rats and mice, Planta medica, 2001, 67(02), 176–179.
[13] Jumpa-ngern P, Kietinun S, Sakpakdeejaroen I, Cheomung A, Na-Bangchang K, Pharmacokinetics of piperine following single dose administration of benjakul formulation in healthy Thai subjects, African Journal of Pharmacy and Pharmacology, 2013, 7(10), 560–566.
[14] Liu J, Bi Y, Luo R, Wu X, Simultaneous UFLC–ESI–MS/MS determination of piperine and *piper longum* in ine in rat plasma after oral administration of alkaloids from piper longum L., application to pharmacokinetic studies in rats, Journal of chromatography B Biomedical applications, 2011, 879(27), 2885–2890.
[15] Bhat BG, Chandrasekhara N, Metabolic disposition of piperine in the rat, Toxicology, 1987, 44(1), 99–106.
[16] Suresh D, Srinivasan K, Tissue distribution & elimination of capsaicin, piperine & curcumin following oral intake in rats, IJMR, 2010, 131(5).
[17] Bano G, Raina RK, Zutshi U, Bedi KL, Johri RK, Sharma SC, Effect of piperine on bioavailability and pharmacokinetics of propranolol and theophylline in healthy volunteers, European Journal of Clinical Pharmacology, 1991, 41(6), 615–617.
[18] Junsaeng D, Anukunwithaya T, Songvut P, Sritularak B, Likhitwitayawuid K, Khemawoot P, Comparative pharmacokinetics of oxyresveratrol alone and in combination with piperine as a bioenhancer in rats, BMC complementary and alternative medicine, 2019, 19(1), 1–10.
[19] Atal S, Atal S, Vyas S, Phadnis P, Bio-enhancing effect of piperine with metformin on lowering blood glucose level in alloxan induced diabetic mice, Pharmacognosy research, 2016, 8(1), 56.
[20] Kasibhatta R, Naidu MU, Influence of piperine on the pharmacokinetics of nevirapine under fasting conditions, Drugs in R & D, 2007, 8(6), 383–391.
[21] Dudhatra GB, Mody SK, Awale MM, Patel HB, Modi CM, Kumar A, Kamani DR, Chauhan BN, A comprehensive review on pharmacotherapeutics of herbal bioenhancers, Scientific World Journal, 2012.
[22] Mujumdar AM, Dhuley JN, Deshmukh VK, Raman PH, Thorat SL, Naik SR, Effect of piperine on pentobarbitone induced hypnosis in rats, Indian Journal of Experimental Biology, 1990, 28(5), 486–487.
[23] Badmaev V, Majeed M, Norkus EP, Piperine, an alkaloid derived from black pepper increases serum response of beta-carotene during 14-days of oral beta-carotene supplementation, Nutrition Research, 1999, 19(3), 381–388.
[24] Shoba G, Joy D, Joseph T, Majeed M, Rajendran R, Srinivas PS, Influence of piperine on the pharmacokinetics of curcumin in animals and human volunteers, Planta medica, 1998, 64, 353–356.
[25] Kapil RS, Zutshi U, Bedi KL, Singh G, Johri RK, Dhar SK, Kaul JL, Sharma SC, Pahwa GS, Kapoor N, Tickoo AK. Pharmaceutical compositions containing piperine and an antituberculosis or antileprosy drug. European Patent Number, EP0650728B1, 2002.
[26] Hiwale AR, Dhuley JN, Naik SR, Effect of co-administration of piperine on pharmacokinetics of β-lactam antibiotics in rats.
[27] Lambert JD, Hong J, Kim DH, Mishin VM, Yang CS, Piperine enhances the bioavailability of the tea polyphenol [–]-epigallocatechin-3-gallate in mice, The Journal of Nutrition, 2004, 134(8), 1948–1952.

[28] Singh M, Varshneya C, Telang RS, Srivastava AK, Alteration of pharmacokinetics of oxytetracycline following oral administration of piper longum in hens, Journal of Veterinary Science, 2005, 6(3).

[29] Balkrishna B, Yogesh P, Influence of co-administration of piperine on pharmacokinetic profile of ciprofloxacin, Indian drugs, 2002, 39(3), 166–168.

[30] Pooja S, Agrawal RP, Nyati P, Savita V, Phadnis P, Analgesic activity of piper nigrum extract per se and its interaction with diclofenac sodium and pentazocine in albino mice, International Journal of Pharmacology, 2007, 5(1), 3.

[31] Janakiraman K, Manavalan R, Compatibility and stability studies of ampicillin trihydrate and piperine mixture, International journal of pharmaceutical sciences research, 2011, 2(5), 1176.

[32] Jin MJ, Han HK, Effect of piperine, a major component of black pepper, on the intestinal absorption of fexofenadine and its implication on food–drug interaction, Journal of Food Science, 2010, 75(3), 93–96.

[33] Amar S, Pawar VK, Vikash J, Parabia MH, Rajendra A, Gaurav S, In-vivo assessment of enhanced bioavailability of metronidazole with piperine in rabbits, Research Journal of Pharmaceutical, Biological and Chemical Sciences, 2010, 1(4), 273–278.

[34] Venkatesh S, Durga KD, Padmavathi Y, Reddy BM, Mullangi R, Influence of piperine on ibuprofen induced antinociception and its pharmacokinetics, Arzneimittelforschung, 2011, 61(09), 506–509.

[35] Majeed M, Badmaev V, Rajendran R, inventors, Sabinsa Corp, assignee. Use of piperine to increase the bioavailability of nutritional compounds. United States patent US 5,536,506. 1996.

[36] Majeed M, Badmaev V, Rajendran R, inventors, Sabinsa Corp, assignee. Use of piperine as a bioavailability enhancer. United States patent US 5,744,161. 1998.

[37] El-Saber Batiha G, Magdy Beshbishy A, G Wasef L, Elewa YH, A Al-Sagan A, El-Hack A, Mohamed E, Taha AE, M Abd-Elhakim Y, Prasad Devkota H, Chemical constituents and pharmacological activities of garlic [Allium sativum L.], A review, Nutrients, 2020, 12(3), 872.

[38] Salehi B, Zucca P, Orhan IE, Azzini E, Adetunji CO, Mohammed SA, Banerjee SK, Sharopov F, Rigano D, Sharifi-Rad J, Armstrong L, Allicin and health, A comprehensive review, Trends Food Sci Technology, 2019, 86, 502–516.

[39] Salehi B, Zucca P, Orhan IE, Azzini E, Adetunji CO, Mohammed SA, Banerjee SK, Sharopov F, Rigano D, Sharifi-Rad J, Armstrong L, Allicin and health, A comprehensive review, Trends Food Sci Technology, 2019, 86, 502–516.

[40] Yang LJ, Fan L, Liu ZQ, Mao YM, Guo D, Liu LH, Tan ZR, Peng L, Han CT, Hu DL, Wang D, Effects of allicin on CYP2C19 and CYP3A4 activity in healthy volunteers with different CYP2C19 genotypes, European Journal of Clinical Pharmacology, 2009, 65(6), 601–608.

[41] Ogita A, Hirooka K, Yamamoto Y, Tsutsui N, Fujita KI, Taniguchi M, Tanaka T, Synergistic fungicidal activity of Cu2+ and allicin, an allyl sulfur compound from garlic, and its relation to the role of alkyl hydroperoxide reductase 1 as a cell surface defense in saccharomyces cerevisiae, Toxicology, 2005, 215(3), 205–213.

[42] Ogita A, Fujita KI, Taniguchi M, Tanaka T, Enhancement of the fungicidal activity of amphotericin B by allicin, an allyl-sulfur compound from garlic, against the yeast *Saccharomyces cerevisiae* as a model system, Planta medica, 2006, 72(13), 1247–1250.

[43] Surjushe A, Vasani R, Saple DG, Aloe vera, a short review, Indian Journal of Dermatology, 2008, 53(4), 163.

[44] Sharma P, Kharkwal AC, Kharkwal H, Abdin MZ, Varma A, A review on pharmacological properties of Aloe vera, International Journal of Pharmaceutical Sciences Review and Research, 2014, 29(2), 31–37.

[45] Vinson JA, Al Kharrat H, Andreoli L, Effect of Aloe vera preparations on the human bioavailability of vitamins C and E, Phytomedicine, 2005, 12(10), 760–765.

[46] Haasbroek A, Willers C, Glyn M, Du Plessis L, Hamman J, Intestinal drug absorption enhancement by Aloe vera gel and whole-leaf extract, in vitro investigations into the mechanisms of action, Pharmaceutics, 2019, 11(1), 36.

[47] Upadhyay HC, Saini DC, Srivastava SK, Phytochemical analysis of *Ammannia multiflora*, Research Journal of Phytochemistry, 2011, 5(3), 170–176.

[48] Upadhyay HC, Dwivedi GR, Darokar MP, Chaturvedi V, Srivastava SK, Bioenhancing and anti-mycobacterial agents from *Ammannia multiflora*, Planta medica, 2012, 78(01), 79–81.

[49] Cho SW, Lee JS, Choi SH, Enhanced oral bioavailability of poorly absorbed drugs. I. Screening of absorption carrier for the ceftriaxone complex, Journal of Pharmaceutical Sciences, 2004, 93(3), 612–620.

[50] Fattori V, Hohmann MS, Rossaneis AC, Pinho-Ribeiro FA, Verri WA, Capsaicin, current understanding of its mechanisms and therapy of pain and other pre-clinical and clinical uses, Molecules, 2016, 21(7), 844.

[51] Bouraoui A, Toumi A, Brazier JL, Effects of capsicum fruit on theophylline absorption and bioavailability in rabbits, Drug-nutrient Interactions, 1988, 5(4), 345–350.

[52] Cruz L, Castañeda-Hernández G, Navarrete A, Ingestion of chilli pepper [Capsicum annuum] reduces salicylate bioavailability after oral aspirin administration in the rat, Canadian Journal of Physiology and Pharmacology, 1999, 77(6), 441–446.

[53] Sumano-López H, Gutiérrez-Olvera L, Aguilera-Jiménez R, Gutiérrez-Olvera C, Jiménez-Gómez F, Administration of ciprofloxacin and capsaicin in rats to achieve higher maximal serum concentrations, Arzneimittelforschung, 2007, 57(05), 286–290.

[54] Khan A, Srivastava VK, Antitoxic and bioenhancing role of kamdhenu ark [cow urine distillate] on fertilitty rate of male mice [Mus musculus] affected by cadmium chloride toxicity, International Journal of Cow Science, 2005, 1(2), 43–46.

[55] Ganaie JA, Gautam V, Shrivastava VK, Effects of kamdhenu ark and active immunization by gonadotropin releasing hormone conjugate [GnRH-BSA] on gonadosomatic indices [GSI] and sperm parameters in male Mus musculus, Journal of Reproduction and Infertility, 2011, 12(1), 3.

[56] Randhawa G, Cow urine distillate as bioenhancer, Journal of Ayurveda and integrative medicine, 2010, 1(4), 240.

[57] Niranjan A, Prakash D, Chemical constituents and biological activities of turmeric [Curcuma longa L.]-A review, Journal of Food Science and Technology, 2008, 45(2), 109.

[58] Zhang W, Tan TM, Lim LY, Impact of curcumin-induced changes in P-glycoprotein and CYP3A expression on the pharmacokinetics of peroral celiprolol and midazolam in rats, Drug Metabolism and Disposition, 2007, 35(1), 110–115.

[59] Alexander A, Qureshi A, Kumari L, Vaishnav P, Sharma M, Saraf S, Saraf S, Role of herbal bioactives as a potential bioavailability enhancer for active pharmaceutical ingredients, Fitoterapia, 2014, 97, 1–4.

[60] Jamwal R, Bioavailable curcumin formulations, A review of pharmacokinetic studies in healthy volunteers, Journal of Integrative Medicine, 2018, 16(6), 367–374.

[61] Lao CD, Ruffin MT, Normolle D, Heath DD, Murray SI, Bailey JM, Boggs ME, Crowell J, Rock CL, Brenner DE, Dose escalation of a curcuminoid formulation, BMC complementary and alternative medicine, 2006, 6(1), 1–4.

[62] Dei Cas M, Ghidoni R, Dietary curcumin, correlation between bioavailability and health potential, Nutrients, 2019, 11(9), 2147.

[63] Garcea G, Jones DJ, Singh R, Dennison AR, Farmer PB, Sharma RA, Steward WP, Gescher AJ, Berry DP, Detection of curcumin and its metabolites in hepatic tissue and portal blood of patients following oral administration Br, Journal of Cancer, 2004, 90(5), 1011–1015.

[64] Garcea G, Berry DP, Jones DJ, Singh R, Dennison AR, Farmer PB, Sharma RA, Steward WP, Gescher AJ, Consumption of the putative chemopreventive agent curcumin by cancer patients, assessment of curcumin levels in the colorectum and their pharmacodynamic consequences, CEBP, 2005, 14(1), 120–125.

[65] Schiborr C, Eckert GP, Rimbach G, Frank J, A validated method for the quantification of curcumin in plasma and brain tissue by fast narrow-bore high-performance liquid chromatography with fluorescence detection anal, Bioanalytical Chemistry, 2010, 397(5), 1917–1925.

[66] Hussaarts KG, Hurkmans DP, Oomen-de Hoop E, van Harten LJ, Berghuis S, van Alphen RJ, Spierings LE, van Rossum-schornagel QC, Vastbinder MB, van Schaik RH, van Gelder T, Impact of curcumin [with or without piperine] on the pharmacokinetics of tamoxifen, Cancers, 2019 Mar, 11(3), 403.

[67] Abo-El-Sooud K, Samar MM, Fahmy MA, Curcumin ameliorates the absolute and relative bioavailabilities of marbofloxacin after oral administrations in broiler chickens, Wulfenia, 2017, 24(3), 284–297.

[68] Li C, Byung-Chul Choi DK, Choi JS, Effects of curcumin on the pharmacokinetics of loratadine in rats, possible role of CYP3A4 and P-glycoprotein inhibition by curcumin, KSAP, 2011, 19(3), 364–370.

[69] He X, Mo L, Li ZY, Tan ZR, Chen Y, Ouyang DS, Effects of curcumin on the pharmacokinetics of talinolol in human with ABCB1 polymorphism, Xenobiotica, 2012, 42(12), 1248–1254.

[70] Rehman T, Fatima Q, Ginger [Zingiber officinale], A mini review, International Journal of Complementary and Alternative Medicine, 2018, 11(2), 88–89.

[71] Qazi G, Bedi K, Johri R, Tikoo M, Tikoo A, Sharma S, Abdullah S, Suri O, Gupta B, Suri K, Satti N, inventors. Bioavailability enhancing activity of Zingiber officinale Linn and its extracts/ fractions thereof. United States patent application US 10/318,314. 2003.

[72] Kesarwani K, Gupta R, Bioavailability enhancers of herbal origin, an overview, Asian Pacific Journal of Tropical Biomedicine, 2013, 3(4), 253–266.

[73] Chivte VK, Tiwari SV, Pratima A, Nikalge G, Bioenhancers, A brief review, Advanced Journal of Pharmacie and Life Science Research, 2017, 2, 1–8, Wikipedia – Glycyrrhizin (Assessed on 17 February 2021, at, https://en.wikipedia.org/wiki/Glycyrrhizin.

[74] Chen L, Yang J, Davey AK, Chen YX, Wang JP, Liu XQ, Effects of diammonium glycyrrhizinate on the pharmacokinetics of aconitine in rats and the potential mechanism, Xenobiotica, 2009, 39(12), 955–963.

[75] Chavhan SA, Shinde SA, Gupta HN, Current trends on natural bioenhancers, a review, International Journal of Pharmacognosy & Chinese Medicine, 2018, 2, 000123.

[76] Khanuja SP, Kumar S, Arya JS, Shasany AK, Singh M, Awasthi S, Gupta SC, Darokar MP, Rahman LU, inventors, Council of Scientific, Industrial Research [CSIR], assignee. Composition comprising pharmaceutical/nutraceutical agent and a bio-enhancer obtained from Glycyrrhiza glabra United States patent US 6,979,471. 2005.

[77] Khanuja S, Arya J, Srivastava S, Shasany A, Kumar TS, Darokar M, Kumar S, inventors, Council of Scientific, Industrial Research [CSIR], assignee. Antibiotic pharmaceutical composition with lysergol as bio-enhancer and method of treatment. United States patent application US 11/395,527. 2007.

[78] Khanuja SS, Arya J, Srivastava S, Shasany A, Kumar TS, Darokar M, Kumar S, inventors, Council of Scientific, Industrial Research [CSIR], assignee. Antibiotic pharmaceutical composition with lysergol as bio-enhancer and method of treatment. United States patent application US 10/103,726. 2003.

[79] Patil S, Dash RP, Anandjiwala S, Nivsarkar M, Simultaneous quantification of berberine and lysergol by HPLC-UV, evidence that lysergol enhances the oral bioavailability of berberine in rats, Biomedical Chromatography, 2012, 26(10), 1170–1175.

[80] https://en.wikipedia.org/wiki/Lysergol (Accessed May 19, 2021)
[81] Shrivastava A, A review on peppermint oil, Asian Journal of Pharmaceutical and Clinical Research, 2009, 2(2), 27–33.
[82] Wacher VJ, Wong S, Wong HT, Peppermint oil enhances cyclosporine oral bioavailability in rats, comparison with d-α-tocopheryl poly [ethylene glycol 1000] succinate [TPGS] and ketoconazole, Journal of Pharmaceutical Sciences, 2002, 91(1), 77–90.
[83] Liu ZQ, Zhou H, Liu L, Jiang ZH, Wong YF, Xie Y, Cai X, Xu HX, Chan K, Influence of co-administrated sinomenine on pharmacokinetic fate of paeoniflorin in unrestrained conscious rats, Journal of Ethnopharmacology, 2005, 99(1), 61–67.
[84] https://en.wikipedia.org/wiki/Sinomenine (Accessed May 19, 2021).
[85] Gokaraju GR, Gokaraju RR, D'souza C, Frank E, inventors, Laila Impex, assignee. Bio-availability/bio-efficacy enhancing activity of Stevia rebaudiana and extracts and fractions and compounds thereof. United States patent application US 12/610,502. 2010.
[86] https://en.wikipedia.org/wiki/Stevia (Accessed June 15, 2021).
[87] Gopalakrishnan L, Doriya K, Kumar DS, *Moringa oleifera*, A review on nutritive importance and its medicinal application, Food Science and Human Wellness, 2016, 5(2), 49–56.
[88] Khanuja SP, Arya JS, Tiruppadiripuliyur RS, Saikia D, Kaur H, Singh M, Gupta SC, Shasany AK, Darokar MP, Srivastava SK, Gupta MM, inventors. Nitrile glycoside useful as a bioenhancer of drugs and nutrients, process of its isolation from *Moringa oleifera*. United States patent US 6,858,588. 2005.
[89] https://en.wikipedia.org/wiki/Niaziridin (Accessed June 15, 2021).
[90] Ghulam Q, Kasturi B, Rakesh J, Manoj T, Ashok T, Subhash S, Tasdaq A, Om S, Bishan G, Krishan S, Naresh S, Ravi K Bioavailability enhancing activity of *Carum carvi* extracts and fractions thereof. US Patent US 20060257505,2006.
[91] Mahboubi M, Caraway as important medicinal plants in management of diseases, Natural Products and Bioprospecting, 2019, 9, 1–11.
[92] Choudhary N, Khajuria V, Gillani ZH, Tandon VR, Arora E, Effect of *Carum carvi*, a herbal bioenhancer on pharmacokinetics of antitubercular drugs, A study in healthy human volunteers, Perspectives in Clinical Research, 2014, 5(2), 80–84.
[93] Rachana B, Lata K, Akshay K, Quercetin as natural bioavailability modulator, an overview, Research Journal of Pharmacy and Technology, 2020, 13(4), 2043–2050.
[94] Su-Lan H, Yu-Chi H, Yao-Horng W, Chih-Wan T, Sheng-Fang S, Pei-Dawn LC, Quercetin significantly decreased cyclosporin oral bioavailability in pigs and rats, Life Sciences, 2002, 72(3), 227–235.
[95] Shin S, Choi J, Li X, Enhanced bioavailability of tamoxifen after oral administration of tamoxifen with quercetin in rats, International Journal of Pharmaceutics, 2006, 313(1–2), 144–149.
[96] Srinivas N, Sali K, Bajoria AA, Therapeutic aspects of Tulsi unraveled, A review, Journal of Indian Academy of Oral Medicine and Radiology, 2016, 28, 17–23.
[97] Pattanayak P, Behera P, Das D, Panda SK, *Ocimum sanctum* Linn. A reservoir plant for therapeutic applications, An overview, Pharmacognosy Reviews, 2010, 4(7), 95–105.
[98] Kokare V, Nagras MA, Patwardhan SK, Bioanalytical method development of metformin co-administered with *ocimum sanctum* for potential bioenhancer activity, International Journal of Pharmaceutical Sciences Research, 2015, 6(5), 2056–2065.
[99] Verma S, Rai S, Bioenhancers from mother nature, a paradigm for modern medicines, Scholars Academic Journal of Pharmacy, 2018(10), 442–451.
[100] Alexander A, Qureshi A, Kumari L, Vaishnav P, Sharma M, Saraf S, Saraf S, Role of herbal bioactives as a potential bioavailability enhancer for active pharmaceutical ingredients, Fitoterapia, 2014, 97, 1–4.
[101] Kesarwani K, Gupta R, Bioavailability enhancers of herbal origin, an overview, Asian Pacific Journal of Tropical Biomedicine, 2013, 3(4), 253–266.

Vaibhav Bhamare, Rakesh Amrutkar, Vinod Patil,
Chandrashekhar Upasani

Chapter 7
Growing impact of herbal bioenhancers in pharmaceutical industries

Abstract: Ayurvedic expertise has made a significant contribution to drug research in the world, with new methods of identifying active compounds. In comparison to modern medicines, herbal medicines have succeeded to emphasize the world for their use, with advantageous therapeutic effects and fewer adverse effects. However, in vitro/in vivo findings for these herbal drugs or extracts are not so impressively correlative. Poor lipid solubility, improper molecular size, and prolonged therapy of phytoconstituents leads to poor absorption, followed by poor bioavailability and treatment expenses. Herbal bioenhancers are nontherapeutic active phytomolecules that, when co–administered at low doses, improve the bioavailability, bioefficacy, and biological activity of different drugs without having a synergistic impact with medication. Since herbal bioenhancers are healthy, nontoxic, inexpensive, easy to obtain, nonaddictive, pharmacologically inert, and nonallergenic, they are enticing pharmaceutical industries as a valuable and most effective means of bioavailability enhancement. Nowadays, industries are focusing on improvement in pharmacokinetic parameters of potent active pharmaceutical ingredients using various bioenhancing mechanisms that can help through alteration in enzyme activity, phytosomal formulation system, escape protein modifications, effects of cholagogic or choleretic agent, and heat production in the organism. An emphasis is made to tackle multidrug resistance in the treatment of infectious diseases. Bioenhancers significantly contribute to the drug development process, with innovative methods for identifying active compounds, using natural drugs and products. Herbal bioenhancers are becoming increasingly popular as a paradigm-shifting technology for improving the bioavailability and bioefficacy of various classes of drugs, nutraceuticals, and veterinary medicines, thanks to recent advances in drug delivery.

Vaibhav Bhamare, Department of Pharmaceutics, K. K. Wagh College of Pharmacy,
Nashik 422006, Maharashtra, India, e-mail: vaibhav.bhamre@gmail.com
Rakesh Amrutkar, K. K. Wagh College of Pharmacy, Nashik 422006, Maharashtra, India
Vinod Patil, SPH College of Pharmacy, Malegaon 423105, Maharashtra, India
Chandrashekhar Upasani, SNJB's SSDJ College of Pharmacy, Chandwad 423101, Maharashtra,
India

https://doi.org/10.1515/9783110746808-007

7.1 Introduction

Bioenhancers are substances that do not possess any pharmacological activity but have the capability to increase the bioavailability and bioefficacy of an active pharmaceutical ingredient (API). Numerous researchers have succeeded in enhancing the bioavailability of antibiotics, anti-inflammatory, drugs acting on the central nervous system, drugs for the treatment of cancer, cardiovascular drugs, vitamins, nutrients, etc. Bioenhancers serve by promoting maximum concentration of the drug reaching the blood stream, thereby augmenting the effectiveness of the drug. Increased bioavailability of API can significantly reduce the dose and the cost of the active drug, benefitting the dose economy toward the poorer segment of the society. A substantial bioefficacy alteration will drive safety, tolerance, compliance, and drug resistance [1].

7.2 History and concept of bioavailability enhancers

In 1929, Bose was the first scientist to document the elevation of antihistaminic effect of *Vasaka* by combining with long pepper [2]. In a research on traditional medicine, Indian scientist Prof. Dr. Atal C. K. (former director of the Indian Institute of Integrative Medicine, previously known as the Regional Research Laboratory, Jammu) proposed the hypotheses of increased drug bioavailability upon co-administration of the principal drug in low doses and, thereafter, in 1979, the term bioavailability enhancers was coined. Following this, Dr. Atal and his team has scientifically established the term "bioenhancers" with numerous researches at RRL Jammu. Piperine was the first bioenhancer in the world, which was discovered and validated for the bioavailability enhancing effects using sparteine and vasicine. Due to the results of the research, sparteine and vasicine became the first experimentally bioenhanced drugs in the world. The world's first bioenhanced antitubercular drug formulation containing rifampicin was the glorious output of research work initiated by Dr. Atal [3]. DCGI approved the bioenhanced rifampicin antitubercular formulation after Phase IIIb clinical trial, which was officially released in 2011 by the Government of India on the occasion of World Tuberculosis Day, and also presented it to Mr. Bill Gates.

The discovery of the most potent, safe, effective, easy to manufacture, and extremely economical bioenhancer piperine has amplified global interest and it has added a new chapter in medical science [4–6].

Ayurvedic expertise has made a significant contribution to the impression of the 'bioavailability enhancers' concept and is being used in medicinal system over the centuries. Therefore, the concept of bioavailability enhancer is immemorial. In 600 BC, a combination of three acrid drugs, namely, *Maricha, Pippali*, and *Shunthi*, was referred to as *Trikatu*, which boosts the bioavailability of drugs, nutrients, vitamins, etc. [7].

The extensive research in the field has triggered the discovery of many other bioenhancers. The novel lipid technology base formulation, where drug is entrapped along with bioenhancers, has proven effective when administered via various routes. Table 7.1 summarizes some of the recent advances associated with bioenhancement.

7.3 Classification of bioenhancers

Bioenhancers can be classified on the basis of their origin and on the basis of the mechanism of action [13, 14].

The first step in the classification of the bioenhancers is dependent on the source from which it originates. Table 7.2 shows the various plant and animal sources that can produce effective bioenhancers.

The second step in the classification of the bioenhancers is the mechanism of action by which they act, as shown in Table 7.3. There are various mechanisms of action, which signal the possible ways by which herbal bioenhancers reflect their utility as bioavailability enhancers.

7.4 Methods used for amplification of absorption of orally administered drugs

Numerous approaches have been used to enhance the absorption of poorly absorbed drugs (Figure 7.1). Particle technology, activation or inactivation of drug molecules, in vivo complex formation, and the use of biomolecule have proven to be effective tools.

7.4.1 Working of bioenhancers

In accordance with the mechanism of action by which bioenhancers work and looking at the significance they can produce, several marked working styles are [15, 16, 23],

- Encouraging gastrointestinal tract absorption of the medicines.
- Barring or lowering the proportion of hepatic or intestinal bioconversion of drugs.
- Immune system modification in order to reduce the overall drug requirement.
- Though pathogens emerge as persistors inside the macrophages, bioenhancers will increase the penetration of the active drug, ensuring desirable killing of the causative pathogens.

Table 7.1: Recent advances of bioenhancers [8–12].

Formulation	Active ingredient	Application	Pharmacology	Method employed	% Entrapment efficiency/size	Route of administration
Liposome	Quercetin	Mean dose, improved penetration in blood–brain barrier	Antioxidant, anticancer	Reverse evaporation technique	60%	Intranasal
Liposome	Silymarin	Improved bioavailability	Hepatoprotective	Reverse evaporation technique	69.2 ± 0.6%	Buccal
microspheres	Rutin	Targeting the cardiovascular and cerebrovascular system	Cardiovascular and cerebrovascular	Complex co-acervation method	165–195 (size in μm)	In vitro
Microspheres	Zedoary	Sustained release and higher bioavailability	Hepatoprotective	Quasi-emulsion solvent diffusion method 100	600 (size in μm)	oral
Nanoparticles	Triptolide	Increased hydration, leading to enhanced drug penetration through stratum corneum	Anti-inflammatory	Emulsification ultrasound		Topical

Table 7.2: Bioenhancers classified on the basis of origin [13, 14].

Plant origin	Animal origin
Examples: capsaicin, ginger, aloe vera, curcumin, *Stevia*, genistein, naringin, caraway, turmeric, pepper, peppermint oil	Example: cow urine distillate (Kamdhenu ark)

Table 7.3: Bioenhancers classified on the basis of mechanism of action [13, 14].

Inhibitors of P-glycoprotein (P-gp) efflux pump and other efflux pump	Suppressors of cytochrome P-450 (CYP-450) enzyme and its isoenzymes	Regulators of gastrointestinal tract characteristic to facilitate higher absorption
Examples: caraway, genistein, sinomenine, black cumin, naringin, quercetin	Examples: naringin, gallic acid and its ester, quercetin	Examples: aloe vera, niaziridin, ginger, liquorice

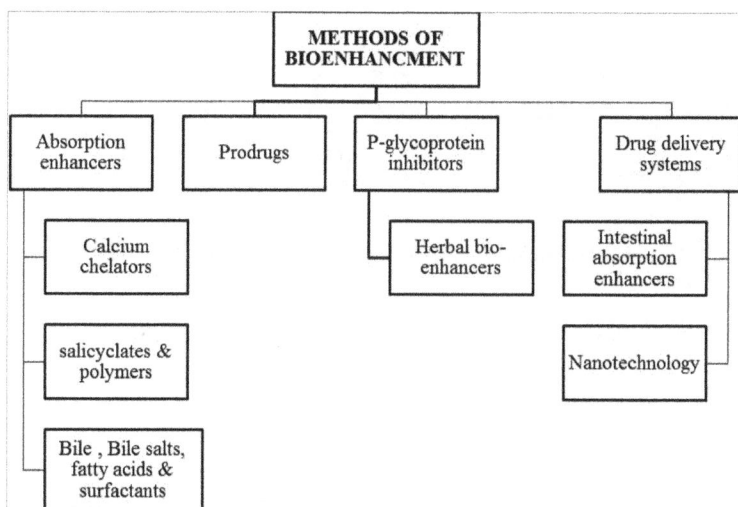

Figure 7.1: Methods of bioenhancement by bioenhancer [15–22].

- Impeding the functionality of the infectious agent or atypical tissue to reject the drug.
- Altering the signaling technique amongst the host and the infectious agent, ensuring improved approachability of the medicines to the infectious agent.
- Potentiating and prolonging the drug effect against the pathogen by enhancing drug binding with the target site.
- Bioenhancers may promote transport of drugs and nutrients across the blood–brain barrier, helping the treatment and control of diseases associated with the central nervous system.

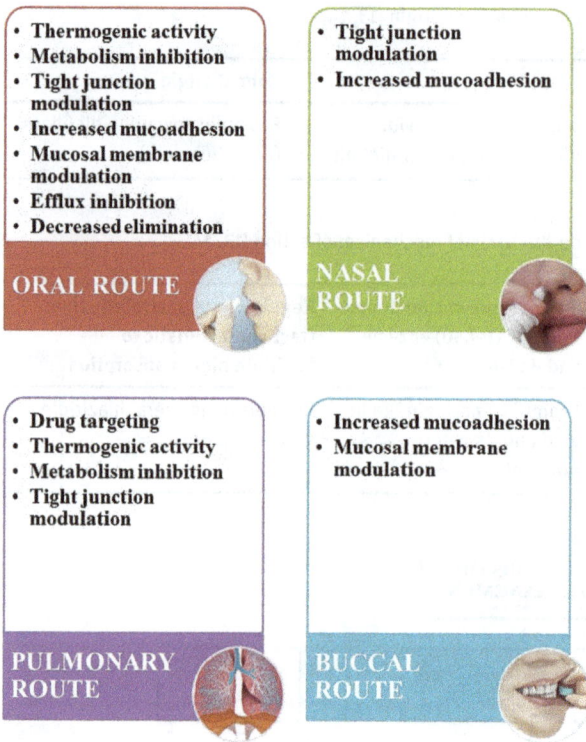

- Thermogenic activity
- Metabolism inhibition
- Tight junction modulation
- Increased mucoadhesion
- Mucosal membrane modulation
- Efflux inhibition
- Decreased elimination

ORAL ROUTE

- Tight junction modulation
- Increased mucoadhesion

NASAL ROUTE

- Drug targeting
- Thermogenic activity
- Metabolism inhibition
- Tight junction modulation

PULMONARY ROUTE

- Increased mucoadhesion
- Mucosal membrane modulation

BUCCAL ROUTE

Figure 7.2: Mechanism of action of bioenhancers in accordance with the route of administration [24].

Figure 7.2 summarizes the influential mechanisms of actions in accordance with the most popular routes through which herbal bioenhancers can be administered. Bioenhancers can potentiate systemic delivery of poorly bioavailable drugs via alternative route of administrations. Herbal bioenhancers are advantageous due to their improved oral bioavailability, being noninvasive, due to the painless local and systemic nasal drug delivery, and due to the copious blood supply in lungs, resulting in rapid drug delivery.

7.5 Herbal bioenhancers and novel pharmaceutical approach

Various updates have been observed with regard to the biopotentiation using natural options. Herbal bioenhancers has proven their value in the enhancement of bioavailability and by virtue of which, pharmaceutical industries have made significant contributions by developing various approaches to synergize the role of herbal bioen-hancers [25, 26].

7.5.1 Nasal spray

Aromatic constituents extracted from natural oil have proven efficient in the treatment of nasal congestion. Nouveau Technologies has come up with an improved herbal-based decongestant and antihistamine nasal spray, which have a combination of triterpenesaponins and aromatic constituents selected form natural oils or extracts. The results were surprisingly curative in the treatment of nasal congestion [27, 28].

7.5.2 Antibody purification

An antibody preparation with reduced host cell protein can be produced using herbal bioenhancers. Host cell protein and antibody blend is subjected to an ion exchange material, wherein ion exchange separation can lead to HCP-reduced antibody preparation [29].

7.5.3 Novel herbal formulation as brain tonic

Seed oil of *Sesamum indicum* and the fresh leaves alcoholic extract of *Centella asiatica* have a potential memory enhancing role [27, 30].

7.5.4 Bioenhancers in pulmonary drug delivery

Bioenhancers can be crucial for improved drug delivery via the pulmonary route of administration by passive paracellular transport where chloride–bicarbonate exchangers are activated and cytochrome P450 isoenzyme plus UDP glucuronyl transferase are used in combination to inhibit metabolism [31].

7.5.5 Nutritional bioenhancers

Various studies have revealed that a nonspecific mechanism and thermogenic properties of some herbal bioenhancers lead to increased bioavailability and retention of nutrients in alimentary canal and serves as nutritional bioenhancers [32, 33].

7.5.6 Cosmeceutical approach

Healthy appearance and skin well-being are the areas that concern consumers. Topical antioxidant supplementation is increasingly becoming popular and has shown improved beneficial effect in the treatment of skin aging. Green extract of *Camellia sinensi* were able to address the skin damage due to the deleterious effects of sunlight [34]. Silibinin, extracted form *Silybum marianum*, has shown potent antioxidant activity against molecular changes caused by skin exposure to xenobiotics [35]. Trans-isomer of resveratrol found in grapes is more stable [36] and a biologically active antioxidant with antiproliferative properties [37]. Several potent antioxidants such as chlorogenic acid, condensed proanthocyanidins, quinic acid, and ferulic acid can be extracted from the berries of *Coffea arabica* [38]. An extract obtained from juice, peel, or seed of the *Punica granatum* fruit is rediscovered as a medicine, which for centuries has been used in various cultures for its many beneficial effects [39].

7.5.7 Bioenhancers with Ayurveda

Various herbs having action as bioenhancers and different strategies have been adopted for bioenhancement using *P. longum*, *Z. officinale*, and *G. glabra* and many other puranaaushadhies. Samshodhana, i.e., biopurification is very much appreciated activity as medications and penetration enhancers. The definite investigation of these ideas clarifies the idea of bioenhancers [40].

7.5.8 Herbal bioenhancers for veterinary

Several examples are available in literature where herbal extracts and phytoactive compounds are proven to have potential in the veterinary medicine. Due to the specific differences in major microsomal enzyme for drug metabolism/detoxification, bioenhancement of veterinary medicine might not show the same enhancement as in humans. Piperine or its blend with rhizomes of ginger, homeopathic tinctures of *Ginkgo biloba*, grapefruit juice, curcumin, *Nigella sativa* Linn methanolic extracts, and many more have resulted in an increase in bioavailability and in modification of pharmacokinetics of veterinary medicines [41].

7.5.9 Bioenhancers for respiratory illness treatment

Nearly 200 antigenic viruses from different virus families cause serious respiratory illnesses in mammals. *Andrographis* extract and turmeric extract, in combination, provides synergistic anticholinergic effect and enhanced immunomudulatory activity against virus. The said combination is beneficial in the prevention and treatment of several respiratory illness as well as existing symptoms [42].

7.5.10 Bioenhancers in a detoxifier herbal formulation

Extract or powder of *Berberis aristata* stem and *Tinospora cordifolia* stem were found beneficial for the bioavailability enhancement of synergistic detoxifier composition. A synergistic herbal detoxifier composition, comprising *Echinacea purpurea* leaves, *Andrographis paniculata* leaves, *Boerhaavia diffusa* whole plant, *Arctium lappa* root, and *Rubia cordifolia* root, is effective in the handling and management of disorders due to the accretion of toxins in the body. An effective amount of the said bioenhancers renders high solubility and high bioavailability of the said active ingredient, in comparison to solubility and bioavailability of said active ingredient alone [43].

7.5.11 Herbal bioactive compounds and metabolic syndromes

Metabolic syndromes play an important role in the spread of noncommunicable diseases. Multiple comorbidities caused by various risk factors and the related healthcare costs show how the population is being affected with this syndrome. A natural and side effect-free characteristic of the herbal bioactive compound is a beneficial tool to tackle the causes of metabolic syndrome. Looking to the effects and the patient compliance associated with the clinical management of metabolic syndrome, herbal bioactive compounds are proving to be valid and influential tools.

Figure 7.3 highlights the pivotal role of herbal bioactive compounds via modified eating habits and pharmacological therapy. Bioactive compounds can delay or alter the comorbidities associated with metabolic syndrome and exert long term effects. Antiobesity action, modulatory functions, hypotensive functions, and lipid metabolism modulation are key roles that bioactive compounds play in the management of metabolic syndromes.

7.5.12 Drugs bioenhanced by various bioenhancers

Capacious research and simultaneous animal or human in vitro studies have proved effective applications of herbal bioenhancers. Their bioenhacing effects along with drugs and vitamins are visible. Examples are drugs such as antiviral, anticancer, penicillin's, anti-inflammatory, antiarthritic, antifungal, cardiovascular, immunosuppressants, antiulcer, antituberculosis, antileprosy, and vitamin A, vitamin E, vitamin C, and folic acid [45–47].

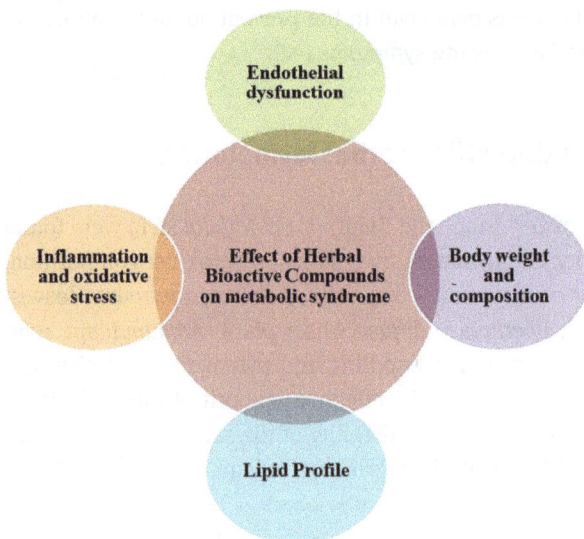

Figure 7.3: Beneficial effects of herbal bioactive compounds on metabolic syndromes [44].

7.6 Patents in the field

Nearly 40% of drugs fail to reach the market as a result of failure to achieve adequate bioavailability and bioefficacy. Herbal bioenhancers play a crucial role in the rectification of this issue and they have become the preferred choice for various researchers throughout the globe. Table 7.4 summarizes the various patents granted for the successful implementations of herbal bioenhancers and bioactive compounds in the field of bioavailability enhancement.

The emerging trend of bioavailability enhancement using herbal bioenhancers has set an impact in the formulation development, and various herbal bioenhancers are under investigation in several vital areas.

7.7 Herbal bioenhancers captivating pharmaceutical industries

Researchers have planned their studies and worked on stepwise protocol, producing significant data and have published in wide-reaching journals. Prolonged and expensive studies have been reported using various herbal compounds sourced from different plants and animal sources. The summary of literature findings are tabulated in Table 7.5.

Table 7.4: Patents granted for various herbal bioenhancers and bioactive compounds.

Title	Active ingredient/ Bioactive fraction	Patent number	Current Assignee	Status	Anticipated expiration
Nitrile glycoside useful as a bioenhancer of drugs and nutrients; process of its isolation from moringa oleifera	Novel nitrile glycoside named NIAZIRIDIN from the pods of *Moringa oleifera*.	US6858588B2	Council of Scientific and Industrial Research CSIR	Active	2023-03-31
Use of herbal agents for potentiation of bioefficacy of anti-infectives	Fraction of *Cuminum cyminum and Piperine*	US7119075B2	Council of Scientific and Industrial Research CSIR	Active	2023-03-31
Novel herbal composition	Extracts of herbal ingredients of *Tribulus terrestris, Withania somnifera, Curculigo orchioides, Mucuna pruriens, Asparagus adscendens, Asteracantha longifolia, Asphaltum,* and optionally the extracts of *Piper longum* and *Anagclus pyrethrum*	US20140294998A1	Enovate Biolife Private Ltd, Enovate Biolife LLC	Active	2031-04-14
Herbal composition for the treatment of kidney stone and other urinary tract disorders	Combination of therapeutically effective amount of Vrun bhavit, Kadalikshar, Apamargkshar, Yavkshar and optionally Gomutrakshar	US9259441B2	Suresh Balkrishna Patankar (inventor)	Active	2032-03-05
Synergistic herbal ophthalmic formulation for lowering the intra ocular pressure in case of glaucoma	Extract of herbs selected from *Ocimum* species, *Curcuma* species, *Solanum nigrum* and *Areca catechu*	US20100112106A1	Sentiss Pharma Private Ltd, Delhi Inst of Pharmaceutical Sciences and Res DIPSAR	Active	2029-11-15

(continued)

Table 7.4 (continued)

Title	Active ingredient/ Bioactive fraction	Patent number	Current Assignee	Status	Anticipated expiration
Pharmaceutical composition containing cow urine distillate and an antibiotic	Cow urine distillate	US6410059B1	Council of Scientific and Industrial Research CSIR	Expired	2020-12-01
Detoxifier herbal formulation	*Berberis aristata* stem and *Tinospora cordifolia* stem, and combinations	US20140147394A1	Manu Chaudhary	Expired – Fee Related	2031-10-25
Use of piperine as a bioavailability enhancer	98% of the alkaloid piperine	US5744161A	Sami Chemicals and Extracts Ltd	Expired	2015-02-24
Use of essential oils to increase the bioavailability of orally administered pharmaceutical compounds	Essential oil comprises All spice Berry, Benzoin Gum, Cananga Extract, Carrot Seed, Cinnamon, Clovebud, Patchouly, Rose Absolute, Celery, Sandalwood Oil, Cumin, Marigold pot, Olibanum oil, Lemongrass etc	US6121234A	Lyotropic Therapeutics Inc	Expired – Fee Related	2015-06-07
Composition comprising pharmaceutical/nutraceutical agent and a bioenhancer obtained from Glycyrrhiza glabra	isolated and purified Glycyrrhizic acid, isolated and purified Glycyrrhizin or mixtures	US6979471B1	Council of Scientific and Industrial Research CSIR	Expired	2020-09-05

Table 7.5: Profile of bioenhancers and their reported drug combinations.

Source	Compound	Biological Source	Family	Geographical Source	Plant Part Used	Effective in Enhancement of
Long Pepper or Black Pepper	Piperine	*Piper Longum* or *Piper Nigrum*	Piperaceae	India	Fruit, Root	Rifampicin, Phenytoin, Sulfadiazine and Propranolol
Aloe	Extract or Gel	*Aloe Barbadensis*	Liliaceae	India	Leaf	Glipizide, Pantoprazole, Vitamin C and E
Turmeric	Curcumin	*Curcuma Longa*	Zingiberaceae	Indian Subcontinent, Southeast Asia	Rhizomes	Celiprolol, Midazolam
Ginger	Gingerol	*Zingiber Officinale*	Zingiberaceae	India, Nepal, China, Indonesia, Thailand	Rhizomes	Azithromycin. Erythromycin, Cephalexin, Cefadroxil, Amoxycillin, Zidovudine, 5-Fluorouracil
Grape Fruit	Naringin	Hybrid originating by cross between the *Citrus Sinensis* and *Citrus Maxima*	Rutaceae	South Asia, East Asia, Southeast Asia, Melanesia, and Australia	Fruits	Diltiazem, Verapamil, Saquinavir and Cyclosporine A
Apples Onions Broccoli	Quercetin	*Malus Domestica* *Allium Cepa L.* *Brassica Oleracea*	Rosaceae Amaryllidaceae Brassicaceae	India, United States, China	Fruit or Leaf	Diltiazem, Digoxin, Verapamil, Paclitaxel, Doxorubicin and Epigallocatechin Gallate

(continued)

Table 7.5 (continued)

Source	Compound	Biological Source	Family	Geographical Source	Plant Part Used	Effective in Enhancement of
Drumstick	Niaziridin	*Moringa Oleifera*	Moringaceae	India Taiwan Oceanian Countries Caribbean Countries	Pods	Rifampicin, Ampicillin, Nalidixic Acid, Clotrimazole
Soyabean Kudzu	Genistein	*Glycine Max PuerariaLobata*	Fabaceae	India United States China Argentina	Bean	Paclitaxel
Liquorice	Glycyrrhizin	*Glycyrrhiza Glabra*	Fabaceae	India, Iran, Italy, Afghanistan, China, Azerbaijan, Turkey	Root and Stolon	Rifampicin, Ampicillin, Tetracycline, Nalidixic Acids, Clotrimazole
Cumin	3′,5dihydroxy Flavone-7-O-B- D-Galactouronide -4′-B-O-D- Glucopyranoside	*Cuminum Cyminum*	Apiaceae	India, China, Mexico	Seeds	Erythromycin, Cephalexin, Amoxycillin, Fluconazole, Ketoconazole, Zidovudine and 5-Fluorouracil
Chilli Pepper	Capsaicin	*Capsicum Annum*	Solanaceae	Spain, United States, China, Mexico, Turkey	Fruit	Theophylline

Garlic	Allicin	Allium Sativum	Amaryllidaceae	India, China, Bangladesh, Egypt, Spain	Bulb	Amphotericin B
Morning Glory Plant	Lysergol (Ergoline)	Rivea Corymbosa, Ipomoea Violacea and Ipomoea Muricata	Convolvulaceae	Australian Bushland, China, United States	Flowers	Berberine. Verapamil, Pantoprazole
Sinomenium	Sinomenine	Sinomenium Acutum Thunb	Menispermaceae	Northern India, Nepal, Japan and China	Root	Paeoniflorin
Honey Leaf	Steviosides	Stevia rebaudiana	Asteraceae	Japan, Southeast Asia, Nepal, United States, Europe	Leaves	Antitubercular, Anti-Leprotic, Anticancer, Antifungal and Antiviral Drugs
Cow Urine Distillate	Kamdhenu Ark					Rifampicin, Tetracycline, Ampicillin, Taxol, Rifampicin

Various drugs and herbal bioenhancer combinations have added astonishing benefit levels. The toxicity profile (if any) must not be overlooked, and research should be carried out in solving the issues and contributing to the unexploited roles of herbal bioenhancers in pharmaceutical industries.

7.8 Conclusion

Bioenhancement is a crucial aspect that the field of pharmacy is continuously dealing with. Various approaches and methods have successfully contributed with appropriate impact on economy. Use of natural resources as bioenhancers has multiplied their significance and applications in formulation development. With the aid of Ayurveda, systematic innovative means have merged in solving the problems of bioavailability, safety, and cost. Compounds derived from nature can augment the therapeutic activity of the active ingredient. A combination of herbal bioenhancers and new medical science can offer a paradigm changing drug delivery system.

List of abbreviations

CYP-450   Cytochrome P450
DCGI   Drugs Controller General of India
P-gp   P-glycoprotein

References

[1] Dudhatra GB, Mody SK, Awale MM, Patel HB, Kronn Y, Cm M, Dr K, Bn C, 2012, A comprehensive review on pharmacotherapeutics of herbal bioenhancers, The Scientific World Journal, 2012, 1(2), 1–33.
[2] Kesarwani K, Gupta R, Mukerjee A, Bioavailability enhancers of herbal origin: An overview, Asian Pacific Journal of Tropical Biomedicine, 2013, 3(4), 253–266.
[3] Zutshi RK, Singh R, Zutshi U, Johri RK, Atal CK, Influence of piperine on rifampicin blood levels in patients of pulmonary tuberculosis, The Journal of the Association of Physicians of India, 1985, 33(3), 223–234.
[4] Chopra B, Dhingra A, Kapoor RP, Prasad DN, Piperine and its various physicochemical and biological aspects: A review, Open Chemistry Journal, 2016, 3, 75–96.
[5] Kirtikar KR, Basu B, Indian medicinal plants, 2nd Dehradun, 1995.
[6] Kesarwani K, Gupta R, Mukerjee A, Bioenhancers of herbal origin, Asian Pacific Journal of Tropical Biomedicine, 2013, 3(4), 253–266.

[7] Muttepawar SS, Jadhav SB, Kankudate AD, Sanghai SD, Usturge DR, A review on bioavailability enhancers of herbal origin, World Journal of Pharmaceutical Science, 2014, 3, 667–677.

[8] Ajazuddin SS, Applications of novel drug delivery system for herbal formulations, Nanomedicine: Nanotechnology, Biology and Medicine, 2008, 4(51), 70–78.

[9] Samaligy MS, Afifi NN, Mahmoud EA, Evaluation of hybrid liposomes-encapsulated silymarin regarding physical stability and in vivo performance, International Journal of Pharmaceutics, 2006, 319(1–2), 121–129.

[10] Xiao L, Zhang YH, Xu JC, Xh J, Preparation of floating rutinalginate- chitosan microcapsule, Chinese Traditional Herbal Drugs, 2008, 2, 209–212.

[11] You J, Cui F, Han X, Wang Y, Yang L, *et al.*, Study of the preparation of sustained-release microspheres containing zedoary turmeric oil by the emulsion solvent-diffusion method and evaluation of the self-emulsification and bioavailability of the oil, Colloids and Surfaces B, 2006, 48(1), 35–41.

[12] Mei Z, Chen H, Weng T, Yang Y, Yang X Solid lipid nanoparticle and microemulsion for topical triptolide, European Journal of Pharmaceutics and Biopharmaceutics, 2003, 56(2), 189–196.

[13] Tatiraju DV, Bagade VB, Karambelkar PJ, Jadhav VM, Natural bio enhancers: An overview, Research & Reviews: Journal of Pharmacognosy and Phytochemistry, 2013, 55–60.

[14] Sindhoora D, Bhattacharjee A, Shabaraya AR, Bioenhancers, A Comprehensive Review, 2020, 60(1), 126–131.

[15] Schipper NGM, Olsson S, Hoogstraate JA, Boer AG, Varum KM, Artursson P, Chitosan an absorption enhancers for poorly absorbable drugs, influence of molecular weight and degree of acetylation on derug transport across human intestinal epithelial(Caco-2) cells, Pharmaceutical Research, 1997, 113, 1686–1692.

[16] Lundin S, Artursson P, Absorption enhancers as an effective method in improving the intestinal absorption, International journal of pharmaceutics, 1990, 64, 181–186.

[17] Buur A, Bundgaard H, Falch E, Prodrugs of 5-fluorouracil. VII. Hydrolysis kinetics and physicochemical properties of N-ethoxyand N-phenoxycarbonyloxymethyl derivatives of 5 fluorouracil, Acta pharmaceutica Suecica, 1986, 23, 205–216.

[18] Saxena V, Singh A, An update on bio-potentiation of drugs using natural options, Asian Journal of Pharmaceutical Sciences, 2020, 13(11), 25–32.

[19] Jain G, Patil UK, Strategies for enhancement of bioavailability of medicinal agents with natural products, IJPSR, 6(12), 5315–5324.

[20] Drabu S, Khatri S, Babu S and Lohani P, Use of herbal bioenhancers to increase the bioavailability of drugs, Research Journal of Pharmaceutical, Biological and Chemical Sciences, 2011, 2(4), 107–119.

[21] Dudhatra GB, Mody SK, Awale M, Patel HB, Modi CM, Kumar A, Kamani DR and Chauhan BN: A comprehensive review on pharmacotherapeutics of herbal bioenhancers, The Scientific World Journal, 2012, 1–33.

[22] Varma MV, Ashokraj Y, Dey CS, Panchagnula R, P – glycoprotein inhibitors and their screening: A perspective from bioavailability enhancement, Pharmacological research: The official journal of the Italian Pharmacological Society, 2003, 48(4), 347–359.

[23] Chivte VK, Tiwari SV, Pratima A, Nikalge G, Bioenhancers: A brief review, Advanced Journal of Pharmacie and Life Science Research, 2017, 2, 1–18.

[24] Peterson B, Weyers M, Steenekamp JH, Steyn JD, Gouws C, Hamman JH, Drug Bioavailability Enhancing Agents of Natural Origin (Bioenhancers) that Modulate Drug Membrane Permeation and Pre-Systemic Metabolism, Pharmaceutics, 2019, 11, 33.

[25] Liversidge GG, Cundy KC, Particle size reduction for improvement of oral bioavailability of hydrophobic drugs: I. Absolute oral bioavailability of nanocrystalline danazol in beagle dogs, International Journal of Pharmaceutics, 1995, 125, 91–97.

[26] Veiga F, Fernandes C, Teixeira F, Oral bioavailability and hypoglycaemic activity of Ttolbutamide/Cyclodextrin inclusion complexes, International Journal of Pharmaceutics, 2000, 202, 165–171.

[27] Goyal A, Kumar S, Nagpal M, Singh I, Arora S, Potential of novel drug delivery systems for herbal drugs, Indian Journal of Pharmaceutical Education and Research, 2011, 45(3), 225–235.

[28] Wiersma JG (1998). Herbal based nasal spray (U.S. Patent No. US 5948414) U.S. Patent and Trademark Office. https://patents.google.com/patent/US5948414A/en

[29] Wan M, Avgerinos G, Zarbis-Papastoitsis G (2006). Antibody purification (U.S. Patent No. US20070292442A1) U.S. Patent and Trademark Office. https://patents.google.com/patent/US20070292442A1/en

[30] Pushpangadan P, Rao CV, Kartik R, et al (2005). Novel herbal formulation as brain tonic (U.S. Patent No. 2005/0142232 A) U.S. Patent and Trademark Office. https://patentimages.storage.googleapis.com/44/be/02/224fd7447e80e6/US20050142232A1.pdf

[31] Peterson B, Weyers M, Steenekamp JH, Steyn JD, Gouws C, Hamman JH, Drug bioavailability enhancing agents of natural origin (bioenhancers) that modulate drug membrane permeation and pre-systemic metabolism, Pharmaceutics, 2019, 11, 33, 1–46.

[32] Badmaev V, Majeed M, Norkus EP, Piperine, an alkaloid derived from black pepper increases serum response of beta-carotene during 14-days of oral betacarotene supplementation, Nutrition Research, 1999, 19, 381–388.

[33] Majeed M, Badmaev V, Rajendran R (1995). Use of piperine to increase the bioavailability of nutritional compounds (U.S. Patent No US5536506A) U.S. Patent and Trademark Office. https://patents.google.com/patent/US5536506A/en

[34] Elmets CA, Singh D, Tubesing K, Matsui M, Katiyar S, Mukhtar H, Cutaneous photoprotection from ultraviolet injury by green tea polyphenols, Journal of the American Academy of Dermatology, 2001, 44, 425–432.

[35] Svobodova A, Psotova J, Walterova D, Natural phenolics in the prevention of UV-induced skin damage. A review, Biomedical papers of the Medical Faculty of the University Palacky, Olomouc, Czechoslovakia, 2003, 147, 137–145.

[36] Orallo F, Comparative studies of the antioxidant effects of cis – And trans -resveratrol, Current Medicinal Chemistry, 2006, 13, 87–98.

[37] Amri A, Chaumeila JC, Sfarb S, Charrueau C, Administration of resveratrol: What formulation solutions to bioavailability limitations?, Journal of Controlled Release, 2012, 158, 182–193.

[38] Farris P, Idebenone, green tea, and Coffeeberry® extract: New and innovative antioxidants, Dermatologic Therapy, 2007, 20, 322–329.

[39] Chidambara Murthy KN, Jayaprakasha GK, Singh RP, Studies on antioxidant activity of pomegranate (*Punica granatum*) peel extract using in vivo models, Journal of Agricultural and Food Chemistry, 2000, 50, 4791–4795.

[40] Raut SV, Nemade LS, Desai MT, Bonde SD, Dongare SU, Chemical penetration enhancers: For transdermal drug delivery systems, International Journal of Pharmaceutical Sciences Review and Research, 2014, 4, 33–40.

[41] Dikmen BY, Turgut Y, Filazi A, Herbal bioenhancers in veterinary phytomedicine, Frontiers in Veterinary Science, 5, 249.

[42] Clymer JW, Ho BY, Jump ML, Walanski AA, Zukowski CK (2006). Compositions and methods useful for treatment of respiratory illness (U.S. Patent No US20110159124A1) U.S. Patent and

Trademark Office. https://patents.google.com/patent/US20110159124?oq=herbal +bioenhancers

[43] Chaudhary M (2010). A detoxifier herbal formulation (Patent No WO2012056476A1) WIPO (PCT). https://patents.google.com/patent/WO2012056476A1/en?oq=herbal+bioenhancers

[44] Noce A, Di Lauro M, Di Daniele F, et al., Natural Bioactive Compounds Useful in Clinical Management of Metabolic Syndrome, Nutrients, 2021, 13(2), 630.

[45] Prasad R, Singh A, Gupta N, Tarke C, Role of bioenhancers in tuberculosis, International Journal of Health Sciences and Research, 2016, 6, 307–313.

[46] Dudhatra GB, Mody SK, Awale MM, et al., Comprehensive review on pharmacotherapeutics of herbal bioenhancers, The Scientific World Journal, 2012, 1, 1–33.

[47] Chavhan SA, Shinde SA, Gupta HN, Current trends on natural bioenhancers: A review, International Journal of Pharmacognosy & Chinese Medicine, 2018, 2, 2576–4772.

Deepa Mandlik, Satish Mandlik, Amarjitsing Rajput

Chapter 8
Herbal bioenhancers and improvement of the bioavailability of drugs

Abstract: Recent research has suggested that several natural substances could increase the bioavailability of therapeutic compounds and phytochemicals. It resulted in excellent effectiveness and decreased dosage regimen. These compounds are bioavailability enhancers and consist of several herbs such as cumin liquorice, aloe, genistein, naringenin, rutin, and *Withania somnifera*. They help increase the bioavailability of the compounds using different mechanisms. In addition, several nanocarrier-loaded bioenhancers also play an essential role in improving the bioavailability of therapeutic actives. Numerous phytoconstituents are being researched and tested for success and better drug delivery for increased bioavailability at the lowest effective dose. Future research could concentrate on marine herbal bioenhancers, obtained from seaweeds and plants, to delve deeper into their new principles and mechanisms. Future novel carriers are in the pipeline, which could potentiate bioavailability for low bioavailable drugs. The herbal bioenhancer-loaded nanocarriers have proven synergistic effects for the improvement of the bioavailability of hydrophobic drugs.

8.1 Bioenhancer

8.1.1 Introduction

Bioenhancers, also called bioavailability enhancers, are substances capable of increasing the absorption and bioavailability of certain compounds. The increase in the effectiveness in the body of these compounds is observed when they were co-administered with bioenhancers [1, 2]. Bioenhancers could reduce the dose of the active compound and also the adverse effects associated with it. However, a combined dose of the drug with a bioenhancer is also cost-effective [3]. Hence, scientists are more interested in identifying novel bioenhancers from herbal origin for drugs suffering from low bioavailability or significant unwanted reactions. Recent research has suggested that several natural substances such as genistein, quercetin,

Deepa Mandlik, Satish Mandlik, Department of Pharmaceutics, Bharti Vidyapeeth Deemed University, Poona College of Pharmacy, Erandwane, Pune 411038, Maharashtra, India
Amarjitsing Rajput, Department of Pharmaceutics, Bharti Vidyapeeth Deemed University, Poona College of Pharmacy, Bharti Vidyapeeth Educational Complex, Erandwane, Pune 411038, Maharashtra, India, e-mail: amarjit.rajput@yahoo.com

https://doi.org/10.1515/9783110746808-008

curcumin, naringenin, sinomenine, piperine, glycyrrhizin, and nitrite glycoside were capable of increasing the bioavailability of therapeutic compounds and phytochemicals [4].

Bioavailability is the rate and extent to which any compound enters the blood circulation and is made available at the desired site of action [5]. The drugs administered by the oral route showed poor bioavailability due to incomplete absorption and first-pass metabolism. Maximum bioavailability of the drugs is achieved via the intravenous route of administration. Such drugs resulted in unwanted effects and drug resistance. Hence, there is a requirement of substances with no therapeutic potential, but when complexed with other compounds/molecules, increases their bioavailability. Most natural substances from natural origin can enhance the bioavailability, when co-administered with other compounds [6]. The process of enhancing the total availability of any chemical substance (drug substance or nutrient) in systemic circulation or biological fluid or systemic circulation is known as "biopotentiation or bioenhancement." The main component's secondary compounds capable of this increment of plasma concentration are known as "bioavailability enhancers or biopotentiers" [7].

For example, the bioavailability of curcumin was enhanced by 154% in rats, and 200% in humans, post co-administration of piperine [8]. The absorption of poorly soluble drugs was increased by developing suitable pharmaceutical formulations, and natural bioenhancers might be helpful for enhancing their bioavailability [1, 9].

8.1.2 Role of bioenhancer in bioavailability enhancement

Several herbal compounds (bioenhancers) are responsible for improving the bioavailability of therapeutic substances and phytopharmaceuticals. These bioenhancers increase the bioavailability through various metabolic enzymes and intestinal absorption [4]. The positive effect resulting in enhancing the bioavailability of therapeutic compounds may ultimately cause excellent effectiveness and decrease the dosage regimen [10]. The various bioenhancers may work in a similar or dissimilar manner. Nutritional bioenhancers function on the gastrointestinal tract to promote absorption. Antimicrobial enhancers principally work on drug metabolism:

1. Enhancement in the gastrointestinal blood supply and decrease in gastric acid secretion.
2. Suppression of emptying time (gastric), gastrointestinal transit, gastrointestinal transit, and intestinal motility.
3. Alteration of GIT epithelial cell penetration potential.
4. Cholagogous impact.
5. Bioenergetics and thermogenic characteristics.
6. Prevention of first-pass effect and inhibition of acids and enzymes responsible for drugs metabolism [11].

In addition to these, biopotentiators might be of clinical significance as they may act by inhibiting the drug degradation enzymes transporter that mediates transcellular transport [1, 9].

The bioenhancer role for drugs administered via various routes of administration for their enhance delivery is as shown in Table 8.1.

8.2 Oral bioavailability improvement by bioenhancer

The drug must be in the form of an aqueous solution at the absorption target site for it to be absorbed. Any compound's aqueous solubility and dissolution rate as well as their impact on oral bioavailability are critical features [22]. A major objective of the formulation development process is to enhance bioavailability for poorly water-soluble compounds. Water solubility is estimated to be poor in 70% of all new chemical entities (NCEs) and is not commercially viable due to their limited bioavailability [23]. To reach blood circulation, drug molecules must pass through the biological layer at the absorption site. As a result, the substance's gastrointestinal penetrability is also a key determinant in estimating the rate and amount of drug absorption [24]. For the Biopharmaceutics Classification System (BCS), solubility and penetration ability have been proposed to be used in bioavailability monitoring. BCS is a scientific classification system for medicinal substances, based on solubility and intestinal permeability, which have proved significant in determining the absorption profile of drug substances when combined with data on the rate of dissolution [25–27].

Based on the BCS classification system, drug substances are classified into four types as follows:
1. BCS class I compound (high solubility, high permeability)
2. BCS class II compounds (low solubility, high permeability)
3. BCS class III compounds (high solubility, low permeability)
4. BCS class IV compounds (low solubility, low permeability) [28]

It is critical to evaluate a compound's BCS class to decide the best strategy to use when increasing a drug's bioavailability. In humans, drug permeability may be directly related to the amount of pharmacological compounds absorbed through the digestive tract. It can also be assessed indirectly using in vitro or in vivo models to determine the mass transfer rate across the intestinal membranes. When a mass balance shows that the systemic bioavailability or extent of absorption in humans is 85% or more of an administered dosage (along with proof indicating drug stability in the gastrointestinal tract) or compared to a reference dose administered via intravenous route, the actives are considered highly permeable [29].

Table 8.1: Bioenhancer role for drugs administered via various routes of administration for their enhance delivery [12].

Name of the bioenhancer	Biological source	Route of administration	Mechanism of action	Studied compound	References
Chitosan–TBA	Deacetylated chitin from crustaceans and fungi	Buccal	Mucoadhesion; mucosal membrane modulation	PACAP: pituitary adenylate cyclase-activating peptide	[13, 14]
Cod-liver oil extract	Animal (cod fish)	Buccal	No mechanism specified	Ergotamine tartrate: ergopeptine alkaloid	[15]
Sodium glycodeoxycholate (bile salt)	Intestinal bacterial by-product	Buccal	No mechanism specified	Dideoxycytidine: nucleoside analog reverse transcriptase inhibitor (NRTI)	[16]
Curcumin (flavonoid)	Plant (turmeric: Curcuma longa)	Oral	Metabolism (UDP-glucuronyl transferase) inhibition	Mycophenolic acid: immunosuppressant	[17]
Cyclosporine A (immunosuppressant)	Fungi (Tolypocladium inflatum Gams)	Oral	Efflux transporter (P-gp) inhibition	Clopidogrel: platelet aggregation inhibitor	[18]
Emodin (anthraquinone derivative)	Plant (senna: Cassia angustifolia, Aloe vera (syn Aloe barbadensis), rhubarb: Rheum officinale)	Oral	Efflux transporter (P-gp) inhibition	Digoxin: digitalis glycoside	[19]
Lanthanum, cerium, gadolinium (lanthanides)	Natural elements	Pulmonary	Drug targeting	Insulin: peptide hormone	[20]
Chitosan (biopolymer)	Chemically modified: deacetylation of chitin animal (crustaceans), fungi	Pulmonary	Tight junction modulation	Octreotide: somatostatin analog	[21]

8.2.1 Techniques for improvement of bioavailability

There are various methods used for bioavailability improvement of orally adminis-
tered compounds such as
Absorption enhancers
Prodrugs
Dosage form and other pharmaceutical modes
P-glycoprotein inhibitors

8.2.1.1 Absorption enhancers

Several absorption enhancers, including fatty acids, bile salts, chelating compounds, salicylates, and polymers, can improve intestinal absorption [30, 31]. Chitosan, specifically, trimethylated chitosan, enhances medication absorption via the
paracellular pathway by redistributing cytoskeletal F-actin, resulting in tight junction opening. Surfactants such as bile, bile salts, and fatty acids help lipophilic
medicines dissolve in water and improve the fluidity of the apical and basolateral
membranes, making them more absorbable [32]. Several substances of herbal origin, comprising naringenin, quercetin, sinomenine, piperine, glycyrrhizin, genistein, and nitrile glycoside have shown potential to improve the bioavailability [2].

8.2.1.2 Prodrug

Prodrugs and more permeable analogs are made by chemically altering active components. It has been intensively investigated by various researchers as a critical technique for improving absorption and bioavailability. One of the most popular examples
of enhancing the lipophilicity of compounds to increase absorption of a polar drug
substance by prodrug approach is different ampicillin derivatives [33]. Because of its
hydrophilic nature, ampicillin absorbs only 30–40% of its dose via the gastrointestinal tract. Hence, prodrugs of ampicillin such as bacampicillin, talampicillin, and pivampicillin were synthesized using esterification of the carboxyl group of ampicillin.
As a result, prodrugs were more lipophilic than the parent molecule and had higher
bioavailability after oral administration than ampicillin [34].

8.2.1.3 Dosage form and other pharmaceutical modes

Aside from chemical alterations, one of the most common methods of increasing the
intestinal absorption of poorly absorbed compounds is to use permeability-improving
dosage forms. Different formulations, such as liposomes, were used to improve the

intestinal absorption of insoluble medicinal compounds [35] and emulsions [36]. In addition to this, particle diameter reduction strategies such as micronization, complexation, nanoparticulate carrier, complexation, and liquid crystalline system need to be studied to elicit absorption [37, 38].

8.2.1.4 P-glycoprotein inhibitors

Many studies have suggested that P-gp inhibitors, which block P-gp-mediated efflux, could be utilized to improve drug transport across epithelia and increase oral bioavailability. P-gp inhibitors may affect the absorption, distribution, metabolism, and elimination of P-gp substrates, when used to influence pharmacokinetics [39].

Verapamil's potential to reverse P-gp-mediated resistance to vincristine and vinblastine was demonstrated in early trials, paving the way for its clinical usage as a P-gp inhibitor [40]. Additionally, orally administered verapamil has been proven to improve doxorubicin distribution, extend elimination half-life, and boost peak plasma levels, following oral intake [41].

8.2.2 Mode of action of herbal bioenhancer

Herbal bioenhancers have many different mechanisms of action. Antimicrobial bioenhancers primarily work on medication metabolic activities, while nutritional bioenhancers promote absorption through acting on the gastrointestinal system. Several distinct modes of action for herbal bioenhancers have been proposed, including
a) Reduction of hydrochloric acid secretion and improvement of gastrointestinal blood supply [42].
b) Inhibition of gastrointestinal transit, gastric emptying time, and intestinal movement [43, 44].
c) Cholagogous impact [45]
d) Bioenergetics and thermogenic characteristics [45, 46]
e) Inhibition of drug-metabolizing enzymes and suppression of first-pass metabolism [46]
f) Stimulation of glutamyl transpeptidase (GGT) activity, which increases amino acid absorption [47]. The mechanism of natural bioenhancer in bioavailability enhancement is as shown in Figure 8.1.

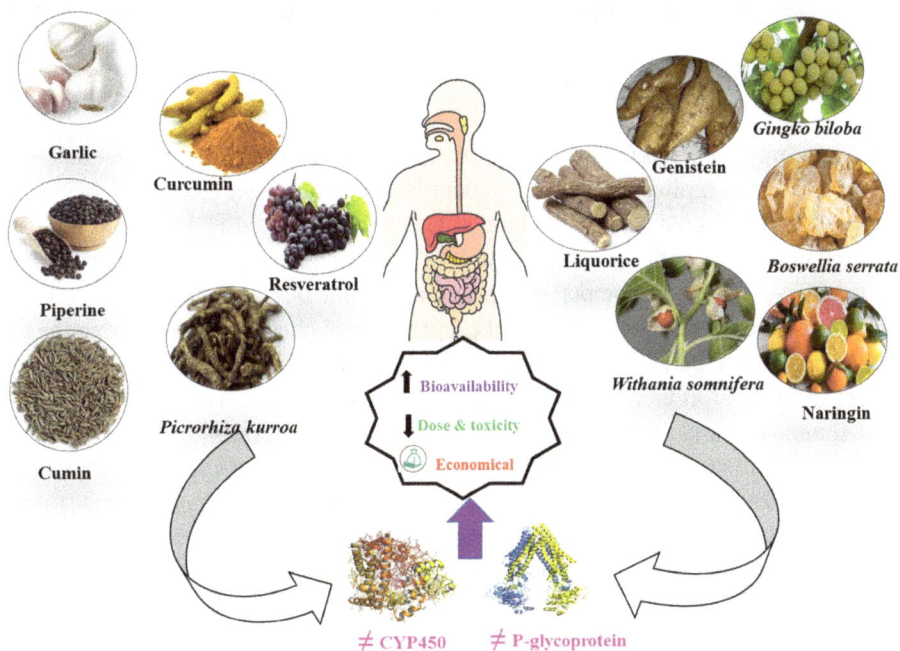

Figure 8.1: Mechanism of natural bioenhancer in bioavailability enhancement.

8.2.3 Need for bioavailability enhancers

The major limiting characteristics for compounds passing through the cellular membrane and being absorbed systemically after oral or topical treatment are lipid solubility and molecular size.

Despite having excellent biological activity in vitro, many plant extracts and phytoconstituents have inadequate lipid solubility, inappropriate molecular size, or both, resulting in poor absorption and bioavailability. When certain elements of a plant are isolated, a reduction of biological activity is always observed. Occasionally, few components are isolated from plant extract are degraded in gastric fluid upon oral consumption. They decrease the dose, decrease the drug duration of therapy, and issues associated with drug resistance. As a result of dosage economy, they reduce drug toxicity and side effects, making treatment more cost-effective [2].

8.2.4 Herbal bioenhancer role in oral bioavailability improvement

8.2.4.1 Cumin (*Cuminum cyminum*)

The main ingredients of *C. cyminum* oil are *p*-mentha-1,4-dien-7-al, cuminaldehyde, γ-terpinene, and β-pinene [48]. *C. cyminum* has several activities such as hypolipidemic [49], estrogenic [50], antimicrobial [51], and antifungal [52], antinociceptive, anti-inflammatory, anticonvulsant, anticancer activity [53], antioxidant activity, and antitussive effect [54]. Its components, including volatile oils, luteolin, and other flavonoids, are thought to have biostimulant properties. Luteolin works by blocking the efflux transporter P-gp [55]. Qazi et al. studied the efficacy of a *C. cyminum* extract or bioactive fraction on increasing the systemic bioavailability of several medicines, essential minerals, and herbal formulations. To investigate the synergistic bioenhancing activity, a small quantity of an extract from *C. cyminum* is mixed with several pharmaceutically acceptable excipients and, sometimes, piperine. *C. cyminum* extract (10–30 mg/kg), while bioactive fraction doses range from 2 to 20 mg/kg. Therefore, the composition comprising *C. cyminum* extract increases the bioavailability by 25–335% [56].

8.2.4.2 Liquorice (glycyrrhizin)

Glycyrrhizin is a saponin glycoside produced by *Glycyrrhiza glabra's* roots and stolon. Because of its expectorant properties, it is commonly used in cough medicines. It is a laxative and diuretic that boosts the immune system and stimulates the adrenal gland. It increases the effectiveness of antimicrobials, such as nalidixic acids, rifampicin, and doxycycline, against Gram-positive bacteria. Glycyrrhizin may work by inhibiting the efflux transporter, P-gp, which is found in the intestinal part [57]. The ability of glycyrrhizin to improve absorption is dependent on its conversion to glycyrrhetic acid by the intestine bacterial enzyme, β-glucuronidase. This bioenhancer's effective concentration ranges from 0.05% to 50%, 0.1% to 10%, and 0.25% to 20% of the overall weight of the nutraceutical, antibacterial, and antifungal compounds, individually [58]. With the administration of glycyrrhizinate with aconitine orally, the peak plasma concentration, area under the curve, and the absolute bioavailability of aconitine increased. At the same time, the half-life and clearance did not change significantly. The pharmacokinetic parameter of aconitine at 0.2 mg/kg dose did not significantly alter the administration of glycyrrhizinate via a hepatic vein and tail vein. The intestinal suppression of efflux transporter, P-gp, may be responsible for the increased absorption [57]. Khanuja et al. have revealed the usefulness of glycoside "glycyrrhizin" in improving the efficacy and bioavailability of several antibiotics, anti-infectives, and anticancer medicines.

Glycyrrhizin enhances the antibiotic and other agent absorption through biological membranes, increasing blood levels, and thus, bioavailability. It boosts medicines' effectiveness, including ampicillin, rifampicin, nalidixic acid, and tetra-cycline, against Gram-negative bacteria such as *E. coli*. It also promotes tetracy-cline, nalidixic acid, and rifampicin bioactivity against gram-positive bacteria such as *Mycobacterium smegmatis* and *Bacillus subtilis*. Glycyrrhizin, similarly, makes antifungal medications such as clotrimazole more effective against *Candida albicans*. Taxol (paclitaxel) was also researched to see whether it improves its anticancer effi-cacy by inhibiting cell division. Taxol works by preventing MCF-74 cancer cells from proliferating and multiplying. Glycyrrhizin increases the activity of taxol by a factor of five. With the administration of taxol (0.01 g/mL) with glycyrrhizin (1 g/mL), the action is elevated than when taxol (0.05 g/mL) is administered alone. Because of its usefulness and harmless nature at minimal concentrations, glycyrrhizin has a lot of po-tential as a bioavailability enhancer for various drugs such as anticancer, antifungal, antibacterial, and nutraceuticals. It also supports minimizing the dose-dependent side effects of chemotherapeutic medicines and developing antimicrobial resistance [59].

8.2.4.3 Aloe

Aloe vera gel and whole-leaf extracts have improved vitamin C and vitamin E absorp-tion in the mouth [60]. The polysaccharide ingredients of these products are assumed to be responsible for their bioenhancing properties. Natural polysaccharides, such as chitosan, operate as penetration enhancers by opening the strong connections be-tween neighboring epithelial cells for a brief period. As a result, they can improve the intestinal absorption of medications given at the same time. The use of *Aloe vera* gel and leaf extract improved the transportation of the peptide drug insulin through monolayers of Caco-2 cell substantially [61]. There is currently limited evidence on the effects of *Aloe vera* extracts on drug absorption.

8.2.4.4 Genistein

Genistein is an isoflavone flavonoid, extracted from *Pueraria lobata* and *Glycine max*, having phytoestrogen anticancer and anti-inflammatory action [62]. It shows its action mostly by the inhibition of P-gp [63] and MRP21 [64], and BCRP2 [65] ef-flux roles. The outcome of orally treated genistein on the pharmacokinetics parame-ters of paclitaxel (anticancer drug) given via oral and intravenous (i.v.) routes was explored by [66]. Paclitaxel was administered orally once (30 mg/kg) or IV (3 mg/kg) alone or 30 min after the oral treatment of genistein (10 mg/kg) in rats. In oral administration of paclitaxel, the inclusion of genistein (10 mg/mL) led to a rise in the area under the curve (AUC) and peak plasma concentration (C_{max}), with an in-crease in the absolute bioavailability and relative bioavailability.

The addition of genistein (10 mg/kg) to intravenous paclitaxel resulted in a significant rise in AUC of 40.5%. The combined suppression of CYP3A and P-gp by genistein may explain the enhanced bioavailability of paclitaxel [66]. As epigallocatechin-3-gallate (EGCG) is combined with genistein in the mice colon cancer cells, the concentration of cytoplasmic EGCG rises, compared to cells treated with EGCG alone. In the treatment of mice with genistein (200 mg/kg) and with EGCG (75 mg/kg), there was an increase in AUC and plasma half-life ($t_{1/2}$). This research shows that genistein can improve the bioavailability of EGCG. In male mice, on the other hand, a combination of EGCG (0.01%), in drinking water, and genistein (0.2%), in the diet increased, intestinal carcinogenesis. As a result, careful administration of this combination is required [67]. Genistein is a dietary supplement that is found in many people's everyday diets. Genistein can be used to improve the bioavailability of anticancer medicines by inhibiting CYP3A and P-gp.

8.2.4.5 Naringin

Grapefruit, apples, onions, and tea contain the flavonoid, glycoside naringin [68]. It has antioxidant, antiallergic, antiulcer, and anticancer properties, and blood cholesterol-lowering properties [69]. Naringin blocks CYP 3A4 [70] and P-gp [71], leading to a decrease in the dose of the drug and a rise in plasma drug concentration of several drugs, such as verapamil, diltiazem, and paclitaxel. Choi et al. studied the activity of naringin on the oral bioavailability of paclitaxel in rats. The rats were orally treated with either paclitaxel (40 mg/kg) alone or in combination with naringin (3 mg/kg). This resulted in higher paclitaxel peak plasma concentrations, C_{max}, and AUC. After coadministration with naringin, the absolute bioavailability of paclitaxel was significantly higher than in the control group. In comparison to the control group, relative bioavailability was also improved. The increased bioavailability of paclitaxel may be due to the inhibition of CYP-450 and P-gp in the gastric mucosa [72]. Oral pretreatment of rats with naringin (5 mg/kg) and with diltiazem (15 mg/kg) determined the absorptive behavior of diltiazem. The AUC and C_{max} of diltiazem were dramatically enhanced two times, although the T_{max} and plasma $t_{1/2}$ of diltiazem did not alter appreciably. C_{max} and AUC were also found to be increased. Furthermore, diltiazem's relative bioavailability and absolute bioavailability were significantly improved. As a result, it may be stated that naringin reduces the dose and side effects of diltiazem by the inhibition of intestinal metabolism of the drug and P-gp [73].

8.2.4.6 Rutin

Rutin, commonly known as quercetin-3-rutinosoid or sophori, is a flavonol glycoside consisting of the flavanol, quercetin, and the disaccharide, rutinose. It can be found in various plants, particularly citrus fruits, and is frequently sold as herbal supplement. Quercetin, the active component of rutin, has been shown to have multiple pharmacological properties, including anti-inflammatory, antidiabetic, cardiovascular effects, and antioxidant. It has been used to treat capillary fragility and various chronic conditions [74–76]. Because of the flavonoids' size and polarity, some medicinal chemistry laws predict that they will have trouble crossing biomembranes. Flavonoids are often poorly absorbed from the small intestine. Their high molecular weight additionally limits rutin's intestinal permeability. According to studies, rutin's trans-epithelial transport is low, relying primarily on passive diffusion [77, 78]. Methods that promote solubility and intestinal permeability are encouraging methodologies to increase bioavailability for this family of drugs. Human investigations have revealed that rutin's bioavailability on oral administration is very low (only 20–30%) [79]. Numerous researches have been published that use various excipients and strategies to improve bioavailability, such as surfactants, cyclodextrin complexation, chitosan, nanoemulsions, and nano-phytosomes [80–82]. The metabolism of rutin in the gut is one factor that contributes to its poor oral absorption. The bacteria hydrolyze rutin in the lower stomach to quercetin and isoquercetin [83]. Heim et al. stated that the enzyme uridine diphosphate-glucoronosyl transferase (UGT) conjugates a portion of the free quercetin with glucuronide, resulting in a bulky and inert adduct that is unabsorbable, lowering quercetin absorption [84]. Since piperine has been found to inhibit UGT, this bioenhancer has been shown to boost the neurocognitive potential of quercetin in mice [85], implying that piperine increases quercetin bioavailability. Piperine has also been proven to boost quercetin's anti-inflammatory and antioxidant activities, and quercetin has been demonstrated to have higher neuroprotective benefits, when combined with piperine [86]. Since rutin and similar chemicals are not substrates for efflux transporters, piperine is unlikely to have increased bioavailability and, thus, the effectiveness of quercetin or rutin by modifying P-gp action [79].

8.2.4.7 Boswellia serrata

Boswellia serrata is commonly known as salai guggul, belonging to the family Burseraceae, predominantly found in Africa and Asia. The dried exudate from this tree's bark is an oleo-gum-resin with a broad variety of pharmacological actions that have been reported to help with bronchitis, cough, asthma, inflammatory diseases, and different digestive disorders [87–89]. Terpenoids make up most gum-resin, with boswellic acid, a pentacyclic triterpene acid, the most physiologically active phytoconstituent [90]. The first to be isolated were α- and β-boswellic acids, pursued by some of its derivatives, such as 3-acetyl-α-boswellic acid, 3-acetyl-11-keto-β-boswellic acid, 11-keto-β-boswellic acid, and 3-acetyl-β-boswellic acid [88]. Furthermore, due to the first-pass metabolic effect, the keto derivatives of boswellic acid promote their low oral bioavailability [91, 92]. Numerous efforts have been made to develop enhanced absorption drug delivery systems, including solid lipid nanoparticles (SLN), loading boswellic acids into liposomes, nano-micelles, niosomes, and phytosomes [93]. As piperine is a CYP3A4 inhibitor, co-administration with it could boost keto-boswellic acid bioavailability. When piperine and boswellia extracts were given together, the bioavailability of the extracts was dramatically increased [94].

8.2.4.8 Gingko biloba

Ginkgo biloba is a huge tree growing all over the world. The leaf extract has been utilized in conventional medication to treat blood ailments, tinnitus, cognitive problems, asthma, and vertigo for ages. It is one of the widely used phytomedicines worldwide [95, 96]. The components of terpenoids, flavonoids, and proanthocyanidins are responsible for this HMP's pharmacological activities [97]. Bilobalide, ginkgolides A, B, C, kaempferol, isorhamnetin, rutin hydrate, and quercetin are the active ingredients [98]. As a result, many techniques for improving the bioavailability of *Ginkgo* extract have been examined, including the usage of solid dispersions and phospholipid complexes [99]. Hydrolysis in the intestinal microflora is the primary pathway for flavonoid metabolism in humans, followed by phase II conjugation reactions, with UGTs playing a pivotal role [100].

They are built on our understanding of piperine's bioenhancing activity and the fact that UGT inhibitor, a concomitant dose of piperine with *Ginkgo* flavanols, is expected to improve oral absorption. Furthermore, investigations have revealed that the flavonols, kaempferol, quercetin, and isorhamnetin, found in the plant are P-gp substrates. This efflux pump supports flavanol bioavailability and is limited [101]. Because piperine inhibits the function of P-gp, it can also improve the bioavailability of *Ginkgo* flavanols by this mechanism. According to research, piperine significantly increased the *Ginkgo* extracts bioavailability, when supplied simultaneously [94]. *Ginkgo* terpene lactones (bilobalide and ginkgolides) have distinct physicochemical features and metabolic fates than flavanols. The intermediate membrane permeability of terpene lactones determines their intestinal absorption [102, 103]. Thus, piperine's process of raising the blood supply in the intestinal channel due to piperine's local vasodilatory impact could plausibly be enhancing the oral absorption of these terpene lactones [104].

8.2.4.9 Withania somnifera

The herb is commonly known as Indian ginseng or ashwagandha, belonging to the family Solanaceae and has long been used in Indian traditional medicine. Its roots are the principal portions of the plant that are used medically. The herb is used to treat several pharmacological activities such as neurological disorders, anxiety, cognitive, hyperlipidemia, Parkinson's disease, and inflammation, and has been supported by preclinical and clinical studies [105]. Immunomodulatory, aphrodisiac, and sedative effects are among its many pharmacological qualities [106]. *Withania somnifera* root extracts are primarily composed of steroidal lactones known as withanolides, which are thought to be responsible for the herbal product's therapeutic qualities [107]. Withanolide A and D and Withaferin A and its derivatives are the most important withanolides. Alkaloids are other physiologically active chemical compounds; withanine as the primary constituent and lesser alkaloids such as somniferine, cuscohygrine, isopelletierine, anaferine, and anahydrine as minor constituents [108]. Because steroidal lactones, such as withanolides, have limited water solubility, they have bioavailability issues.

Several techniques have been established to enhance this high melting phenylephrine oxazolidine's bioavailability and, thus, its pharmacological action. These include the creation of polymeric nanoparticles [109], phytosomes [110, 111], and enteric-coated dosage forms of herbal extracts that preserve the content from hydrolysis in the stomach's acidic environment, hence improving absorption [112]. When ginger is added to the extract of *Withania somnifera*, it enhances absorption by 64% [1]. A well-known activity of ginger mediates this impact as a bioavailability enhancer that works by regulating the function of the intestine and increasing GIT medication absorption [48].

8.2.4.10 Picrorhiza kurroa

Picrorhiza kurroa is commonly called Kutki, belonging to the family *scrophulariaceae*. This plant's roots and rhizomes are used in traditional medicine to treat liver and upper respiratory tract diseases, diarrhoea, dyspepsia, and dysentery, among other ailments [113]. Other pharmacological effects of the HMP have been discovered, including antioxidant and anti-inflammatory properties [114, 115]. The main bioactive ingredients responsible for its therapeutic actions are irioid glycosides [116]. Due to the hydrophilic nature of picrosides, they have low intestinal penetrability. Because of its low absorption, it is suggested to be given in larger doses, resulting in higher costs of therapy [117].

A review of the literature reveals that only a few techniques have been devised to increase the bioavailability of these biomolecules, one of which is a plant extract nano-encapsulation formulation [118]. Another study reported the formation of phytosomes of picroside II with phospholipid, which dramatically enhances picroside II bioavailability [119]. The bioavailability of the *Picrorhiza kurroa* product was raised by as much as 56% when the ginger extract was added to the formulation, and this was improved even more, to 87%, when administered with piperine [48]. Both are bioenhancers, although they work in distinct ways. The various herbs explored in bioavailability enhancement are as shown in Figure 8.2.

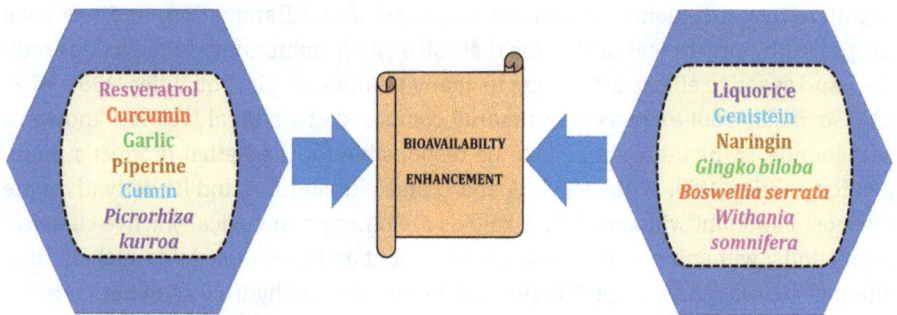

Figure 8.2: Representation of compounds explored in bioavailability enhancement.

8.3 Nanocarrier-loaded bioenhancer in bioavailability improvement

8.3.1 Nanoparticles

Bioenhancers, along with nanoparticles, were found to be an effective and synergistic combination for bioavailability improvement of drugs. The piperine enhanced the bioavailability of nisoldipine 4.9-fold in poly(lactic-co-glycolic acid) polymeric nanoparticles [120]. Rathee P et al. evaluated the pharmacokinetic profiles of nisoldipine nanoparticles and conventional nisoldipine formulation. Nanoparticles showed a significantly larger magnitude of absorption (255.545.92 g/mL/h) than conventional formulation (52.093.76 g/mL/h).

Kumar et al. developed eudragit-coated SLNs (coated SLN) and lipoid (glycerol monostearate and soy lecithin) containing isradipine, with rutin as a bioenhancing agent. When compared to a coated formulation of isradipine without rutin and conventional drug suspension, the pharmacokinetic study revealed a 3.2- to 4.7-fold increase in oral bioavailability of coated SLN of isradipine with rutin. In vivo investigations demonstrated a considerable increase in oral absorption, showing a high degree of entrapment (97.85%) that could boost biological activity against hypertension. As a result, isradipine-loaded nanoparticles against hypertension have been successfully developed, resulting in a reduction of dose with improved bioavailability [121].

The bioavailability of curcumin is enhanced with SLNs. The studies carried out by Gupta et al. revealed that the relative bioavailability of free curcumin was increased 70 times in the presence of SLNs [122].

8.3.2 Liposomes

Liposomes are biocompatible, spherical, lipid bilayers vesicles assembled with an inner core comprising an aqueous phase. Liposomes showed significant advantages, such as efficacy, affordability, accessibility, and economical tuberculosis treatment [123]. Various herbal bioenhancers, along with liposomes, have demonstrated the improvement of bioavailability for drugs. The reported bioenhancers-loaded liposomes, namely, piperine [124], genistein, resveratrol [125], quercetin [126], thymoquinone [127] have shown significant improvement in bioavailability.

Takahashi et al. developed and evaluated the curcumin-encapsulated liposome using lecithin. The animal studies using Sprague–Dawley (SD) rats confirmed the high bioavailability of curcumin when given orally. A curcumin liposome also showed faster absorption rate *vis a vis* other forms [128]. Hong et al. demonstrated improved bioavailability of curcumin- and catechin-loaded liposomes, which were prepared using microfluidic technique [129].

8.3.3 Phytosomes

Phytosomes are the advanced phytolipidic carriers embedded with plant extracts and phytoconstituents. The phytosomes significantly increase the absorption rate compared to the plain phytoconstituents, thus enhancing bioavailability [130, 131]. The bioavailability of many poorly absorbed polyphenols, namely terpenoids, flavonoids, and phenolics, were found to be enhanced with their phytosomal carriers [131]. The Meriva, a trademark product of Thorne contains curcumin phytosomes which have proven a 29-fold increase in absorption than curcumin alone.

8.3.4 Transfersomes

Transfersomes are ultra-deformable elastic vesicular carriers composed of phospholipids and an edge activator. Edge activators, namely sodium deoxycholate, Tweens, and Spans, intercalate with phospholipid bilayer and are responsible for the vesicle's elasticity [132, 133]. Transfersomes can transport both low and high molecular weight drug molecules. With more than 50% transport efficiency, transfersomes transfer hydrophilic and lipophilic molecules through the skin [134]. Sana et al. developed and characterized curcumin-loaded transfersomes. The curcumin transfersomes were found to be therapeutically more efficacious for rheumatoid arthritis treatment when studied on in vivo arthritic animal models [135].

8.3.5 Microspheres

Microspheres are potential drug carriers for enhancing the bioavailability of poorly bioavailable drugs. The combined effect of penetration enhancers with polymeric microspheres could enhance bioavailability synergistically [136].

The drug targeting, as well as bioavailability of acyclovir, was enhanced with piperine containing floating microspheres. The added piperine showed enhanced bioavailability of acyclovir in the microspheres [137].

The 4-fold increase in area under the curve was observed when Boddupalli et al. evaluated the omeprazole microspheres, along with piperine, compared to omeprazole alone. The pharmacokinetic parameters and bioavailability were significantly enhanced with piperine in microspheres [138].

The bioavailability of curcumin microspheres was improved with the addition of ascorbic acid, when studied on Wistar rats. The colon-targeting and bioavailability of curcumin were observed due to the antioxidant effect of ascorbic acid [139].

The bioenhancers (hydroalcoholic extracts of *Carum carvi* and *Ocimum sanctum*) were used to formulate antitubercular microspheres. These bioenhancers potentiated isoniazid's bioavailability and rifampicin-loaded microspheres [140].

8.3.6 Self-nanoemulsifying drug delivery system (SNEDDS)

Self-nanoemulsifying drug delivery systems (SNEDDS) are homogenous mixtures consisting of oil, surfactant, drug, and co-emulsifier, which spontaneously form aqueous nanoemulsion (< 200 nm) upon dilution with water and with stirring [141]. The SNEDDS are commonly used to improve the oral bioavailability of hydrophobic drugs by various mechanisms.

Kazi et al. developed SNEDDS containing piperine, curcumin, and black seed oil. The SNEDDS, containing black seed oil, performed well in self-emulsification, droplet size, and bioavailability [142].

Kanwal et al. effectively incorporated permeation enhancers and curcumin in the development of SNEDDS. Curcumin anticancer activity and cellular uptake in colon cancer cells were increased by including an absorption enhancer in the SNEDDS. Similarly, when drugs were administered orally in developed SNEDDS, their bioavailability increased significantly [143].

In another case study, Usmani et al. developed SNEDDS of doxorubicin and *Nigella sativa* oil for hepatic carcinoma. An optimized formulation showed improved efficacy in HepG2 cells, by cytotoxicity and IC50 value of 2.5 µg/mL. In human hepatocellular cancer, the delivery of doxorubicin and *Nigella sativa* oil in SNEDDS proved to be an effective carrier for oral delivery [144].

The list of nanocarriers loaded bioenhancers in bioavailability improvement is mentioned in Table 8.2.

Table 8.2: List of nanocarriers loaded bioenhancer in bioavailability improvement.

Nanocarriers	Drug/herbal bioenhancers	Key observations	References
Solid lipid nanoparticles	Isradipine/rutin	Enhanced oral bioavailability (4.7 times) with rutin	[145]
Nanostructured lipid carriers	Curcumin/ hydrolyzed ginsenoside	Higher drug release flux with bioenhancer	[146]
Chitosan nanoparticles	Piperine	High cytotoxicity on human brain cancer cell line (Hs683)	[147]
Guar gum nanoparticles	Amphotericin B/ piperine	Nanoparticles revealed good efficacy in golden hamster *L. donovani* model, high bioavailability	[148]
Poly(lactic-co-glycolic acid) nanoparticles	Nislodipine/ piperine	Pharmacokinetic studies showed a 4.9-fold increase in oral bioavailability and a >28.376 ± 1.32% reduction in systemic blood pressure by using nanoparticles as compared to control (nisoldipine suspension)	[120]
SMEDDS	Curcumin/ curcumin oleoresins	29 times higher bioavailability was obtained with bioenhancers	[149]
Liposomes	Curcumin/ piperine	Enhanced bioavailability	[150]
Liposomes	Silymarin	Improved bioavailability	[151]
Niosomes	Resveratrol	Improved bioavailability of the drug	[152]
Niosomes	Methotrexate/ piperine and curcumin	Improved bioavailability	[153]
Phytosomes	Domperidone/ piperine	Increased bioavailability of domperidone	[154]
Ethosomes	Curcumin	Improved transdermal delivery	[155]
Transethosomes	Resveratrol	Enhanced permeation	[156]
Nano-transferosomes	Naringenin/ pravastatin	Reduced side effects and improved bioavailability	[157]

8.4 Conclusion and future perspectives

In pharmacotherapy, phytomedicine has played a significant part. However, numerous medicinal plant extracts and phytoconstituents obtained from them have been reported to have limited bioavailability, leading to diminished in vivo activity. Enhancing the bioavailability of these herbal medicinal products (HMP) will go a long way toward improving the products' efficacies. Although much has been researched to improve the oral bioavailability of inadequately absorbed traditional drugs, this review found that HMPs have received significantly less attention in this area. The purpose of this review was to highlight the several ways of increasing HMP bioavailability, with a focus on the use of herbs-derived bioenhancers. It was determined that identifying the bioactive phytoconstituents of HMPs was necessary so that information on their metabolic fate, physicochemical properties, and factors affecting their intestinal absorption could be observed and used as a guide for improving the bioavailability of other HMPs with low bioavailability of active constituent. Before choosing a suitable method or bioenhancer, it is critical to understand why bioactive moieties have poor bioavailability. The advantages of incorporating appropriate bioenhancers into drug products with low bioavailability cannot be overstated since this is linked to a decrease in dose, medication toxicity, and overall therapy cost.

As a result, more significant research efforts on combining HMPs with appropriate bioenhancers are needed, as this can improve efficacy while also lowering medication treatment costs. Numerous phytoconstituents are being researched and tested for success and better drug delivery, with increased bioavailability at the lowest effective dose. Future research could concentrate on marine herbal bioenhancers obtained from seaweeds and plants, with the goal of delving deeper into their new principles and mechanisms. Combinatory bioenhancers and their bioenhancing potential, mechanism, and toxicity can be investigated soon for significant improvement in bioavailability. Plants, such as *Artemisia anuua*, *Rauwolfia serpentine*, *Taxus baccata*, *Catharanths roseus*, and *Bacopa monnierie*, can be studied for their bioenhancing properties in the upcoming years. Future novel carriers are in the pipeline, which could potentiate bioavailability for low bioavailable drugs. Herbal bioenhancer-loaded nanocarriers have proven synergistic effects for the improvement of the bioavailability of hydrophobic drugs.

Conflict of interest

Authors have no conflict of interest.

List of abbreviations

BCS	Biopharmaceutics Classification Systems
SNEDDS	Self-nanoemulsifying drug delivery system
nm	Nanometer
HMP	Herbal medicinal products
i.v.	Intravenous
P-gp	P-glycoprotein
EGCG	Epigallocatechin-3-gallate
AUC	Area under the curve
C_{max}	Peak plasma concentration
$t_{1/2}$	Plasma half-life
UGT	Uridine diphosphate-glucoronosyl transferase

References

[1]   Dudhatra GB, Mody SK, Awale MM, Patel HB, Modi CM, Kumar A, Kamani DR, Chauhan BN, A comprehensive review on pharmacotherapeutics of herbal bioenhancers, The Scientific World Journal, 2012, 2012.

[2]   Kesarwani K, Gupta R, Bioavailability enhancers of herbal origin: An overview, Asian Pacific Journal of Tropical Biomedicine, 2013, 3, 253–266.

[3]   Atal N, Bedi K, Bioenhancers: Revolutionary concept to market, Journal of Ayurveda and integrative medicine, 2010, 1, 96.

[4]   Cho H-J, Yoon I-S, Pharmacokinetic interactions of herbs with cytochrome p450 and p-glycoprotein, Evidence-Based Complementary and Alternative Medicine, 2015, 2015.

[5]   Brahmankar D, Jaiswal SB, Biopharmaceutics and pharmacokinetics-a treatise Vallabh Prakashan, New Delhi, 1995.

[6]   Gopal V, Prakash Yoganandam G, Velvizhi Thilagam T, Bio-enhancer: A pharmacognostic perspective, European Journal of Molecular Biology and Biochemistry, 2016, 3, 33–38.

[7]   Jhanwar B, Gupta S, Biopotentiation using herbs: Novel technique for poor bioavailable drugs, International Journal of Pharmtech Research, 2014, 6, 443–454.

[8]   Shoba G, Joy D, Joseph T, Majeed M, Rajendran R, Srinivas PSSR, Influence of piperine on the pharmacokinetics of curcumin in animals and human volunteers, Planta medica, 1998, 64, 353–356.

[9]   Kesarwani K, Gupta R, Bioavailability enhancers of herbal origin: An overview, Asian Pacific Journal of Tropical Biomedicine, 2013, 3, 253–266.

[10] Ajazuddin AA, Qureshi A, Kumari L, Vaishnav P, Sharma M, Saraf S, Saraf S, Role of herbal bioactives as a potential bioavailability enhancer for active pharmaceutical ingredients, Fitoterapia, 2014, 97, 1–14.

[11] VTatiraju D, Bagade VB, JKarambelkar P, MJadhav V, Kadam V, Natural bioenhancers. An overview, Journal of Pharmacognosy and Phytochemistry, 2013, 2.

[12] Peterson B, Weyers M, Steenekamp JH, Steyn JD, Gouws C, Hamman JH. Drug bioavailability enhancing agents of natural origin (Bioenhancers) that modulate drug membrane permeation and pre-systemic metabolism, Pharmaceutics, 2019, 11, 33.

[13] Langoth N, Bernkop-Schnürch A, Kurka P, In vitro evaluation of various buccal permeation enhancing systems for PACAP (pituitary adenylate cyclase-activating polypeptide), Pharmaceutical Research, 2005, 22, 2045–2050.

[14] Langoth N, Kahlbacher H, Schöffmann G, Schmerold I, Schuh M, Franz S, Kurka P, Bernkop-Schnürch A, Thiolated chitosans: Design and in vivo evaluation of a mucoadhesive buccal peptide drug delivery system, Pharmaceutical Research, 2006, 23, 573–579.

[15] Tsutsumi K, Obata Y, Takayama K, Loftsson T, Nagai T, Effect of cod-liver oil extract on the buccal permeation of ergotamine tartrate, Drug Development and Industrial Pharmacy, 1998, 24, 757–762.

[16] Xiang J, Fang X, Li X, Transbuccal delivery of 2', 3'-dideoxycytidine: In vitro permeation study and histological investigation, International Journal of Pharmaceutics, 2002, 231, 57–66.

[17] Basu NK, Kole L, Kubota S, Owens IS, Human UDP-glucuronosyltransferases show atypical metabolism of mycophenolic acid and inhibition by curcumin, Drug Metabolism and Disposition, 2004, 32, 768–773.

[18] Lee JH, Shin Y-J, Oh J-H, Lee Y-J, Pharmacokinetic interactions of clopidogrel with quercetin, telmisartan, and cyclosporine A in rats and dogs, Archives of Pharmacal Research, 2012, 35, 1831–1837.

[19] Li X, Hu J, Wang B, Sheng L, Liu Z, Yang S, Li Y, Inhibitory effects of herbal constituents on P-glycoprotein in vitro and in vivo: Herb–drug interactions mediated via P-gp, Toxicology and Applied Pharmacology, 2014, 275, 163–175.

[20] Shen Z-C, Cheng Y, Zhang Q, Wei S-L, Li R-C, Wang K, Lanthanides enhance pulmonary absorption of insulin, Biological Trace Element Research, 2000, 75, 215–225.

[21] Florea BI, Thanou M, Junginger HE, Borchard G, Enhancement of bronchial octreotide absorption by chitosan and N-trimethyl chitosan shows linear in vitro/in vivo correlation, Journal of Controlled Release, 2006, 110, 353–361.

[22] Kumar S, Dilbaghi N, Rani R, Bhanjana G, Umar A, Novel approaches for enhancement of drug bioavailability, Reviews in Advanced Sciences and Engineering, 2013, 2, 133–154.

[23] Kawabata Y, Wada K, Nakatani M, Yamada S, Onoue S, Formulation design for poorly water-soluble drugs based on biopharmaceutics classification system: Basic approaches and practical applications, International Journal of Pharmaceutics, 2011, 420, 1–10.

[24] Khadka P, Ro J, Kim H, Kim I, Kim JT, Kim H, Cho JM, Yun G, Lee J, Pharmaceutical particle technologies: An approach to improve drug solubility, dissolution and bioavailability, Asian Journal of Pharmaceutical Sciences, 2014, 9, 304–316.

[25] Roy K, Mao H-Q, Huang S-K, Leong KW, Oral gene delivery with chitosan–DNA nanoparticles generates immunologic protection in a murine model of peanut allergy, Nature Medicine, 1999, 5, 387–391.

[26] Yu LX, Amidon GL, Polli JE, Zhao H, Mehta MU, Conner DP, Shah VP, Lesko LJ, Chen M-L, Lee VH, Biopharmaceutics classification system: The scientific basis for biowaiver extensions, Pharmaceutical Research, 2002, 19, 921–925.

[27] Wagh MP, Patel JS, Biopharmaceutical classification system: Scientific basis for biowaiver extensions, International Journal of Pharmacy and Pharmaceutical Sciences, 2010, 2, 12–19.

[28] Oladimeji FA, Adegbola AJ, Onyeji CO, Appraisal of bioenhancers in improving oral bioavailability: Applications to herbal medicinal products, Journal of Pharmaceutical Research International, 2018, 1–23.

[29] USFDA. Waiver of In Vivo Bioavailability and Bioequivalence Studies for Immediate-Release Solid Oral Dosage Forms Based on a Biopharmaceutics Classification System Guidance for Industry, 2017.

[30] Aungst BJ, Blake JA, Hussain MA, An in vitro evaluation of metabolism and poor membrane permeation impeding intestinal absorption of leucine enkephalin, and methods to increase absorption, Journal of Pharmacology and Experimental Therapeutics, 1991, 259, 139–145.

[31] Lundin S, Artursson P, Absorption enhancers as an effective method in improving the intestinal absorption, International Journal of Pharmaceutics, 1990, 64, 181–186.

[32] Schipper NG, Olsson S, Hoogstraate JA, deBoer AG, Vårum KM, Artursson P, Chitosans as absorption enhancers for poorly absorbable drugs 2: Mechanism of absorption enhancement, Pharmaceutical Research, 1997, 14, 923–929.

[33] Buur A, Bundgared H, Falch E, Prodrugs of 5-fluorouracil. VII: Hydrolysis kinetics and physicochemical properties of N-ethoxy-and N-phenoxycarbonyloxymethyl derivatives of 5-fluorouracil, Acta Pharmaceutica Suecica, 1986, 23, 205–216.

[34] Kang+ MJ, Cho+ JY, Shim BH, Kim DK, Lee J, Bioavailability enhancing activities of natural compounds from medicinal plants, Journal of Medicinal Plants Research, 2009, 3, 1204–1211.

[35] He H, Lu Y, Qi J, Zhu Q, Chen Z, Wu W, Adapting liposomes for oral drug delivery, Acta pharmaceutica Sinica. B, 2019, 9, 36–48.

[36] Sato Y, Yokoyama S, Yamaki Y, Nishimura Y, Miyashita M, Maruyama S, Takekuma Y, Sugawara M, Enhancement of intestinal absorption of coenzyme Q10 using emulsions containing oleyl polyethylene acetic acids, European Journal of Pharmaceutical Sciences: Official Journal of the European Federation for Pharmaceutical Sciences, 2020, 142, 105144.

[37] Liversidge GG, Cundy KC, Particle size reduction for improvement of oral bioavailability of hydrophobic drugs: I, Absolute oral bioavailability of nanocrystalline danazol in beagle dogs, International journal of pharmaceutics, 1995, 125, 91–97.

[38] Veiga F, Fernandes C, Teixeira F, Oral bioavailability and hypoglycaemic activity of tolbutamide/cyclodextrin inclusion complexes, International Journal of Pharmaceutics, 2000, 202, 165–171.

[39] Varma MV, Ashokraj Y, Dey CS, Panchagnula R, P-glycoprotein inhibitors and their screening: A perspective from bioavailability enhancement, Pharmacological Research : The Official Journal of the Italian Pharmacological Society, 2003, 48, 347–359.

[40] Tsuruo T, Iida H, Tsukagoshi S, Sakurai Y, Overcoming of vincristine resistance in P388 leukemia in vivo and in vitro through enhanced cytotoxicity of vincristine and vinblastine by verapamil, Cancer Research, 1981, 41, 1967–1972.

[41] Kerr D, Graham J, Cummings J, Morrison J, Thompson G, Brodie M, Kaye S, The effect of verapamil on the pharmacokinetics of adriamycin, Cancer Chemotherapy and Pharmacology, 1986, 18, 239–242.

[42] Annamalai A, Manavalan R, Effect of "Trikatu" and its individual components and piperine on gastrointestinal tracts, Indian Drugs, 1990, 27, 595–604.

[43] Majeed M, Badmaev V, Rajendran R, Use of piperine to increase the bioavailability of nutritional compounds, in, Google Patents, 1996.

[44] Bajad S, Bedi K, Singla A, Johri R, Piperine inhibits gastric emptying and gastrointestinal transit in rats and mice, Planta Medica, 2001, 67, 176–179.

[45] Khajuria A, Thusu N, Zutshi U, Piperine modulates permeability characteristics of intestine by inducing alterations in membrane dynamics: Influence on brush border membrane fluidity, ultrastructure and enzyme kinetics, Phytomedicine, 2002, 9, 224–231.

[46] Atal C, Dubey RK, Singh J, Biochemical basis of enhanced drug bioavailability by piperine: Evidence that piperine is a potent inhibitor of drug metabolism, Journal of Pharmacology and Experimental Therapeutics, 1985, 232, 258–262.

[47] Johri RK, Thusu N, Khajuria A, Zutshi U, Piperine-mediated changes in the permeability of rat intestinal epithelial cells: The status of γ-glutamyl transpeptidase activity, uptake of amino acids and lipid peroxidation, Biochemical Pharmacology, 1992, 43, 1401–1407.

[48] Qazi G, Bedi K, Johri R, Tikoo M, Tikoo A, Sharma S, Abdullah S, Suri O, Gupta B, Suri K. Bioavailability enchancing activity of Zingiber officinale Linn and its extracts/fractions thereof, in, Google Patents, 2003.

[49] Dhandapani S, Subramanian VR, Rajagopal S, Namasivayam N, Hypolipidemic effect of cuminum cyminum L. on alloxan-induced diabetic rats, Pharmacological Research, 2002, 46, 251–255.

[50] Malini T, Vanithakumari G, Estrogenic activity of cuminum cyminum in rats, Indian Journal of Experimental Biology, 1987, 25, 442–444.

[51] Gachkar L, Yadegari D, Rezaei MB, Taghizadeh M, Astaneh SA, Rasooli I, Chemical and biological characteristics of Cuminum cyminum and Rosmarinus officinalis essential oils, Food Chemistry, 2007, 102, 898–904.

[52] Pai MB, Prashant G, Murlikrishna K, Shivakumar K, Chandu G, Antifungal efficacy of Punica granatum, Acacia nilotica, Cuminum cyminum and Foeniculum vulgare on Candida albicans: An in vitro study, Indian Journal of Dental Research, 2010, 21, 334.

[53] Gagandeep SD, Mendiz E, Rao AR, Kale RK, Chemopreventive effects of Cuminum cyminum in chemically induced forestomach and uterine cervix tumors in murine model systems, Nutrition and Cancer, 2003, 47, 171–180.

[54] El-Ghorab AH, Nauman M, Anjum FM, Hussain S, Nadeem M, A comparative study on chemical composition and antioxidant activity of ginger (Zingiber officinale) and cumin (Cuminum cyminum), Journal of Agricultural and Food Chemistry, 2010, 58, 8231–8237.

[55] Boumendjel A, Di Pietro A, Dumontet C, Barron D, Recent advances in the discovery of flavonoids and analogs with high-affinity binding to P-glycoprotein responsible for cancer cell multidrug resistance, Medicinal Research Reviews, 2002, 22, 512–529.

[56] Qazi G, Bedi K, Johri R, Tikoo M, Tikoo A, Sharma S et al. Bioavailability enhancing activity of Carum carvi extracts and fractions thereof, US2007002034, 2007.

[57] Chen L, Yang J, Davey A, Chen Y-X, Wang J-P, Liu X-Q, Effects of diammonium glycyrrhizinate on the pharmacokinetics of aconitine in rats and the potential mechanism, Xenobiotica, 2009, 39, 955–963.

[58] Imai T, Sakai M, Ohtake H, Azuma H, Otagiri M, Absorption-enhancing effect of glycyrrhizin induced in the presence of capric acid, International Journal of Pharmaceutics, 2005, 294, 11–21.

[59] Khanuja S, Arya J, Srivastava S, Shasany A, Kumar TS, Darokar M, Kumar S. Antibiotic pharmaceutical composition with lysergol as bio-enhancer and method of treatment, in, Google Patents, 2007.

[60] Vinson JA, Al Kharrat H, Andreoli L, Effect of aloe vera preparations on the human bioavailability of vitamins C and E, Phytomedicine, 2005, 12, 760–765.

[61] Sharma K, Mittal A, Chauhan N, Aloe vera as penetration enhancer, International Journal of Drug Development and Research, 2015, 7, 31–43.

[62] Kurzer MS, Xu X, Dietary phytoestrogens, Annual Review of Nutrition, 1997, 17, 353–381.

[63] Sparreboom A, Van Asperen J, Mayer U, Schinkel AH, Smit JW, Meijer DK, Borst P, Nooijen WJ, Beijnen JH, Van Tellingen O, Limited oral bioavailability and active epithelial excretion of paclitaxel (Taxol) caused by P-glycoprotein in the intestine, Proceedings of the National Academy of Sciences, 1997, 94, 2031–2035.

[64] Huisman MT, Chhatta AA, van Tellingen O, Beijnen JH, Schinkel AH, MRP2 (ABCC2) transports taxanes and confers paclitaxel resistance and both processes are stimulated by probenecid, International Journal of Cancer, 2005, 116, 824–829.

[65] Doyle LA, Ross DD, Multidrug resistance mediated by the breast cancer resistance protein BCRP (ABCG2), Oncogene, 2003, 22, 7340–7358.

[66] Li X, Choi J-S, Effect of genistein on the pharmacokinetics of paclitaxel administered orally or intravenously in rats, International Journal of Pharmaceutics, 2007, 337, 188–193.

[67] Lambert JD, Kwon S-J, Ju J, Bose M, Lee M-J, Hong J, Hao X, Yang CS, Effect of genistein on the bioavailability and intestinal cancer chemopreventive activity of (-)-epigallocatechin-3-gallate, Carcinogenesis, 2008, 29, 2019–2024.

[68] Dixon RA, Steele CL, Flavonoids and isoflavonoids—a gold mine for metabolic engineering, Trends in Plant Science, 1999, 4, 394–400.

[69] Nijveldt RJ, Van Nood E, Van Hoorn DE, Boelens PG, Van Norren K, Van Leeuwen PA, Flavonoids: A review of probable mechanisms of action and potential applications, The American Journal of Clinical Nutrition, 2001, 74, 418–425.

[70] Zhang H, Wong C, Coville P, Wanwimolruk S, Effect of the grapefruit flavonoid naringin on pharmacokinetics of quinine in rats, Drug Metabolism and Drug Interactions, 2000, 17, 351–364.

[71] Lim SC, Choi JS, Effects of naringin on the pharmacokinetics of intravenous paclitaxel in rats, Biopharmaceutics & Drug Disposition, 2006, 27, 443–447.

[72] Choi J-S, Shin S-C, Enhanced paclitaxel bioavailability after oral coadministration of paclitaxel prodrug with naringin to rats, International Journal of Pharmaceutics, 2005, 292, 149–156.

[73] Choi J-S, Han H-K, Enhanced oral exposure of diltiazem by the concomitant use of naringin in rats, International Journal of Pharmaceutics, 2005, 305, 122–128.

[74] Al-Dhabi NA, Arasu MV, Park CH, Park SU, An up-to-date review of rutin and its biological and pharmacological activities, EXCLI J, 2015, 14, 59.

[75] Kim HP, Son KH, Chang HW, Kang SS, Anti-inflammatory plant flavonoids and cellular action mechanisms, Journal of Pharmacological Sciences, 2004, 0411110005–0411110005.

[76] Pu F, Mishima K, Irie K, Motohashi K, Tanaka Y, Orito K, Egawa T, Kitamura Y, Egashira N, Iwasaki K, Neuroprotective effects of quercetin and rutin on spatial memory impairment in an 8-arm radial maze task and neuronal death induced by repeated cerebral ischemia in rats, Journal of Pharmacological Sciences, 2007, 0707310004–0707310004.

[77] Rastogi H, Jana S, Evaluation of physicochemical properties and intestinal permeability of six dietary polyphenols in human intestinal colon adenocarcinoma Caco-2 cells, European Journal of Drug Metabolism and Pharmacokinetics, 2016, 41, 33–43.

[78] Scalbert A, Williamson G, Dietary intake and bioavailability of polyphenols, The Journal of Nutrition, 2000, 130, 2073S–2085S.

[79] Hollman PC, Van Trijp JM, Buysman MN, van der Gaag MS, Mengelers MJ, De Vries JH, Katan MB, Relative bioavailability of the antioxidant flavonoid quercetin from various foods in man, FEBS letters, 1997, 418, 152–156.

[80] Hooresfand Z, Ghanbarzadeh S, Hamishehkar H, Preparation and characterization of rutin-loaded nanophytosomes, Pharmaceutical Sciences, 2015, 21, 145–151.

[81] Macedo AS, Quelhas S, Silva AM, Souto EB, Nanoemulsions for delivery of flavonoids: Formulation and in vitro release of rutin as model drug, Pharmaceutical Development and Technology, 2014, 19, 677–680.

[82] Miyake K, Arima H, Hirayama F, Yamamoto M, Horikawa T, Sumiyoshi H, Noda S, Uekama K, Improvement of solubility and oral bioavailability of rutin by complexation with 2-hydroxypropyl-β-cyclodextrin, Pharmaceutical Development and Technology, 2000, 5, 399–407.

[83] Scholz Interactions affecting the bioavailability of dietary polyphenols in vivo, International journal for vitamin and nutrition research, 2007, 77, 224–235.

[84] Heim KE, Tagliaferro AR, Bobilya DJ, Flavonoid antioxidants: Chemistry, metabolism and structure-activity relationships, The Journal of Nutritional Biochemistry, 2002, 13, 572–584.

[85] Rinwa P, Kumar A, Quercetin along with piperine prevents cognitive dysfunction, oxidative stress and neuro-inflammation associated with mouse model of chronic unpredictable stress, Archives of Pharmacal Research, 2017, 40, 1166–1175.

[86] Singh S, Jamwal S, Kumar P, Neuroprotective potential of Quercetin in combination with piperine against 1-methyl-4-phenyl-1, 2, 3, 6-tetrahydropyridine-induced neurotoxicity, Neural Regeneration Research, 2017, 12, 1137.

[87] Alam M, Khan H, Samiullah L, Siddique K, A review on phytochemical and pharmacological studies of kundur (Boswellia serrata Roxb ex Colebr.)-A Unani drug, Journal of Applied Pharmaceutical, 2012, 2, 148–156.

[88] Hussain H, Al-Harrasi A, Csuk R, Shamraiz U, Green IR, Ahmed I, Khan IA, Ali Z, Therapeutic potential of boswellic acids: A patent review (1990–2015), Expert opinion on therapeutic patents, 2017, 27, 81–90.

[89] Upaganlawar A, Ghule B, Pharmacological activities of Boswellia serrata Roxb.-mini review, Ethnobotanical Leaflets, 2009, 2009, 10.

[90] Al-Harrasi A, Ali L, Rehman NU, Hussain H, Hussain J, Al-Rawahi A, Langley GJ, Wells NJ, Abbas G, Nine triterpenes from Boswellia sacra Flückiger and their chemotaxonomic importance, Biochemical Systematics and Ecology, 2013, 113–116.

[91] Barthe L, Woodley J, Houin G, Gastrointestinal absorption of drugs: Methods and studies, Fundamental & clinical pharmacology, 1999, 13, 154–168.

[92] Reising K, Meins J, Bastian B, Eckert G, Mueller WE, Schubert-Zsilavecz M, Abdel-Tawab M, Determination of boswellic acids in brain and plasma by high-performance liquid chromatography/tandem mass spectrometry, Analytical Chemistry, 2005, 77, 6640–6645.

[93] Mehta M, Satija S, Nanda A, Garg M, Nanotechnologies for boswellic acids, American Journal of Drug Discovery and Development, 2014, 4, 1–11.

[94] Majeed M, Badmaev V, Rajendran R. Use of piperine as a bioavailability enhancer, in, Google Patents, 1998.

[95] Ernst E, The risk–benefit profile of commonly used herbal therapies: Ginkgo, St. John's Wort, Ginseng, Echinacea, Saw Palmetto, and Kava, Annals of Internal Medicine, 2002, 136, 42–53.

[96] Kennedy DO, Wightman EL, Herbal extracts and phytochemicals: Plant secondary metabolites and the enhancement of human brain function, Advances in Nutrition, 2011, 2, 32–50.

[97] Brondino N, De Silvestri A, Re S, Lanati N, Thiemann P, Verna A, Emanuele E, Politi P, A systematic review and meta-analysis of Ginkgo biloba in neuropsychiatric disorders: From ancient tradition to modern-day medicine, Evidence-Based Complementary and Alternative Medicine, 2013, 2013.

[98] Ding S, Dudley E, Plummer S, Tang J, Newton RP, Brenton AG, Quantitative determination of major active components in Ginkgo biloba dietary supplements by liquid chromatography/mass spectrometry, Rapid Communications in Mass Spectrometry: An International Journal Devoted to the Rapid Dissemination of Up-to-the-Minute Research in Mass Spectrometry, 2006, 20, 2753–2760.

[99] Chen Z-P, Sun J, Chen H-X, Xiao Y-Y, Liu D, Chen J, Cai H, Cai B-C, Comparative pharmacokinetics and bioavailability studies of quercetin, kaempferol and isorhamnetin after oral administration of Ginkgo biloba extracts, Ginkgo biloba extract phospholipid complexes and Ginkgo biloba extract solid dispersions in rats, Fitoterapia, 2010, 81, 1045–1052.

[100] Chen Z, Zheng S, Li L, Jiang H, Metabolism of flavonoids in human: A comprehensive review, Current Drug Metabolism, 2014, 15, 48–61.

[101] Wang Y, Cao J, Zeng S, Involvement of P-glycoprotein in regulating cellular levels of Ginkgo flavonols: Quercetin, kaempferol, and isorhamnetin, Journal of Pharmacy and Pharmacology, 2005, 57, 751–758.

[102] Gaudineau C, Beckerman R, Welbourn S, Auclair K, Inhibition of human P450 enzymes by multiple constituents of the Ginkgo biloba extract, Biochemical and Biophysical Research Communications, 2004, 318, 1072–1078.

[103] Li L, Zhao Y, Du F, Yang J, Xu F, Niu W, Ren Y, Li C, Intestinal absorption and presystemic elimination of various chemical constituents present in GBE50 extract, a standardized extract of Ginkgo biloba leaves, Current drug metabolism, 2012, 13, 494–509.

[104] Bhardwaj RK, Glaeser H, Becquemont L, Klotz U, Gupta SK, Fromm MF, Piperine, a major constituent of black pepper, inhibits human P-glycoprotein and CYP3A4, Journal of Pharmacology and Experimental Therapeutics, 2002, 302, 645–650.

[105] Gupta GL, Rana A, Withania somnifera (Ashwagandha): A review, Pharmacognosy Reviews, 2007, 1.

[106] Uddin Q, Samiulla L, Singh V, Jamil S, Phytochemical and pharmacological profile of Withania somnifera dunal: A review, J Appl Pharm Sci, 2012, 2, 170–175.

[107] Mishra L-C, Singh BB, Dagenais S, Scientific basis for the therapeutic use of Withania somnifera (ashwagandha): A review, Alternative medicine review, 2000, 5, 334–346.

[108] Narinderpal K, Junaid N, Raman B, A review on pharmacological profile of Withania somnifera (Ashwagandha), Research and reviews: Journal of botanical sciences, 2013, 2, 6–14.

[109] Haripriya S, Vadivel E, Ilamurugu K, Venkatachalam R, Preparation of Withaferin-A loaded PLGA nanoparticles by modified emulsion diffusion evaporation technique, International Journal of Nanoparticles, 2010, 4, 163–168.

[110] Ahmad H, Arya A, Agrawal S, Samuel SS, Singh SK, Valicherla GR, Sangwan N, Mitra K, Gayen JR, Paliwal S, Phospholipid complexation of NMITLI118RT+: Way to a prudent therapeutic approach for beneficial outcomes in ischemic stroke in rats, Drug delivery, 2016, 23, 3606–3618.

[111] Keerthi B, Pingali PS, Srinivas P, Formulation and evaluation of capsules of ashwagandha phytosomes, International Journal of Pharmaceutical Sciences Review and Research, 2014, 29, 138–142.

[112] Antony B. Process to enhance the bioactivity of Ashwagandha extracts, in, Google Patents, 2018.

[113] Bhattacharjee S, Bhattacharya S, Jana S, Baghel D, A review on medicinally important species of Picrorhiza, International Journal of Pharma and Bio Sciences, 2013, 2, 1–16.

[114] Chander R, Kapoor NK, Dhawan B, Picroliv, picroside-I and kutkoside from Picrorhiza kurrooa are scavengers of superoxide anions, Biochemical pharmacology, 1992, 44, 180–183.

[115] Singh G, Bani S, Singh S, Khajuria A, Sharma M, Gupta B, Banerjee S, Antiinflammatory activity of the iridoids kutkin, picroside-1 and kutkoside from Picrorhiza kurrooa, Phytotherapy Research, 1993, 7, 402–407.

[116] Tiwari SS, Pandey MM, Srivastava S, Rawat A, TLC densitometric quantification of picrosides (picroside-I and picroside-II) in Picrorhiza kurroa and its substitute Picrorhiza scrophulariiflora and their antioxidant studies, Biomedical Chromatography, 2012, 26, 61–68.

[117] Hussain A, Shadma W, Maksood A, Ansari SH, Protective effects of Picrorhiza kurroa on cyclophosphamide-induced immunosuppression in mice, Pharmacognosy research, 2013, 5, 30.

[118] Jia D, Barwal I, Thakur S, Yadav SC, Methodology to nanoencapsulate hepatoprotective components from Picrorhiza kurroa as food supplement, Food bioscience, 2015, 9, 28–35.

[119] Shuxiang FYZ. Picroside II phospholipid composition and preparation method., in, USA,2010.

[120] Rathee P, Kamboj A, Sidhu S, Enhanced oral bioavailability of nisoldipine-piperine-loaded poly-lactic-co-glycolic acid nanoparticles, Nanotechnology Reviews, 2017, 6, 517–526.

[121] Kumar V, Chaudhary H, Kamboj A, Development and evaluation of isradipine via rutin-loaded coated solid–lipid nanoparticles, Interventional Medicine and Applied Science, 2018, 10, 236–246.

[122] Gupta T, Singh J, Kaur S, Sandhu S, Singh G, Kaur IP, Enhancing bioavailability and stability of curcumin using solid lipid nanoparticles (CLEN): A covenant for its effectiveness, Frontiers in bioengineering and biotechnology, 2020, 8, 879.

[123] Bozzuto G, Molinari A, Liposomes as nanomedical devices, International journal of nanomedicine, 2015, 10, 975.

[124] Burande AS, Viswanadh MK, Jha A, Mehata AK, Shaik A, Agrawal N, Poddar S, Mahto SK, Muthu MS, EGFR targeted paclitaxel and piperine co-loaded liposomes for the treatment of triple negative breast cancer, AAPS PharmSciTech, 2020, 21, 1–12.

[125] Thornthwaite JT, Shah HR, England SR, Roland LH, Thibado SP, Ballard TK, Goodman BT, Anticancer effects of curcumin, artemisinin, genistein, and resveratrol, and vitamin C: Free versus liposomal forms, Advances in Biological Chemistry, 2017, 7, 27–41.

[126] Toniazzo T, Peres MS, Ramos AP, Pinho SC, Encapsulation of quercetin in liposomes by ethanol injection and physicochemical characterization of dispersions and lyophilized vesicles, Food bioscience, 2017, 19, 17–25.

[127] Mohammadabadi M, Mozafari M, Enhanced efficacy and bioavailability of thymoquinone using nanoliposomal dosage form, Journal of Drug Delivery Science and Technology, 2018, 47, 445–453.

[128] Takahashi M, Uechi S, Takara K, Asikin Y, Wada K, Evaluation of an oral carrier system in rats: Bioavailability and antioxidant properties of liposome-encapsulated curcumin, Journal of agricultural and food chemistry, 2009, 57, 9141–9146.

[129] Hong S-C, Park K-M, Hong CR, Kim J-C, Yang S-H, Yu H-S, Paik H-D, Pan C-H, Chang P-S, Microfluidic assembly of liposomes dual-loaded with catechin and curcumin for enhancing bioavailability, Colloids and surfaces. A, Physicochemical and engineering aspects, 2020, 594, 124670.

[130] Acharya N, Parihar G, Acharya S, Phytosomes: Novel approach for delivering herbal extract with improved bioavailability, Pharma Science Monitor, 2011, 2, 144–160.

[131] Lu M, Qiu Q, Luo X, Liu X, Sun J, Wang C, Lin X, Deng Y, Song Y, Phyto-phospholipid complexes (phytosomes): A novel strategy to improve the bioavailability of active constituents, Asian Journal of Pharmaceutical Sciences, 2019, 14, 265–274.

[132] Fernández-García R, Lalatsa A, Statts L, Bolás-Fernández F, Ballesteros MP, Serrano DR, Transferosomes as nanocarriers for drugs across the skin: Quality by design from lab to industrial scale, International Journal of Pharmaceutics, 2020, 573, 118817.

[133] Rajan R, Jose S, Mukund VB, Vasudevan DT, Transferosomes-A vesicular transdermal delivery system for enhanced drug permeation, Journal of Advanced Pharmaceutical Technology & Research, 2011, 2, 138.

[134] Cevc G. Transfersomes, liposomes and other lipid suspensions on the skin: permeation enhancement, vesicle penetration, and transdermal drug delivery, Critical reviews™ in therapeutic drug carrier systems, 13, 1996.

[135] Sana E, Zeeshan M, Ain QU, Khan AU, Hussain I, Khan S, Lepeltier E, Ali H, Topical delivery of curcumin-loaded transfersomes gel ameliorated rheumatoid arthritis by inhibiting NF-κβ pathway, Nanomedicine, 2021, 16, 819–837.

[136] Rassu G, Ferraro L, Pavan B, Giunchedi P, Gavini E, Dalpiaz A, The role of combined penetration enhancers in nasal microspheres on in vivo drug bioavailability, Pharmaceutics, 2018, 10, 206.

[137] Khatri S, Awasthi R, Piperine containing floating microspheres: An approach for drug targeting to the upper gastrointestinal tract, Drug Delivery and Translational Research, 2016, 6, 299–307.

[138] Boddupalli BM, Anisetti RN, Ramani R, Malothu N, Enhanced pharmacokinetics of omeprazole when formulated as gastroretentive microspheres along with piperine, Asian Pacific Journal of Tropical Disease, 2014, 4, S129–S133.

[139] Karade PG, Jadhav NR, Colon targeted curcumin microspheres laden with ascorbic acid for bioavailability enhancement, Journal of microencapsulation, 2018, 35, 372–380.

[140] Pingale P, Pandharinath R, Shrotriya P, Study of herbal bioenhancers on various characteristics of isoniazid and rifampicin microspheres, International Journal of Infectious Diseases, 2014, 21, 235.

[141] Date AA, Desai N, Dixit R, Nagarsenker M, Self-nanoemulsifying drug delivery systems: Formulation insights, applications and advances, Nanomedicine, 2010, 5, 1595–1616.

[142] Kazi M, Shahba AA, Alrashoud S, Alwadei M, Sherif AY, Alanazi FK, Bioactive self-nanoemulsifying drug delivery systems (Bio-SNEDDS) for combined oral delivery of curcumin and piperine, Molecules, 2020, 25, 1703.

[143] Kanwal T, Saifullah S, Ur Rehman J, Kawish M, Razzak A, Maharjan R, Imran M, Ali I, Roome T, Simjee SU, Design of absorption enhancer containing self-nanoemulsifying drug delivery system (SNEDDS) for curcumin improved anti-cancer activity and oral bioavailability, Journal of Molecular Liquids, 2021, 324, 114774.

[144] Usmani A, Mishra A, Arshad M, Jafri A, Development and evaluation of doxorubicin self nanoemulsifying drug delivery system with Nigella Sativa oil against human hepatocellular carcinoma, Artificial Cells, Nanomedicine and Biotechnology, 2019, 47, 933–944.

[145] Kumar V, Kharb R, Chaudhary H, Optimization & design of isradipine loaded solid lipid nanobioparticles using rutin by Taguchi methodology, International Journal of Biological Macromolecules, 2016, 92, 338–346.

[146] Selvaraj K, Yoo B-K, Curcumin-loaded nanostructured lipid carrier modified with partially hydrolyzed ginsenoside, AAPS Pharmaceutical Science Technology, 2019, 20, 1–9.

[147] Sedeky AS, Khalil IA, Hefnawy A, El-Sherbiny IM, Development of core-shell nanocarrier system for augmenting piperine cytotoxic activity against human brain cancer cell line, European Journal of Pharmaceutical Sciences, 2018, 118, 103–112.

[148] Ray L, Karthik R, Srivastava V, Singh SP, Pant A, Goyal N, Gupta KC, Efficient antileishmanial activity of amphotericin B and piperine entrapped in enteric coated guar gum nanoparticles, Drug delivery and translational research, 2021, 11, 118–130.

[149] Vijayan UK, Varakumar S, Sole S, Singhal RS, Enhancement of loading and oral bioavailability of curcumin loaded self-microemulsifying lipid carriers using Curcuma oleoresins, Drug development and industrial pharmacy, 2020, 46, 889–898.

[150] Verma N, Saraf S, Development and optimization of mannosylated naringenin loaded transfersomes using response surface methodology for skin carcinoma, International Journal of Applied Pharmaceutics, 2021, 235–241.

[151] El-Samaligy M, Afifi N, Mahmoud E, Increasing bioavailability of silymarin using a buccal liposomal delivery system: Preparation and experimental design investigation, International journal of pharmaceutics, 2006, 308, 140–148.

[152] Schlich M, Lai F, Pireddu R, Pini E, Ailuno G, Fadda A, Valenti D, Sinico C, Resveratrol proniosomes as a convenient nanoingredient for functional food, Food chemistry, 2020, 310, 125950.

[153] Jose J, Priya S, Shastry C, Influence of bioenhancers on the release pattern of niosomes containing methotrexate, Journal of Health and Allied Sciences NU, 2012, 2, 36–40.

[154] Islam N, Irfan M, Hussain T, Mushtaq M, Khan IU, Yousaf AM, Ghori MU, Shahzad Y, Piperine phytosomes for bioavailability enhancement of domperidone, Journal of Liposome Research, 2021, 1–9.

[155] Li Y, Xu F, Li X, Chen S-Y, Huang L-Y, Bian -Y-Y, Wang J, Shu Y-T, Yan G-J, Dong J, Development of curcumin-loaded composite phospholipid ethosomes for enhanced skin permeability and vesicle stability, International Journal of Pharmaceutics, 2021, 592, 119936.

[156] Wu P-S, Li Y-S, Kuo Y-C, Tsai S-J-J, Lin -C-C, Preparation and evaluation of novel transfersomes combined with the natural antioxidant resveratrol, Molecules, 2019, 24, 600.

[157] Hosny KM, Alharbi WS, Almehmady AM, Bakhaidar RB, Alkhalidi HM, Sindi AM, Hariri AH, Shadab M, Zaki RM, Preparation and optimization of pravastatin-naringenin nanotransfersomes to enhance bioavailability and reduce hepatic side effects, Journal of Drug Delivery Science and Technology, 2020, 57, 101746.

Dattatraya M. Shinkar, Sunil V. Amrutkar, Prashant L. Pingale

Chapter 9
Case study: Indian herbal bioenhancers

Abstract: Herbal bioenhancers have a herbal origin, and the concept of herbal bio-enhancer is mentioned in Ayurveda as "Yogvahi," which acts as a "carrier" for the macromolecules of bulk drugs to transport them to the site of action, without changing their inherent properties, while also being a catalyst to increase the rate of a chemical reaction, without undergoing any change during the process. Yogvahi is used to increase the effect of drugs given orally by increasing their bioavailability, decreasing side effects, and avoiding the parenteral route of drug administration. Herbal medicines have been used and accepted worldwide because of their good therapeutic benefits and lesser toxic effects than modern medicines. In developing countries like India, the major concern for formulation scientists is the overall treatment cost of formulations, so it becomes necessary to develop a modern system of medicine by systematic innovative means that can help reduce the treatment cost, by using natural bioenhancers. In India, in identifying a new compound, a key contribution in the drug development and discovery process was made by Ayurveda. Recent development in bioavailability augmentation of many medicines by herbal bioenhancers has been an innovation in use of drugs in the treatment of many diseases. The world's first bioavailability enhancer, piperine, was discovered and authorized by scientists at the Indian Institute of Integrative Medicine, Jammu, in 1979. Bioavailability and bio efficacy of many drugs like antineoplastic agents, antihypertensive, antiviral, antitubercular, and antifungal were increased at low doses by the use of herbal bioenhancers. They enhanced the oral absorption and bioavailability of a few herbal and essential nutraceutical compounds. Piperine, ginger, garlic, stevia, ghee, and cow distillate urine, etc. are used as herbal bioenhancers in the preparation of medicines. Nowadays, with the advancement in technology, herbal bioenhancers are a new approach used in the development of modern system of drug delivery systems to improve the bioavailability of many drugs. Herbal bioenhancers are economical, safe, easily available, decrease drug toxicity, lower drug resistance problems, and reduce the treatment period in therapy. This chapter presents a detailed study about various aspects of Indian herbal bioenhancers, collects information of all herbal bioenhancers with their mechanism of action and studies conducted on them to improve the drug bioavailability.

Dattatraya M. Shinkar, GES's Sir Dr. M. S. Gosavi College of Pharmaceutical Education and Research, Nashik 425005, Maharashtra, India, e-mail: drdmshinkar@gmail.com
Sunil V. Amrutkar, Prashant L. Pingale, GES's Sir Dr. M. S. Gosavi College of Pharmaceutical Education and Research, Nashik 425005, Maharashtra, India

https://doi.org/10.1515/9783110746808-009

Keywords: bioavailability, herbal, bioenhancers, medicines, Yogvahi

9.1 Introduction

Many medicinal plants are used as a main component in formulation of all indigenous or alternative systems of medicines like Ayurveda, homeopathy, naturopathy,
Siddha, and Unani. The demand and acceptance of herbal drug and natural plant-
based medicines has increased throughout the world, owing to their good therapeutic
efficacy, lesser side effects, and low cost, compared to modern system of medicines.
Many herbal and synthetic origin of drugs have low oral bioavailability because of
poor membrane permeability, low lipophilicity, ionic characteristics, and poor water
solubility [1–2].

Bioavailability is the rate at which an amount of drug reaches systemic circulation in its unchanged form and becomes bioavailable at the site of action. Drugs
administered via intravenous route achieve higher bioavailability, compared to
drugs administered orally, as they are poorly bioavailable due to first-pass effect,
erratic, and incomplete absorption. Poorly absorbed drugs in the body may lead to
adverse drug reactions and resistance to pharmacological activity. Therefore, there
is need for use of bioenhancers with drugs or molecules to augment the bioavailability, without showing therapeutic activity of their own [3]. Coadministration of
many herbal compounds obtained from a variety of medicinal plants have shown
enhancement of bioavailability of drugs. The phenomenon of increasing the total
systemic bioavailability of any drug is called biopotentiation effect, and the agents
responsible for increase in plasma concentration of drug are bioenhancers [4].

Bioavailability enhancers act as drug facilitators; these are the substances that
do not show any pharmacological activity when used in combination with drug or
nutrient at the dose used; they increase the bioavailability of drug across the biological membrane, through conformational interaction due to potentiation of a
drug molecule, acting as a receptor and also making target cells more approachable
to the drug molecules [3]. This concept is somewhat new to the modern system of
medicines; many drug studies have shown improvement in bioavailability when coadministered with herbal bioenhancers. This development has led to new promising approaches, such as developments in prodrugs, micronization, delayed-release
products, sustained-release capsules, bioavailability enhancers in transferosomes,
ethosomes, emulsions, and liposomal formulations [4].

9.2 History

In 1929, the use of herbal bioenhancer was studied for the first time by Bose, a well-known author of the book *Pharmacographia Indica*; he reported that vasaka (*Adhatoda vasica*), when coadministered with long pepper showed enhancement in anti-asthmatic activity of Ayurvedic formulation. In 1979, piperine was the world's first exposed herbal bioenhancer that was scientifically discovered at the Regional Research laboratory (RRL), Jammu, by Indian research scientists. Development and isolation of plant-based herbal molecules like piperine and quercetin, is considered a scientific revolution in the modern system of medicine. Risorine, a fixed drug combination of rifampicin, isoniazid, and piperine, is the result of this research. It prevents the resistance of rifampicin and also decreases its dose by almost 60% because of increase in bioavailability [5].

9.2.1 Bioenhancers in Ayurveda

In Ayurveda, the idea of bioavailability enhancement called "Yogvahi" was used for many years. It intended to use medicinal herbs like piperine to increase the plasma concentration of drug in biological fluid. The thought of herbal bioavailability enhancers from plant origin can be traced back to the Ayurvedic system of medicine. "Trikatu" was a very famous word in a Sanskrit meaning three acrids, used in the seventh century B.C. and the sixth century A.D. in many ancient Ayurvedic formulations to treat many diseases. Trikatu was a mixture of piperine, black pepper, and ginger, with piperine as an active component that has the potential to augment bioavailability of various drugs, vitamins, and nutrients. They are used in many formulations as ingredients, even though many times, they do not have a direct role to play in the disease indicated [6–9].

   The Director of RRL Jammu, C. K. Atal, examined and studied 370 Ayurvedic formulations of Indian origin used in the treatment of many diseases and reported that Trikatu or mainly *Piper longum* Linn was used in about 210 formulations. Piperine, the active constituent was extracted from the *P. longum* and its use as an oral herbal bioenhancer was established in research carried out on various drugs like antifungal, antitubercular, antibiotics, NSAIDs, and cardiovascular showing similar bioenhancement effects. It has shown enhancement of 20–200% in bioavailability of almost all drugs. In the current modern system of medicine, the importance of use of different herbal bioenhancers like genistein, sinomenine, and curcumin, has been extended to improve the pharmacokinetic and pharmacodynamic profile of drugs with narrow therapeutic indices [10–12].

9.2.2 Bioenhancers in modern system of medicine

The term bioavailability enhancer is defined as "a substance when combined and administered with a drug or nutrient at low dose improves amount of drug in the body resulted in reduction of the consumption of both with enhanced therapeutic efficacy." To maximize bioavailability of therapeutically administered drug is important, because the amount of drug bioavailable in systemic circulation directly affects drug concentration in plasma and, thus, its therapeutic effect [5]. Bioavailability augmentation can reduce the toxic effects of drugs and makes them more affordable by reducing cost.

The great importance of the use of bioenhancers to advance the bioavailability of drugs is due to their
– poor oral bioavailability
– use for long periods in treatment
– toxicity and
– cost

Therefore, many herbal drugs and extracts have shown excellent in vitro effect and very poor in vivo effect, because of poor lipid solubility or improper molecular size, resulting in low systemic availability of the drug. Since traditional times, bioavailability enrichment has gained a wide popularity, using secondary agent as a supplement along with the main active ingredient. Ayurvedic literature surveys have revealed the use of herbal bioenhancers as a new approach to improve the absorption of poorly bioavailable drugs. Currently, the development of expertise in novel drug delivery system (NDDS) has opened the doors to use of bioenhancers in herbal drug delivery systems. This concept was witnessed in the last decade for many potent drugs that require immediate effects using novel carriers like liposomal preparations, microspheres, nanoparticles, transferosomes, ethosomes, and lipid-based formulations for successful modified delivery [5].

The herbal bioenhancers used should possess following ideal characteristics:
– It should be nontoxic, nonirritating and nonallergic.
– It should be easy to formulate into a various dosage form.
– It should be pharmacologically inert.
– It should be unidirectional with rapid onset of action.
– It should be well-suited with drugs.
– It should be stable.
– It should be easily available and cost effective.

9.3 Barriers of drug absorption

Orally administered drug crosses the biological membrane of the gut wall mucosae during its passage through the lumen into the systemic circulation, to produce its pharmacological action. There are several biological and structural barriers like presence of lipid bilayers at cell surroundings, aqueous pore channels, and P-glyco-protein (P-gp) efflux pump system in the gastrointestinal tract (GIT), which stick around the passage of aqueous stagnant layer shown in Figure 9.1. Penetration of the drug through the epithelial membrane, acts as a barrier for the passage of drug from the GIT to the systemic circulation. A drug reaches systemic circulation by traveling through the GIT through the epithelial membrane wall and proceeding via hepatic metabolism. Due to the hydrophilic nature, the aqueous stagnant layer present in the epithelium membrane serves as a possible barrier for the absorption of many drugs. The transport of drugs through the biological membrane takes place either by passive or carrier-mediated diffusion and active transport mechanism, which involves the expenditure of energy. Small water-soluble molecules like etha-nol are absorbed through the aqueous pore channels present in the proteins of lipid bilayers. Drug molecules larger than 0.4 nm in size will not pass easily through these aqueous pore channels.

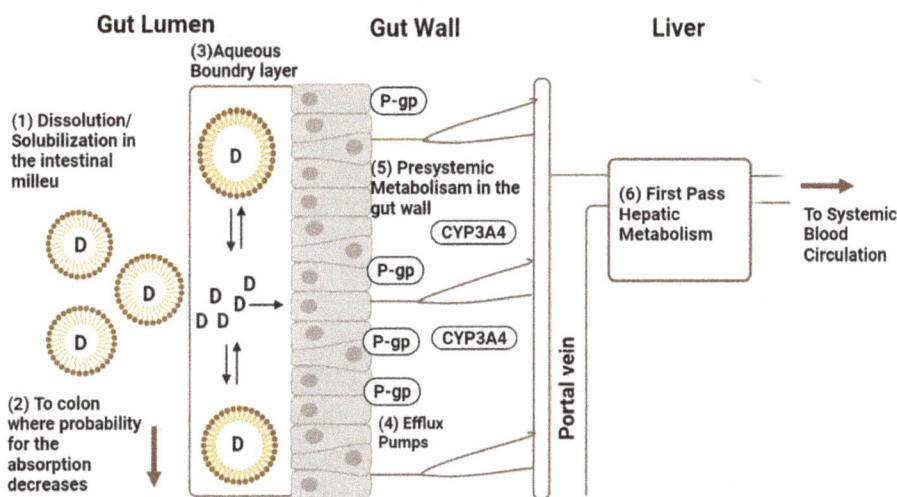

Figure 9.1: Drug absorption barriers.

Current research work has shown that P-gp is an ATPase energy-dependent trans-membrane drug efflux system mechanism that belongs to the members of ABC transporters and plays a vital role in the inhibition of drug entry into the systemic circulation. It has molecular weight of – 170 kDa in the presence of 1,280 amino acid residues; therefore, it has gained importance in absorption enhancement of many drugs. More research work has been done to study the variation of the efflux pump system, owing to its substrate selectivity and distribution at the absorption site. Many herbal bioenhancers work by inhibiting this efflux pump system to augment the absorption of drugs administered orally [3–4, 13].

9.4 Methods to enhance the bioavailability of drugs

The various approaches and methods that have been used to improve intestinal absorption of drugs are as follows.

9.4.1 Absorption enhancers

Bile salts, surface-active agents, fatty acids, chelating agents, salicylates, and polymers are used as effective enhancers to increase the intestinal absorption of poorly soluble drugs. Chitosan, chiefly trimethylated chitosan, increases the drug absorption of many drugs through the paracellular route due to reformation of the cytoskeletal F-actin responsible for opening the tight junctions. The absorption of lipophilic drugs was improved by surfactants, bile salts, and fatty acids by enhancing the solubility and increasing their fluidity in the aqueous layer, apical, and basolateral membranes, respectively. Calcium chelators such as ethylene glycol tetraacetic acid and ethylenediaminetetraacetic acid, increase the absorption of drugs by decreasing the cellular concentration of calcium, which causes disruption in cell–cell contacts [13–15].

9.4.2 Prodrugs

Chemical alteration of drug into a more permeable analogue prodrug is commonly studied and available as an alternative method for increasing the absorption and bioavailability of many drugs. Ampicillin derivatives are examples of augmenting the lipophilicity of agents in the form of prodrug to increase absorption of drug. Due to hydrophilicity of ampicillin, its oral bioavailability is only 30–40% from the GIT. Prodrugs are synthesized by esterification of carboxyl group of ampicillin, pivampicillin, bacampicillin, and talampicillin. Prodrugs have higher lipophilicity than the parent drug ampicillin, provided better oral administration, and showed higher absorption and bioavailability.

9.4.3 Different dosage forms and other pharmaceutical approaches

Of late, the greatest important practical approach is to increase the intestinal absorption of poorly absorbed drugs by formulating them into dosage forms like liposomes and emulsions. Also, the other approaches that include reduction of drug particle size by micronization, nanoparticular carriers, complexation, and liquid crystalline phases are used to enhance bioavailability [16].

9.4.4 P-glycoprotein inhibitors

Many research findings have confirmed the use of P-gp inhibitors in enhancing the absorption of many drugs through peroral delivery. They mediate increase in oral bioavailability through reverse P-gp efflux of drug transport across the epithelium membrane (Figure 9.2). They also influence the overall systemic availability of drug, distribution, biotransformation, and elimination of P-gp substrates in the process of controlling pharmacokinetics of drugs [17].

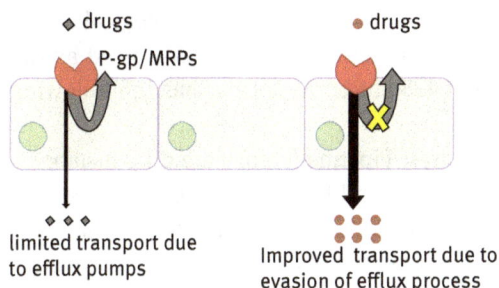

Figure 9.2: P-glycoprotein inhibition process.

9.5 Mechanism of action of herbal bioenhancers

Herbal bioenhancers exert their properties of increasing bioavailability of the drug molecule by several mechanisms of actions. Depending on the type of enhancer selected for the purpose, they show similar or different mechanism of action. Nutritive and antimicrobial bioenhancers improve the bioavailability of drugs by acting on GIT and drug metabolizing processes, respectively.

Amongst the several mechanisms of action suggested for most of the herbal bioenhancing agents, some general mechanisms are as follows:

- Encouraging the absorption of drugs administered orally from GIT through increasing supply of blood.
- Alterations in the permeability of biological membrane of GIT.
- Decrease in hydrochloric acid secretion.
- Inhibition of gastrointestinal transit time and gastric emptying time of drug through improvement in intestinal motility.
- Increasing the bile secretion termed as cholagogous effect.
- Bioenergetics and thermo genic properties.
- Inhibiting the total systemic clearance by prevention of glomerular filtration rate and active tubular secretion due to inhibition P-gp efflux that facilitates passive tubular reabsorption.
- Modification of the immune system to reduce the total necessity of the drug.
- Increasing the permeability or improving the accessibility to the active drug due to improved killing of organisms.
- Increasing the availability of the drugs to the pathogens by modification in the signaling process among host and pathogen.
- Improving the requisite binding capacity drug and targeting sites like receptors, proteins, deoxyribonucleic acid (DNA), and RNA that potentiates and prolongs antibiotic activity against pathogens.
- Herbal bioenhancers are the agents used to improve the pharmacological action of many active pharmaceutical ingredients. Piperine, which was used as the first herbal bioenhancer in the form of Trikatu by Ayurveda, has the following mechanisms of action [18].
- **Inhibits the drug metabolizing pathways:** Piperine inhibits the drug metabolizing enzymes and produces drug degradation and biotransformation during their passage through the liver. It hinders the P-gp efflux system and cytochrome P450 (CYP) enzyme. Piperine also inhibits or induces some of the other CYP class metabolizing enzymes present in GIT like CYP3A4, CYP1A1, CYP1B1, CYP1B2, or CYP2E1, which facilitates overcoming the first-pass effect of the administered drug.
- **Modulating effect of P-glycoproteins:** P-gp is a major efflux pump of the biological membrane, throws out the administered drug, and avoids it to reach at the target site to exert its pharmacological action. In such a situation, herbal bioenhancers act by preventing the P-gp efflux system and enhance the bioavailability of drugs like antimalarial or antineoplastic agents.
- **Inhibits presence of glucuronic acid:** It inhibits the glucuronidation present in the GIT and lowers the content of the UDP-glucuronic acid. Experimental studies carried out on rats using piperine have shown the strong inhibition of UDP-glucoronyl transferase by inhibiting the transferase activity.
- **Stimulation effect on gamma glutamyl transpeptidase (GGT):** Stimulation of GGT raises the absorption of drug conjugates with amino acids due to increase in uptake of amino acids.

In addition to these, the bioenhancers may be beneficial in improving the transport of drugs and nutrients that cross the blood–brain barrier, which plays a vital role in controlling epilepsy, cerebral infections, and central nervous system (CNS)-related disorders. Therefore, metabolism of drugs by hepatic enzymes will be influenced in terms of increase in their bioavailability by use of bioenhancers. A few other proposed mechanisms such as making target receptor more receptive to drugs, vasodilation of GIT, and changes in cell membrane dynamics enhance the passage of drugs through cell membranes to increase the drug absorption and bioavailability [19–20].

9.6 Classification of herbal bioenhancers

Classification of herbal bioenhancers based on their origin and mechanism of action is represented in Table 9.1.

Table 9.1: Classification of herbal bioenhancers [21].

Classification based on source		
Plant-based bioenhancers		Animal-based bioenhancers
Piperine, glycyrrhizin, cumin, ginger, niaziridin. allicin, caraway, curcumin, *Stevia*, *Ammannia multiflora*, gallic acid, *Callistemon rigidus*, capsaicin, lysergol, *Aloe vera*, sinomenine, *Stevia*, capmul, quercetin, genistein, naringin, and peppermint oil.		Cow urine distillate, ghee, and honey
Classification based on mechanisms of action		
P-glycoprotein efflux pump and other efflux pump inhibitors	Cytochrome P450 and its isozyme suppressors	Regulators to facilitate GIT function for enhanced absorption of drugs
Piperine, curcumin, genistein, *Cuminum cyminum*, naringin, quercetin, etc.	Piperine, naringin, quercetin, and gallic acid.	Aloe, drumstick pods, liquorice, and *Zingiber officinale* (ginger)

9.7 Herbal bioenhancers

9.7.1 Piperine

Piperine is regarded as the world's first herbal bioenhancer and as a milestone in the field of biopotentiation of poorly absorbed drugs. It is a major alkaloid component present in plants of long pepper and black pepper belonging to family Piperaceae. It is obtained from the stem, leaf, or pods of the plant. It is the most widely studied herbal bioenhancer and known to augment the bioavailability of many drugs. It is approved by FDA as a generally recognized as safe material (GRAS). Piperine comprises two ingredients of Trikatu out of three. Piperine also shows various pharmacological effects like anti-inflammatory, antipyretic, antifungal, antidiarrheal, antioxidant, antithyroid, antidepressant, analgesic activities, etc. The coupled aromatic bonds present in side chain structure of piperine are mainly responsible for its activity. The daily dose of piperine is 15–20 mg/kg. Piperine alone or with a mixture of other herbal components, helps augment the bioavailability of various drugs by increasing the level of drug in blood. It also helps increase the therapeutic efficiency of many drugs like sparteine, rifampicin, phenytoin, sulfadiazine, propranolol, and drugs containing components of leaves of vasaka, vasicine, etc. It was found to increase the bioavailability for various classes of drugs like NSAIDS, cardiovascular system, CNS antitubercular, antileprosy, and antibiotics in the range of 30–200% [22–23].

Piperine shows its biopotentiation action by encouraging rapid absorption of many drugs and nutrients and also inhibits the enzymes that are responsible for metabolism of drugs during their passage through the liver. It interacts with microsomal enzyme system (aryl hydrocarbon hydroxylase enzymes) present in the gastrointestinal tract and enzymes like uridine diphosphate glucose dehydrogenase (UDP-GD), uridine diphosphate glucoronyl transferase (UDP-GT), 5-lipoxegenase (5-LOX), cyclo-oxygenase-I (COX-I), ethyl morphine-N demethylase, and 7-ethoxycoumarin-O-de-ethylase and inhibits the in vitro and in vivo metabolism and degradation of drugs. Many research findings prove that it is a nonspecific inhibition of the drug metabolism process [24]. It also inhibits the potent P-gp efflux transport system present in the gastrointestinal wall. Binding to the DNA receptor and change in cell signal transduction are the other actions performed by piperine. Its oral administration inhibits the hepatic aryl hydrocarbon hydroxylase (AHH) and UDP-glucoronyl transferase activities [24–25] in rats.

Piperine showed enhancement in bioavailability of many classes of drugs including antimicrobial, antiprotozoal, antihelminthic, cardiovascular, CNS, NSAIDs, vaccines, vitamins, herbal compounds, antihistaminic, respiratory, anticancer, corticosteroids, muscle relaxants, and hormones. Coadministration of piperine as an herbal bioenhancer has shown improvement of bioavailability of many drugs as shown in studies mentioned below [26–27].

9.7.1.1 Piperine and rifampicin

Rifampicin is used as first-line remedy for tuberculosis and leprosy. It has shown a very huge effect on this drug, and hence is thought to be more useful in lowering the dose and reducing the time period of treatment. Piperine acts directly on RNA polymerase enzyme present in human cells by inhibiting the transcription of polymerase catalyzed by *Mycobacterium smegmatis* and increases the activity of rifampicin several folds. Similarly, it stimulates the binding capacity of resistant strains of rifampicin to RNA polymerase. During tuberculosis treatment, when used in the weight ratio of 24:1 (drug: piperine), it has shown great effect. It augments rifampicin bioavailability by 60% and reduces the dose from 450 to 200 mg. At higher doses, piperine shows antagonistic effect, and it decreases the plasma concentration of rifampin [28].

9.7.1.2 Piperine and β-lactam antibiotics

Hiwale and his coworkers coadministered piperine with β-lactam antibiotics like amoxicillin trihydrate and cefotaxime sodium to rats to study the effect of bioenhancer on bioavailability of both drugs. The absorption and bioavailability of both drugs was enhanced due to inhibition of the microsomal metabolizing enzyme system. Pharmacokinetic parameters such as peak plasma concentration, elimination half-life, and AUC of the drug were studied to determine bioavailability [29].

9.7.1.3 Piperine and ciprofloxacin

Studies on coadministration of piperine with ciprofloxacin in rabbits confirmed the increase in bioavailability due to inhibition of the metabolism and changes in the absorption kinetics of drug [30].

The biopotentiation effect of piperine in combination with ciprofloxacin against *Staphylococcus aureus* ATCC 29213 was studied by Khan et al. It significantly reduces the mutation inhibition concentration of drug against *S. aureus* and their methicillin-resistant strain. Piperine, at a concentration of 100 μg/mL, did not show any antibacterial activity. MIC of ciprofloxacin was reduced twofold for *Staphylococcus aureus*, when verified at concentration of 12.5 and 25 μg/mL. If given at 50 μg/mL, it reduced MIC 4-fold due to inhibition of the efflux of drug ciprofloxacin and ethidium bromide from bacterial cells and hence improves the bioavailability of drug [31].

9.7.1.4 Piperine and oxytetracycline

Singh et al. evaluated the influence of piperine on bioavailability of orally administered oxytetracycline on White Leghorn birds weighing in the range of 2.0–2.8 kg. The pharmacokinetic data of the study has shown significant increase in absorption of drug and residence time in animals treated with piperine. This increase in bioavailability of drug declined the initial and maintenance dose of oxytetracycline by 33.3% and 39%, respectively [32].

9.7.1.5 Piperine with nevirapine

The effect of piperine on nevirapine pharmacokinetic parameters in fasting conditions using eight healthy human male volunteers of age between 20 and 40 years by crossover, placebo-controlled design was studied by Kasibhatta et al. All the selected volunteers were randomly on therapy with placebo, and in the morning, were given piperine 20 mg every day, for 6 days, coadministered with piperine (20 mg) and nevirapine (200 mg), or nevirapine plus placebo on day 7, using crossover design. Required amount of blood samples were drawn and collected for 1–144 h, after administering the dose of drug to calculate pharmacokinetic parameters. The results of the study showed evidence of enhanced bioavailability of nevirapine due to piperine with increase in maximum plasma concentration, with AUC_t, AUC_0, and C_{last} values at about 120%, 167%, 170%, and 146%, respectively [33].

9.7.1.6 Piperine and analgesic drug

Interaction of piperine with analgesic drugs such as pentazocine and diclofenac sodium was studied in albino mice. The study revealed that acetic acid-induced writhing reduced to 54.90% and 78.43%, when diclofenac sodium was administered alone and coadministered with piperine, respectively. Similarly, increase in tail flick latency was shown compared to pentazocine administered alone. The outcomes of the study showed that the extract of piperine coadministered with both drugs showed significant increase in activity [34].

9.7.1.7 Piperine with ampicillin and norfloxacin

Coadministration of piperine (20 mg/kg) with ampicillin and norfloxacin in rabbits showed increased oral bioavailability of both drugs in animal model [35].

9.7.1.8 Piperine and fexofenadine

Jin et al. studied the influence of coadministration of piperine with oral fexofenadine in rats. Results of pharmacokinetic study in rats concluded that piperine augmented the AUC of drug by 180–190%, compared to the control group. About 2-fold increase in bioavailability of fexofenadine was seen due to inhibition of P-gp-produced cellular efflux system mechanism by piperine, throughout its intestinal absorption [36].

9.7.1.9 Piperine and metronidazole

A. Singh et al. studied the combination of piperine with metronidazole in healthy male New Zealand white rabbits, and results of all pharmacokinetic parameters revealed that bioavailability of metronidazole was increased, compared to drug administered alone [37].

9.7.1.10 Piperine and ibuprofen

The ibuprofen-induced antinociception activity was evaluated after coadministration with piperine by Venkatesh et al. The results of pharmacokinetic parameters using acetic acid writhing and formalin test methods showed that antinociceptive activity of drug was increased significantly. The increase in plasma concentration of drug was due to its synergistic antinociception activity with piperine as a herbal bioenhancer [38].

9.7.1.11 Piperine with losartan potassium

Losartan potassium was coadministered with piperine to healthy male white albino rats. All the pharmacokinetic parameters were evaluated and compared to study the efficacy of losartan potassium (100 mg/kg) administered alone, and that when combined with 10 mg/kg piperine, to the rats. The results of study showed that losartan potassium bioavailability was increased significantly due to the effect of piperine as an herbal bioenhancer [39].

9.7.1.12 Piperine and propranolol HCl

Bano et al. studied the pharmacokinetics of propranolol and theophylline on healthy human volunteers by oral administration of 40 mg propranolol or 150 mg theophylline alone and combined with 20 mg piperine divided into each group of six for 7 days. The pharmacokinetics study concluded improvement in oral bioavailability of both drugs due to beneficial effects of piperine in attaining better therapeutic efficacy and patient compliance [40].

9.7.1.13 Piperine with nimesulide

Coadministration of piperine with nimesulide at dose of 10 mg/kg enhanced the blood plasma concentration of drug from 8.03 ± 0.99 µg/mL to 11.9 ± 0.23 µg/mL, in mice. This increase in concentration of nimesulide was due to inhibition of metabolism of drug [41].

9.7.1.14 Piperine and curcumin

Shoba and coworkers evaluated the combined influence of piperine and curcumin, on inhibition of hepatic and intestinal glucuronidation in rats and healthy human subjects. The investigations in the study revealed that piperine increases the systemic bioavailability of curcumin [42].

9.7.2 Ginger

The rhizome extract, a major component of *Zingiber officinale*, belonging to family Zingiberaceae, contains gingerols as active constituent transformed to shogaols, zingerone, and paradol. The content of volatile oil varies from 1% to 3% and decides the odor of ginger. The gingerol showed increase in the motility of gastrointestinal tract when administered to animals; it also acts as an analgesic, antipyretic, antibacterial, and sedative agent. 6-Gingerol is the main component of ginger, and its chemopreventive potential presents a promising and cost-effective substitute for the toxic therapeutic agents. Ginger shows powerful action on GIT mucous membrane. Ginger plays an important role as a bioenhancer in regulating the intestinal absorption of drug for better therapeutic effect and provides increase in bioavailability by 30–75%. It has significantly increased the bioavailability of drugs, mainly azithromycin, erythromycin, cephalexin, cefadroxil, amoxycillin, and cloxacillin [43–45].

9.7.3 Niaziridin

A nitrile glycoside, niaziridin is obtained from the leaves, pods, and bark of the drumstick (*Moringa oleifera*) plant. It also acts as antimicrobial, diuretic, anticancer, antifertility agent, analgesic antifungal, antiulcer, antioxidant, hepatoprotective, hypolipidemic, antiteratogenic, and antiarthritic agent. It augments the absorption and bioavailability of the antibiotics rifampicin, ampicillin, and tetracycline. The bioavailability of nalidixic acid was enhanced 1.2–19-fold against gram-negative *E. coli* and gram-positive bacteria *Bacillus subtilis* and *Mycobacterium smegmatis* by niaziridin. The antifungal activity of clotrimazole was augmented 5–6-fold, if used at high concentration (10 µg/mL) against *Candida albicans*. Niaziridin, in combination with nutrients like vitamin B12, helps to facilitate its uptake in the GIT, as a function of a bioavailability enhancer [46–57].

9.7.4 Glycyrrhizin

A plant glycoside, glycyrrhizin, is gained from the root, stolon of *Glycyrrhiza glabra* L. (family: Fabaceae). It is used as an expectorant in the treatment of bronchitis and also used to treat inflammation, allergy, peptic ulcer, rheumatoid arthritis, and throat infection. It acts a detoxifier for the drugs in the treatment of liver disorders. It stimulates the function of the adrenal gland and supports as an immune modulator. It is nearly 50 times sweeter compared to sugar. Glycyrrhizin also exhibits activities like antihepatotoxic, analgesic, antineoplastic, and antiviral agent. It acts as bioenhancer in the concentration range from 0.05% to 50% w/w and 0.10% to 10% w/w of the antibacterial and nutraceutical compounds, respectively. It shows improvement in bioavailability of antifungal agents, when used in concentration ranges 0.25–20% w/w.

Coadministration of glycyrrhizin (1 µg/mL) with "Taxol" (0.01 µg/mL), anticancer agent-induced inhibitory action on cancerous cell line MCF-7 markedly enhanced fivefold, compared to Taxol administered alone at dose of 0.05 µg/mL. It enhanced the absorption and bioavailability of nalidixic acid, rifampicin, ampicillin, tetracycline, and clotrimazole against the *Candida albicans* species [58–65].

9.7.5 Cumin

Cumin aldehyde obtained from *C. cyminum* oil belonging to family Apiaceae has β-pinene, *p*-mentha-1,4-dien-7-al, and γ-terpinene as the main constituents. The doses of its parts like C. cyminum extract or its fractions at 0.5–25 mg/kg body weight enhanced bioavailability of many drugs by 25–335%. Combination of polar and nonpolar extracts of cumin and piperine showed augmented bioavailability between 25% and 435% [66, 67].

9.7.6 Caraway

Caraway oil (carvone) and limonene are the main constituents acquired from dry and crushed seeds of *Carum carvi* L., belonging to family Apiaceae. The color and flavor of caraway are mainly attributed to its seeds [68]. It is effective as a bioenhancer at a dose of 5–100 mg/kg and 1–55 mg/kg w/w of total body weight, for the bioenhancer extract and the dose of bioactive fraction, respectively. It augments the bioavailability of drugs like antibiotics, antifungal, antiviral, and antineoplastic agents. If cumin is combined with *Z. officinale* and piperine, it enhanced bioactivity in the range of 20–110% and 25–95%, respectively [69,70].

9.7.7 Garlic (allicin)

It holds the allyl sulfur compound derived from the garlic (*Allium sativum* L.) belonging to family Amaryllidaceae. Cupric ions (Cu^{2+}) showed tremendous improvement in fungicidal activity, in the presence of allicin against *Saccharomyces cerevisiae* cells [71].

Allicin when used in combination with amphotericin B enhanced the fungicidal activity. It acts as a cellular protective mechanism against amphotericin-produced vacuole disruption in *S. cerevisiae* and also inhibits transport of ergosterol from the plasma membrane to the cytoplasmic reticulum. Therefore, it increases the effect of amphotericin B as a fungicidal agent in controlling serious fungal infections, against *Candida albicans* and *Aspergillus fumigatus* species [72, 73].

9.7.8 Lysergol

Lysergol is an ergoline alkaloid, obtained from morning glory plants (*Ipomoea* sp.), belonging to family Convolvulaceae which improves the pharmacological activity of antibiotics on bacteria and is used an effective herbal bioenhancer. Lysergol is also obtained from mind-blowing seeds of plants ololiuhqui (*Rivea corymbose*), Hawaiian baby woodrose, and *Ipomoea violacea* belonging to family Convolvulaceae. Lysergol presents a minor component of species of fungi *Claviceps*. It is effective as a bioenhancer at dose of 1–10 µg/mL and augments the absorption and bioavailability of rifampicin, tetracycline, and ampicillin [74].

9.7.9 Aloe vera gel

Aloe vera gel is obtained from the whole leaf extracts of plant *Aloe vera* L. belonging to family Asphodelaceae. It has shown enhancement in plasma concentration of ascorbic acid (vitamin C) and tocopherols (vitamin E), and significantly improved the bioavailability. It prolongs the residence time of ascorbate in plasma, even after 24 h of overnight fast and hence increased the absorption of ascorbate. Due to its unique capability to augment the absorption and bioavailability of both vitamins, it should find its use as a dietary herbal bioenhancer [75] in the days to come.

9.7.10 Curcumin

It is a key component of the Indian spice, turmeric, i.e., *Curcuma longa* L. belonging to family Zingiberaceae. Curcumin shows bioenhancing properties similar to piperine. It enhances the systemic availability of celiprolol and midazolam in rats due to suppression of CYP3A4 drug metabolizing enzyme in the liver and inhibition of P-gp efflux mechanism. Curcumin increases the absorption of many drugs due to modification in the physiological activity of the gastrointestinal tract and suppression of amount of UDP-glucuronyl transferase in the intestine and hepatic tissue [76,77].

9.7.11 Capmul

Capmul is obtained from edible fats and oils and contains two medium chain fatty acids, primarily, caprylic and capric. It increases the bioavailability of ceftriaxone in rats in the range of 55–79% [78].

9.7.12 *Callistemon rigidus*

Callistemon rigidus R.Br. is a crude extract of leaves belonging to family Myrtaceae. At a very low concentration of 39.06 µg/mL, it was active against ciprofloxacin-resistant *Staphylococcus aureus*, when combined with drug. It produced synergistic effect with resilient drug ciprofloxacin and showed improvement in in vitro activity against control as well as mutant strain of *S. aureus* [79].

9.7.13 *Ammannia multiflora*

Ammaniol is an innovative compound of *Ammannia multiflora* along with other compounds belonging to family Lythraceae. Its methanolic extract has shown increased bioenhancing activity against the two strains of *Escherichia coli*, namely CA8000 and DH5α, for the antibiotic nalidixic acid [80].

9.7.14 *Stevia*

Stevia rebaudiana is called stevioside, and is the main component of the glycoside stevia, and is also present in constituents like steviol, austroinulin, rebaudioside, and dulcoside; it is 200 times sweeter than sucrose. In South America, it is used as a sweetener called honey leaf. If used in combination with piperine at doses of

0.01–50 and 0.01–12 mg/kg w/w of total body weight, respectively, it showed increase in bioavailability of nutraceutical formulations, drugs, and herbal formulations. *Stevia* and piperine are bioactive at doses of 30 and 8 mg/kg body weight, respectively, and are used as bioenhancers. *Stevia* increases the absorption and bioavailability of many classes of drugs, mainly antineoplastic agents, anti-inflammatory, antibiotics, antidiabetic, antifungal, antiviral, anticancer, cardiovascular, analgesic, antiarthritic agents, antitubercular drugs, antihelminthic, antiulcer drugs, and herbal formulations [81].

9.7.15 Genistein

It is an isoflavone obtained from alimentary plants such as soybean (*Glycine max*) and kudzu (*Pueraria lobata*). It is used as a well-known phytoestrogen and shows many beneficial effects, like antineoplastic and analgesic activities [82–83].

9.7.15.1 Genistein and paclitaxel

Genistein showed improvement in the absorption and bioavailability of paclitaxel administered orally and intravenously in rats. The occurrence of genistein as herbal bioenhancer showed enhancement in bioavailability of paclitaxel in the study [84].

9.7.15.2 Genistein and epigallocatechin gallate (EGCC)

If genistein is coadministered with EGCG, it increases the concentration of HT-29 human colon cancer cells 2- to 5-fold, compared to treatment with EGCG alone. When coadministered with genistein at 200 mg/kg (75 mg/kg) and EGCG 75 mg/kg to mice, it increases the biological half-life and duration of action of drug. It increases the maximum peak plasma concentration of EGCC in the small intestine 2.0- to 4.7-fold and half-life 1.4-fold. It proves that genistein can be used to enhance the oral bioavailability of drug [85].

9.7.16 Capsaicin

Capsaicin is a main component obtained from chili peppers, *Capsicum annum* L. belonging to family Solanaceae. It produces burning sensation to mammals when it comes in contact with any tissue. Its biopotentiation effect was examined after oral administration of fruit suspension, with and without a ground capsicum in 10 male rabbits, and showed improvement in bioavailability of theophylline from sustained release gelatin capsule [86].

9.7.17 Peppermint oil

It is obtained from *Mentha piperita* L. belonging to family Lamiaceae. Coadministration of peppermint oil (100 mg/kg) augmented the plasma concentration and area under the curve of drug cyclosporine 3-fold [29]. Studies concluded that due to inhibition of cytochrome CYP3A enzyme, peppermint oil increases the oral bioavailability of cyclosporine [87].

9.7.18 Sinomenine

It is an alkaloid found from *Sinomenium acutum*, family Menispermaceae, commonly used to treat rheumatoid arthritis in China and Japan. Studies of administration of sinomenine alone and combined with paeoniflorin (150 mg/kg) through gastric gavage in rats were conducted by Chan et al. All the pharmacokinetics parameters results have shown significant improvement in the bioavailability of paeoniflorin [88]. The proposed mechanism of action involved in the enhancement of paeoniflorin's bioavailability was, inhibition of P-gp efflux pumps [89].

9.7.19 Gallic acid

Gallic acid is a main component present in gallnut, sumac, witch hazel, tea leaves, and oak bark. Esters of gallic acid showed the enhancement of bioavailability of drug nifedipine [90]. Coadministration of gallic acid with piperine exerts a synergistic effect and produces a greater therapeutic potential in reduction of beryllium-induced hepatorenal dysfunction and the moments of oxidative stress [91].

9.8 Bioflavonoids used as herbal bioenhancers

Bioflavonoids are the biologically active components of the group flavonoids and were first discovered in 1936, by Nobel Prize winner and pioneer of vitamin C research, Albert Szent – Gyorgi. The prominent bioflavonoids, quercetin, genistein, and naringin are presented abundantly in paste and wools of citrus fruits and foods such as soybeans and root vegetables containing vitamin C. They are used to enhance the biological activity of certain drugs.

9.8.1 Quercetin

It is a plant-based flavonoid present in fruits, vegetables, leaves, and grains and has shown effects as an analgesic, antineoplastic agent, antioxidant, radical scavenging, and antiviral agent. It exhibits action by inhibition of CYP3A4 enzyme and P-gp efflux pump. If it is coadministered with drugs such as diltiazem, digoxin, doxorubicin, epigallocatechin gallate, fexofenadine, verapamil, etoposide, tamoxifen, and paclitaxel, it showed an increase in bioavailability of drugs [92–96].

9.8.1.1 Quercetin with paclitaxel

J. S. Choi studied the effect of administration of drug paclitaxel 40 mg/kg alone and its prodrug with dose of at 280 mg /kg to the pre-treated rats administered with quercetin (2, 10, 20 mg/kg) shown increase in bioavailability of drug paclitaxel [97].

9.8.1.2 Quercetin and verapamil

Choi studied the effect of coadministration of quercetin and norverapamil in rabbits. Results showed that oral absorption of verapamil was significantly increased when pretreated with quercetin. The study concludes that pretreatment of quercetin enhanced the absolute and relative bioavailability of verapamil [98].

9.8.1.3 Quercetin and diltiazem

J. S. Choi studied the influence of quercetin pretreatment on the absorption of diltiazem in rabbits, following oral administration. Results indicated that only the concentration of diltiazem in the rabbit plasma was increased when pre-treated with quercetin, compared to the control, whereas coadministration with quercetin was not enhanced significantly [93].

9.8.1.4 Quercetin with tamoxifen

Quercetin increases the relative bioavailability of tamoxifen by 1.20–1.61 times higher than the control group after its oral administration at dose of 10 mg/kg. Coadministration of quercetin promotes the duodenal absorption, reduces the first-pass hepatic metabolism of drug, and leads to increase in absorption and bioavailability [94].

9.8.1.5 Quercetin with fexofenadine

Kim and his coworkers studied the influence of quercetin on absorption and bio-availability of fexofenadine given orally at a dose of 60 mg in combination with quercetin to 12 healthy human volunteers. The results of the study showed increase in plasma concentration of fexofenadine in humans [99].

9.8.1.6 Quercetin and doxorubicin

Choi et al. studied the influence of coadministration of quercetin in rats with 50 mg/kg oral dose or 10 mg/kg by intravenous administration. The results of pharmacokinetic parameters showed increase in absolute bioavailability of drug compared to control, and the relative bioavailability of doxorubicin enhanced 1.32- to 2.36-fold. However, drug administered intravenously could not affect any pharmacokinetic parameters of drug due to coadministration of quercetin [95].

9.8.2 Naringin

It is the main glycoside of the flavonoid class existing naturally in citrus fruit, grapefruit and also in apple, onion, and tea. It shows many pharmacological effects like antioxidant, antiulcer, antiallergic, anticancer, and blood lipid lowering activity. It exhibits bioenhancer effect by inhibition of CYP3A4 enzyme and acts as a P-gp modulator [100–103].

9.8.2.1 Naringin and diltiazem

Choi et al. studied the effect of coadministration of naringin with diltiazem at a dose of 15 mg/kg and drug alone in rats. Study concluded that absolute and relative bioavailability of diltiazem was enhanced with increase in peak plasma concentration and duration of action of drug 2-fold when pretreated with bioenhancer [104].

9.8.2.2 Naringin and paclitaxel

Lim et al. reviewed the influence of naringin given orally 30 min before the intravenous administration of paclitaxel at 3 mg/kg body weight on the pharmacokinetic parameters using rats. AUC was found to be significantly more than control group after intravenous administration of paclitaxel, due to inhibition of CYP3A1/2 drug metabolizing enzyme, due to oral gavage of naringin. Oral naringin inhibits the hepatic metabolism by P-gp and increases the AUC of intravenous paclitaxel [105].

9.8.2.3 Naringin and verapamil

Yeum and his coworkers studied the absorption of verapamil in rabbits pretreated with herbal bioenhancer, naringin, at concentration of 1.5, 7.5, and 15 mg/kg. The study concluded that absorption and bioavailability of verapamil increased significantly in rabbits after oral administration. Thus, pretreatment of bioenhancer, naringin, augmented the absolute bioavailability and relative bioavailability 1.26- to 1.69-fold in a dose-dependent manner [106].

9.8.3 Genistein

It is also a flavonoid phytoestrogen that belongs to the isoflavones class present in a variety of nutritional plants such as lupin (*Lupinus albus* L.), fava beans (*Vicia faba* L.), kudzu, and soybean. It shows its biopotentiation activity by acting on P-gp and breast cancer resistant protein efflux pump inhibitor. Coadministration of bioenhancer genistein with paclitaxel concurrently enhanced the drug bioavailability by increasing the amount of drug in systemic circulation by 54.7% and decreasing the total plasma clearance by 35.2%. It also enhanced the cytosolic epigallocatechin gallate 2- to 5-fold compared with epigallocatechin gallate alone, when coadministered with HT-29 human colon cancer cells [107].

9.9 Bioenhancers of animal origin

The following are a few bioenhancers of animal origin useful in enhancing the bioavailability of drugs due its own incredible pharmacologic effects.

9.9.1 Cow urine

It is responsible for modifying various body functions, including immunity. Cow urine distillate was found to be more effective than cow urine, when used as a bioenhancer. It enhances the absorption of many drugs like antimicrobial, antifungal, and anticancer due to its probable action by enhancing the permeation through the gastrointestinal tract membrane as well as across artificial membrane.

The bioavailability of rifampicin was enhanced against *Escherichia coli* and Gram-positive bacteria 5- to 7-fold and 3- to 11-fold, respectively, by cow urine.

Cow urine increases B- and T-lymphocyte blastogenesis effect and immuno-globulin antibody titers in mice. Mature male mice showed increase in fertility rate, 90–100% feasibility and lactation index, when they were exposed to cow urine distillate with zinc sulfate and cadmium chloride, compared to 0% fertility rate if cadmium chloride administered alone. Animals given treatment with cadmium chloride and cow urine showed increase in Fertility index. Therefore, the study concludes that cow urine distillate shows antitoxicity and is used as a bio-enhancer for zinc against cadmium chloride. It has been observed that potency of "Taxol" (paclitaxel) was increased against MCF-7, a human breast cancer cell line, due to administration of cow urine distillate.

The immunomodulatory properties of cow urine enhance the concentration of generative hormones and estrous cycle of mice like gonadotropin-releasing hormone conjugates. There was significant improvement observed in sperm motility rate, sperm count, gonadosomatic index, and morphology of sperm for male mice in 90–120 days for the treated group of animals ($P < 0.05$) due to effect of gonadotropin-releasing hormone by administration of cow urine distillate [108, 109].

9.9.2 Ghee

Ghee is also called as *tup*, *ghi*, *ghio*, or *neyy*. Cow ghee is mostly used in Ayurveda; it is satvik in nature, i.e., *sattva guni*. It shows its biopotentiation effect as bioenhancer in many Ayurvedic formulations such as *Brahmi ghrita* and *Trikatrayadi Lauha*.

9.9.3 Honey

It is also called *madhu*. Honey is a sweet food obtained from bees using nectar of flowers. In modern practice of medicines, it is used in crystallized, pasteurized, raw, strained, filtered, ultrasonicated, creamed, or dried forms; nowadays honey decoctions are also available. It is mainly used a bioenhancer in Trikatrayadi Lauha [110].

9.10 Recently introduced bioenhancers

9.10.1 Resveratrol

It is a stilbenoid (3, 5, 4′-trihydroxy-*trans*-stilbene) produced as a result of an attack or response to injury on plants by pathogens like bacteria and fungi that originate from the coating of grape, blueberry, raspberry, mulberry etc. If it is coadministered with apigenin (a natural anti-inflammatory agent), it exhibited increase in plasma drug levels by 2.39 times, compared to administration of apigenin alone [42]. This result indicated that resveratrol helps enhance the absorption of apigenin to bypass the hepatic metabolism [111].

9.10.2 Lutein

Lutein is a xanthophyll, plentifully found in green vegetables such as spinach, kale, and yellow carrots. It shows opthalmoprotective, anti-atherogenic, and anti-carcinogenic activities [112].

9.10.3 Friedelin

Friedelin is a biologically active constituent obtained from plant species such as *Azima tetracantha* Lam. (Salvadoraceae), *Orostachys japonica* A. Berger (Crassula-ceae), and *Quercus stenophylla* Blume. (Fagaceae). If friedelin (50 mg/kg) is coadministered with apigenin orally in rats, it enhances the absorption of apigenin. The suppression of ATPase enzyme activity of P-gp indicates increase in bioavailability of apigenin due to inhibition of P-gp [113].

9.11 Applications of herbal bioenhancers

Recently many researchers have studied bioenhancers of herbal origin by using various lipid-based technologies such as microspheres, liposomes, transferosomes, ethosomes, nanoemulsions, microemulsions, other vesicular herbal formulations, and polymeric micelle formulations. These techniques of bioenhancement are basically targeted for the group of toxic drugs, expensive drugs, drugs with poor bioavailability, and drugs used in chronic therapy for long durations. Piperine is marketed in India as the first mono-preparation bioenhancer and as a constituent of nutritional inert material that contains different vitamins, curcumin resveratrol, or coenzymes. Piperine, used as an herbal bioenhancer in humans, was first approved in India as a fixed dose combination product to treat tuberculosis with 200 mg of rifampicin IP, 300 mg of isoniazid IP, and 10 mg of piperine, instead of high amount of rifampicin 450 mg and isoniazid 300 mg. Herbal bioenhancers reduce the dose and cost of expensive medication, while making treatment safer. The recent advancements in formulation development as a part of innovation and use of bioavailability enhancers of herbal origin have led to several NDDS and some of the patented formulations represented in Tables 9.2 and 9.3.

Table 9.2: Some scientifically documented formulations using herbal bioenhancers.

Type of dosage forms	Drug	Use or application of bioenhancer	Pharmacological effect	Administration route	References
Quercetin liposome	Quercetin	Reduction in dose, enhanced permeability in blood–brain barrier	Antioxidant, anti cancer	Intranasal	[114]
Liposome	Silymarin	Enhances bioavailability	Hepatoprotective	Buccal	[115]
Alginate chitosan microspheres	Rutin	Target to cardiovascular and cerebrovascular system	Cardiovascular and cerebrovascular	In vitro	[116]
Microspheres	Zedoary oil	Prolongs release of drug and Improves bioavailability	Hepatoprotective	Oral	[117]
Triptolide nanoparticles	Triptolide	Increases the permeability of drug through skin	Anti-inflammatory	Topical	[118]
Nanoparticles	Radix salvia	Increases the bioavailability	Antihypertensive effect	In-vitro	[119]
Transferosomes	Capsaicin	Increases skin permeability of drug	Analgesic	Topical	[120]
Transferosomes	Colchicine	Increases skin permeability of drug	Antigout	In-vitro	[121]
Ginseng lipid-based preparations	Flavonoids	Increases absorption	Nutraceutical immune modulator	Oral	[122]
Green tea lipid-based formulation	Ginsenoside	Improves absorption and bioavailability	Nutraceutical, antioxidant, and anticancer agent	Oral	[122]

Table 9.3: Patented formulations with herbal bioenhancers [123].

US patent no.	Drug	Type of formulation
5948414	Opioid analgesic and aloe	Nasal spray
6340478 B1	Ginsenoside	Microencapsulated DDS
6890561 B1	Isoflavone	Microcapsules
6896898 B1	Alkaloids of aconitum specie	Transdermal patch
2005/0142232 A	Oleaginous oil of *Sesamum indicum* and alcoholic extract of *Centella asiatica*	Brain tonic
2007/0042062 A1	Glycine max containing 7s globulin Protein extract, curcumin, *Zingiber officinale*	Herbal tablet
2007/0077284A1	Opioid analgesics	Transdermal patch
7569236132	Flavonoids – quercetin and terpene	Microgranules

9.12 Hurdles with herbal bioenhancers

Bioenhancers have made a great revolution in the field of drug delivery systems, but then there are also various hurdles that need to be overcome, in order to meet with success. Although, some of the challenges have to be modified or solved, which includes improvement in the properties of drug preparations like blood circulation, increase in effective surface area, avoiding the degradation of encapsulated drug, improvement of drug permeability through biological membranes, and site-specific targeting of the drug. To scale up techniques and pilot techniques for commercialization of all, the development of formulations at large-scale productions is also another challenge before the scientist. Large-scale production is another problem associated with use of herbal bioenhancers, which includes low concentration of nanomaterials, glomeration, and chemistry processes, which can be easily modified at laboratory level. Present advancements in herbal bioenhancers have brought in a new challenge for regulatory control of formulations, because there are no regulatory standards available, yet, for determination and evaluation of physical, chemical, and pharmacokinetic parameters of drug products containing herbal bioenhancers by US-FDA and EMEA. Therefore, regulatory authorities should take the initiative to set standards for drug products containing herbal bioenhancers [124].

9.13 Future scope

The concept of "Trikatu" in Ayurveda used an herbal bioenhancer has successfully taken the lead in improving bioavailability of many modern medicines. The researchers now aim at using bioenhancers along with the main pharmacological active ingredient as a method for reduction of the dose and hence low treatment cost for financially challenged people in society. Various research works related to the development of bioenhancers of herbal and nonherbal origin are in process. Nowadays, various research studies are going on in pain management, stress, obesity, anticancer, etc., related to development of bioavailability enhancers of herbal and nonherbal origin. Unexplored areas of traditional system of medicines and Ayurveda could be an additional area where new breakthroughs can be expected. Studies on all categories of drugs along with knowledge of the exact mechanism of bioenhancers could be the next step in research. Modern dosage forms can be prepared by using Yogvahi, like conventional dosage forms.

9.14 Conclusion

The innovative concept of use of herbal bioenhancers in traditional system of medicines has resulted in a great revolution in the field of medicines. However, the knowledge of bioenhancers as an integral part of drug design and development process in Ayurveda is somewhat novel to Western medicine. They are safe, effective, easily available, nonaddictive, and have an effect on many classes of drugs. This will lead to reduction in drug dosage, toxicity, and cost and benefit the national economy. Developing countries like India have a major concern regarding the use of modern medicines – the overall cost of disease treatment. A major concern of a new drug development process is the economics of drug development. Therefore, the scientific community has its eyes on development of effective and optimized formulations to meet the requisite pharmacokinetic parameters of dietary components to expand in vivo absorption and exploit their use as an effective bioenhancers. This strategic development would lead to the reduction in cost of dosage, and hence the entire allopathic treatment must be made inexpensive and cost effective to each and every section of the society, including the financially challenged.

The oral bioavailability of many poorly bioavailable drugs was found to be enhanced due to biopotentiation effect of nature originated agents. Various research works have been done using piperine derivatives and other novel bioenhancers. The existing scientific research on use of herbal bioenhancers like piperine, ginger, etc. has shown significant improvement in the bioavailability of many drugs and nutraceuticals when coadministered or pre-treated with them. Animal-originated cow urine, ghee, and honey are also breakthroughs in the medical field. Herbal bioenhancers decrease the normal dose of potent drugs and nutraceuticals, leading to reduction in drug resistance and toxicity, and cuts the treatment period. This combination of traditional system of medicine and technology satisfies all essential criteria for safe, ideal, and effective drug delivery system. Hence, there is a need for more multicenter trials with a combination of new drugs and natural bioenhancers, in order to take advantage of higher potency with a lower drug toxicity.

List of abbreviations

A.D.	Anno Domini
B.C.	Before Christ
CNS	Central nervous system
CYP	Cytochrome P450
DNA	Deoxyribonucleic acid
FDA	Food Drug Administration
GGT	Gamma glutamyl transpeptidase
GIT	Gastrointestinal tract
GRAS	Generally recognized as safe material
NDDS	Novel drug delivery system
NSAIDS	Nonsteroidal anti-inflammatory drugs
P-gp	P-glycoprotein
RRL	Regional research laboratory
RNA	Ribonucleic acid
UDP	Uridine diphosphate

References

[1]   Kalia AN, Textbook of industrial pharmacognosy, 1st CBS publishers, 2006, 1–3.
[2]   Jhanwar B, Gupta S, 2014, Biopotentiation using Herbs: Novel Technique for Poor Bioavailable Drugs, International Journal of PharmTech Research, 2014, 6(2), 443–454.
[3]   Brahmankar DB, Jaiswal S, Biopharmaceutics and pharmacokinetics: A treatise, 1st (Edn.) Vallabh Prakashan, 1995, 24–26.
[4]   Atal N, Bedi KL, Bioenhancers: Revolutionary concept to market, Journal of Ayurveda and integrative medicine, 2005, 1(2), 96–99.
[5]   Sathyanarayana B, Ayurvedic bioenhancers: A classical and contemporary review, Natural Products Chemistry & Research, 2015, 3, 6.
[6]   Randhawa GK, Kullar JS, Rajkumar, Bioenhancers from mother nature and their applicability in modern medicine, International Journal of Applied and Basic Medical Research, 2011, 1(1), 5–10.
[7]   Jain G, Patil UK. Strategies for enhancement of bioavailability of medicinal agents with natural products.2015, IJPSR 6(12): 5315–5324.
[8]   Gopal V, Yoganandam P, Bio-enhancer: A pharmacognostic perspective, European Journal of Molecular Biology and Biochemistry, 2016, 3(1), 33–38.

[9]   Shanmugam S, Natural Bioenhancers: Current Outlook, Clinical Pharmacology & Biopharmaceutics, 2015, 4(2), 1–3.

[10] Atal CK, A breakthrough in drug bioavailability-a clue from age old wisdom of Ayurveda, IMDA Bulletin, 1979, 10, 483–484.

[11] Johri RK, Zutshi U, An Ayurvedic formulation "Trikatu" and its constituents, Journal of Ethnopharmacology, 1992, 37:85-91.

[12] Annamalai AR, Manavalan R, Effects of "Trikatu" and its individual components and piperine on gastro intestinal tracts: Trikatu: A bioavailable enhancer, Indian Drugs, 1990, 27(12), 595–604.

[13] Lundin S, Artursson P, Absorption enhancers as an effective method in improving the intestinal absorption, International Journal of Pharmaceutics, 1990, 64, 181–186.

[14] Bhavsar KS, Singh R, Devi S, Indian Herbal Bioenhancers: A Review, Pharmacognosy Reviews, 2009, 3(5), 80–82.

[15] Kang MJ, Cho JY, Shim BH, Kim DK, Lee J, Bioavailability enhancing activities of natural compounds from medicinal plants, Journal of Medicinal Plant Research, 2009, 3(13), 1204–1211.

[16] Dudhatra GB, Mody SK, Awale MM, Patel HB, Modi CM, Kumar A, Kamani DR, Chauhan BN, A comprehensive review on pharmacotherapeutics of herbal bioenhancers, The Scientific World Journal, 2012, 63, 53–79.

[17] Varma MV, Ashokraj Y, Dey CS, Panchagnula R, P – glycoprotein inhibitors and their screening: A perspective from bioavailability enhancement, Pharmacological Research: The Official Journal of the Italian Pharmacological Society, 2003, 48(4), 347–359.

[18] Randhawa GK, Kullar JS, Rajkumar, Bioenhancers from Mother Nature and their applicability in modern medicine, International Journal of Applied and Basic Medical Research, 2011, 1(1), 5–10.

[19] Ratndeep S, Devi Sarita B, Patel Jatin H, Patel Urvesh D, Indian herbal bio-enhancers: A review, Pharmacognosy Reviews, 2009, 3, 80-2.

[20] Atal CK, Zutshi U, Rao PG, Scientific evidence on the role of Ayurvedic herbals on bioavailability of drugs, Journal of Ethnopharmacology, 1981, 4:229-32.

[21] Kesarwani K, Gupta R, Mukerjee A, Bioavailability enhancers of herbal origin: An overview, Asian Pacific Journal of Tropical Biomedicine, 2013, 3, 253-66.

[22] Bano G, Amla V, Raina RK, Zutshi U, Chopra CL, The effect of piperine on pharmacokinetics of phenytoin in healthy volunteers, Planta Medica, 1987, 53, 568-9.

[23] Bano G, Raina RK, Zutshi U, Bedi KL, Johri RK, Sharma SC, Effect of piperine on bioavailability and pharmacokinetics of propranolol and theophylline in healthy volunteers, European Journal of Clinical Pharmacology, 1991, 41, 615-7.

[24] Bhardwaj RK, Glaeser H, Becquemont L, Klotz U, Gupta SK, Fromm MF, Piperine, a major constituent of black pepper, inhibits human P-glycoprotein and CYP3A4, Journal of Pharmacology and Experimental Therapeutics, 2002, 302(2), 645–650.

[25] Kesarwani K, Gupta R, Bioavailability enhancers of herbal origin: An overview, Asian Pacific Journal of Tropical Biomedicine, 2013, 3(4), 253–266.

[26] Muttepawar SS, Jadhav SB, Kankudate AD, Sanghai SD, Usturge DR, Chavare SS, A review on bioavailability enhancers of herbal origin, World Journal of Pharmacy and Pharmaceutical Sciences, 2014, 3(3), 667–677.

[27] Ajazuddin AA, Qureshi A, Kumari L, Vaishnav P, Sharma M, Saraf S, Saraf S, Role of herbal bioactives as a potential bioavailability enhancer for Active Pharmaceutical Ingredients, Fitoterapia, 2014, 97, 1–14.

[28] Balakrishnan V, Varma S, Chatterji D, Piperine augments transcription inhibitory activity of rifampicin by several fold in Mycobacterium smegmatis, Current Science, 2001, 80(10), 1302–1305.

[29] Hiwale AR, Dhuley JN, Naik SR, Effect of co-administration of piperine on pharmacokinetics of β-lactam antibiotics in rats, Indian Journal of Experimental Biology, 2002, 40(3), 277–281.

[30] Balkrishna BS, Yogesh PV, Influence of co-administration of piperine on pharmacokinetic profile of ciprofloxacin, Indian Drugs, 2002, 39(3), 166–168.

[31] Ia K, Zm M, Kumar A, Verma V, Qazi GN, "Piperine, a phytochemical potentiator of ciprofloxacin against Staphylococcus aureus, Antimicrobial Agents and Chemotherapy, 2006, 50(2), 810–812.

[32] Singh M, Varshneya C, Telang RS, Srivastava AK, Alteration of pharmacokinetics of oxytetracycline following oral administration of Piper longum in hens, Journal of veterinary science, 2005, 6(3), 197–200.

[33] Kasibhatta R, Naidu MUR, Influence of piperine on the pharmacokinetics of nevirapine under fasting conditions: A randomised, crossover, placebo-controlled study, Drugs in R and D, 2007, 8(6), 383–391.

[34] Pooja S, Agrawal RP, Nyati P, Savita V, Phadnis P, Analgesic activity of Piper nigrumextract per se and its interaction with diclofenac sodium and pentazocine in albino mice, International Journal of Pharmacology, 2007, 5(1), 3–9.

[35] Janakiraman K, Manavalan R, Compatibility and stability studies of ampicillin trihydrate and piperine mixture, International Journal of Pharmaceutical Sciences and Research, 2008, 2, 1176–1181.

[36] Jin MJ, Han HK, Effect of piperine, a major component of black pepper, on the intestinal absorption of fexofenadine and its implication on food-drug interaction, Journal of Food Science, 2010, 75(3), H93–H96.

[37] Singh A, Vk P, Mh J, Ra P, Sharma G, In vivo assessment of enhanced bioavailability of metronidazole with piperine in rabbits, Research Journal of Pharmaceutical, Biological and Chemical Sciences, 2010, 1(4), 273–278.

[38] Venkatesh S, Durga KD, Padmavathi Y, Reddy BM, Mullangi R, Influence of piperine on ibuprofen induced antinociception and its pharmacokinetics, Drug Research, 2011, 61, 506–509.

[39] Singh A, Jain DA, Kumar S, Jaiswal N, Singh RK, Patel PS, Enhanced bioavailability of losartan potassium with piperine in rats, International Journal of Pharmacy Research & Technology, 2012, 2, 34–36.

[40] Bano G, Raina RK, Zutshi U, Bedi KL, Johri RK, Sharma SC, Effect of piperine on bioavailability and pharmacokinetics of propranolol and theophylline in healthy volunteers, European Journal of Clinical Pharmacology, 1991, 41(6), 615–617.

[41] Gupta SK, Velpandian T, Sengupta S, Mathur P, Sapra P, Influence of piperine on nimesulide induced anti nociception, Phytotherapy Research, 1998, 12, 266–269.

[42] Shoba G, Joy D, Joseph T, Majeed M, Rajendran R, Srinivas PS, Influence of piperine on the pharmacokinetics of curcumin in animals and human volunteers, Planta Medica, 1998, 64(4), 353–356.

[43] Govindarajan VS, Ginger-chemistry, technology, and quality evaluation: Part 2, Critical Reviews in Food Science and Nutrition, 1982, 17(3), 189–258.

[44] Jolad SD, Lantz RC, Solyom AM, Chen GJ, Bates RB, Timmermann BN, Fresh organically grown ginger (Zingiber officinale): Composition and effects on LPS-induced PGE2 production, Phytochemistry, 2004, 65(13), 1937–1954.

[45] Evans WC, Ginger. Trease and Evans Pharmacognosy, WB Saunders,, Edinburgh, UK, 15th, 2002.

[46] Aruna K, Sivaramakrishnan VM, Anticarcinogenic effects of some Indian plant products, Food and Chemical Toxicology, 1992, 30(11), 953–956.

[47] Guevara AP, Vargas C, Anti-inflammatory and antitumor activities of seed extracts of malunggay, Moringa oleifera L (Moringaceae)., Philippine Journal of Science, 1996, 125, 175–184.

[48] Gilani AH, Aftab K, Suria A, et al., Pharmacological studies on hypotensive and spasmolytic activities of pure compounds from Moringa oleifera, Phytotherapy Research, 1994, 8(2), 87–91.

[49] Nwosu MO, Okafor JI, Preliminary studies of the antifungal activities of some medicinal plants against Basid iobolus and some other pathogenic fungi, Mycoses, 1995, 38(5–6), 191–195.

[50] Pal SK, Mukherjee PK, Saha BP, Studies on the antiulcer activity of Moringa oleifera leaf extract on gastric ulcer models in rats, Phytotherapy Research, 1995, 9(6), 463–465.

[51] Siddhuraju P, Becker K, Antioxidant properties of various solvent extracts of total phenolic constituents from three different agroclimatic origins of drumstick tree (Moringa oleifera Lam.) leaves, Journal of Agricultural and Food Chemistry, 2003, 51(8), 2144–2155.

[52] Oinam N, Urooj A, Phillips PP, Niranjan NP, Effect of dietary lipids and drumstick leaves (Moringa oleifera) on lipid profile and antioxidant parameters in rats, Food and Nutrition Sciences, 2012, 3, 141–145.

[53] Pari L, Kumar NA, Hepatoprotective activity of Moringa oleifera on antitubercular drug-induced liver damage in rats, Journal of Medicinal Food, 2002, 5(3), 171–177.

[54] Mehta LK, Balaraman R, Amin AH, Bafna PA, Gulati OD, Effect of fruits of Moringa oleifera on the lipid profile of normal and hypercholesterolaemic rabbits, Journal of Ethnopharmacology, 2003, 86(2–3), 191–195.

[55] Saravillo KB, Herrera AA, Biological activity of Moringa oleifera Lam (Malunggay) crude seed extract, Philippine Agricultural Scientist, 2004, 87(1), 96–100.

[56] Mahajan SG, Mali RG, Mehta AA, Protective effect of ethanolic extract of seeds of Moringa oleifera Lam against inflammation associated with development of arthritis in rat, Journal of Immunotoxicology, 2007, 4(1), 39–47.

[57] Khanuja SPS, Arya JS, Ranganathan T et al. Nitrile glycoside useful as a bioenhancer of drugs and nutrients, process of its isolation from Moringa oleifera. United States Patent Number, US006858588B2, 2005.

[58] Kiso Y, Tohkin M, Hikino H, Mechanism of anti- hepatotoxic activity of glycyrrhizin, I: Effect on free radical generation and lipid peroxidation, Planta Medica, 1984, 50(4), 298–302.

[59] Nose M, Ito M, Kamimura K, Shimizu M, Ogihara Y, A comparison of the antihepatotoxic activity between glycyrrhizin and glycyrrhetinic acid, Planta Medica, 1994, 60(2), 136–139.

[60] Akamatsu H, Komura J, Asada JY, Niwa Y, Mechanism of anti-inflammatory action of glycyrrhizin: Effect on neutrophil functions including reactive oxygen species generation, Planta Medica, 1991, 57(2), 119–121.

[61] Fujisawa Y, Sakamoto M, Matsushita M, Fujita T, Nishioka K, Glycyrrhizin inhibits the lytic pathway of complement possible mechanism of its anti-inflammatory effect on liver cells in viral hepatitis, Microbiology and Immunology, 2000, 44((9)), 799–804.

[62] Shibata S, Antitumor promoting and anti-inflammatory activities of liquorice principles and their modified compounds, Food Phytochemicals for Cancer Prevention-II, 1994, 31, 308–321.

[63] Utsunomiya T, Kobayashi M, Pollard RB, Suzuki F, Glycyrrhizin, an active component of liquorice roots, reduces morbidity and mortality of mice infected with lethal doses of influenza virus, Antimicrobial Agents and Chemotherapy, 1997, 41(3), 551–556.

[64] Khanuja SPS, Kumar S, Arya JS et al., "Composition comprising pharmaceutical/nutraceutical agent and a bioenhancer obtained from Glycyrrhiza glabra," United States Patent Number, US006979471B1, 2005.

[65] Khanuja SPS, Kumar S, Arya JS et al. Composition comprising pharmaceutical/nutraceutical agent and a bioenhancer obtained from Glycyrrhiza glabra. United States Patent Number, 2006, US20060057234A1.

[66] Qazi GN, Bedi KL, Rk J et al., Bioavailability enhancing activity of Cuminum cyminum extracts and fractions thereof. World Intellectual Property Organization, International Publication Number, 2003, WO03075685A2.

[67] Qazi GN, Bedi KL, Johri RK et al. Bioavailability/bio efficacy enhancing activity of Cuminum cyminum and extracts and fractions thereof. United States Patent Number, 2009, US007514105B2.

[68] Toxopeus H, Bouwmeester HJ, Improvement of caraway essential oil and carvone production in The Netherlands, Industrial Crops and Products, 1992, 1(2–4), 295–301.

[69] Qazi GN, Bedi KL, Johri RK et al. Bioavailability enhancing activity of Carum carvi extracts and fractions thereof. United States Patent Number, 2007, US20070020347A1.

[70] Qazi GN, Bedi KL, Johri RK et al. Bioavailability enhancing activity of Carum carvi extracts and fractions thereof," United States Patent Number, 2003, US20030228381A1.

[71] Ogita A, Hirooka K, Yamamoto Y, et al., Synergistic fungicidal activity of Cu2+ and allicin, an allyl sulfur compound from garlic, and its relation to the role of alkyl hydroperoxide reductase 1 as a cell surface defense in Saccharomyces cerevisiae, Toxicology, 2005, 215(3), 205–213.

[72] Ogita A, Fujita KI, Tanaka T, Enhancement of the fungicidal activity of amphotericin B by allicin: Effects on intracellular ergosterol trafficking, Planta Medica, 2009, 75(3), 222–226.

[73] Ogita A, Fujita KI, Tanaka T, Dependence of vacuole disruption and independence of potassium ion efflux in fungicidal activity induced by combination of amphotericin B and allicin against Saccharomyces cerevisiae, The Journal of Antibiotics, 2010, 63(12), 689–692.

[74] Patil S, Dash RP, Anandjiwala S, Nivsarkar M, Simultaneous quantification of berberine and lysergol by HPLC- UV: Evidence that lysergol enhances the oral bioavailability of berberine in rats, Biomedical Chromatography, 2012, 26(10), 1170–1175.

[75] Vinson JA, Kharrat HA, Andreoli L, Effect of Aloe vera preparations on the human bioavailability of vitamins C and E, Phytomedicine, 2005, 12(10), 760–765.

[76] Zhang W, Tan TMC, Lim LY, Impact of curcumin- induced changes in P-glycoprotein and CYP3A expression on the pharmacokinetics of peroral celiprolol and midazolam in rats, Drug Metabolism and Disposition, 2007, 35(1), 110–115.

[77] Pavithra BH, Prakash N, Jayakumar K, Modification of pharmacokinetics of norfloxacin following oral administration of curcumin in rabbits, Journal of Veterinary Science, 2009, 10 (4), 293–297.

[78] Cho SW, Lee SH, Choi SH, Enhanced oral bioavailability of poorly absorbed drugs. I. Screening of absorption carrier for the ceftriaxone complex, Journal of pharmaceutical sciences, 2004, 93(3), 612–620.

[79] Singh G, Screening of Herbal Fractions for Antibiotic Drug Resistance Reversal. In: Thesis. Thapar Institute of Engineering and Technology, Patiala, India, 2005, 44.

[80] Upadhyay HC, Dwivedi GR, Darokar MP, Chaturvedi V, Srivastava SK, Bioenhancing and antimycobacterial agents from Ammannia multiflora, Planta medica, 2012, 78(1), 79–81.

[81] Gokaraju GR, Gokaraju RR Bioavailability/bio-efficacy enhancing activity of Stevia rebaudiana and extracts and fractions and compounds thereof. United States Patent Number, 2010,US2010011 2101A1.

[82] Lambert JD, Hong J, Yang GY, Liao J, Yang CS, Inhibition of carcinogenesis by polyphenols: Evidence from laboratory investigations, The American Journal of Clinical Nutrition, 2005, 81 (1), 284S–291S.

[83] Kurzer MS, Xu X, Dietary phytoestrogens, Annual Review of Nutrition, 1997, 17, 353–381.

[84] Li X, Choi JS, Effect of genistein on the pharmacokinetics of paclitaxel administered orally or intravenously in rats, International Journal of Pharmaceutics, 2007, 337(1–2), 188–193.

[85] Lambert JD, Kwon SJ, Ju J, et al., Effect of genistein on the bioavailability and intestinal cancer chemopreventive activity of (-) epigallocatechin-3-gallate, Carcinogenesis, 2008, 29 (10), 2019–2024.

[86] Bouraoui A, Toumi A, Mustapha BH, Brazier JL, Effects of capsicum fruit on theophylline absorption and bioavailability in rabbits, Drug-nutrient interactions, 1988, 5(4), 345–350.

[87] Wacher VJ, Wong S, Wong HT, Peppermint oil enhances cyclosporine oral bioavailability in rats: Comparison with d – Alpha tocopheryl poly (ethylene glycol 1000) succinate (TPGS) and ketoconazole, Journal of Pharmaceutical Sciences, 2002, 91(1), 77–90.

[88] Pavithra BH, Prakash N, Jayakumar K, Modification of pharmacokinetics of norfloxacin following oral administration of curcumin in rabbits, Journal of Veterinary Science, 2009, 10 (4), 293–297.

[89] Liu ZQ, Zhou HL, Liu, et al Influence of co-administrated sinomenine on pharmacokinetic fate of paeoniflorin in unrestrained conscious rats, Journal of ethnopharmacology, 2005, 99(1), 61–67.

[90] Wacher VJ, Benet ZL Use of gallic acid esters to increase bioavailability of orally administered pharmaceutical compounds. US Patent, 2001,US6180666 B.

[91] Zhao JQ, Du GZ, Xiong YC, Wen YF, Bhadauria M, Nirala SK, Attenuation of beryllium induced hepatorenal dysfunction and oxidative stress in rodents by combined effect of gallic acid and piperine, Archives of pharmacal research, 2007, 30(12), 1575–1583.

[92] Hsiu SL, Hou YC, Wang YH, Tsao CW, Su SF, Chao PDL, Quercetin significantly decreased cyclosporin oral bioavailability in pigs and rats, Life Sciences, 2002, 72(3), 227–235.

[93] Choi JS, Li X, Enhanced diltiazem bioavailability after oral administration of diltiazem with quercetin to rabbits, International Journal of Pharmaceutics, 2005, 297(1–2), 1–8.

[94] Shin SC, Choi JS, Li X, Enhanced bioavailability of tamoxifen after oral administration of tamoxifen with quercetin in rats, International Journal of Pharmaceutics, 2006, 313(1–2), 144–149.

[95] Choi JS, Piao YJ, Kang KW, Effects of quercetin on the bioavailability of doxorubicin in rats: Role of CYP3A4 and P-gp inhibition by quercetin, Archives of Pharmacal Research, 2011, 34 (4), 607–613.

[96] Wang YH, Chao PDL, Hsiu SL, Wen KC, Hou YC, Lethal quercetin-digoxin interaction in pigs, Life Sciences, 2004, 74(10), 1191–1197.

[97] Choi JS, Jo BW, Kim YC, Enhanced paclitaxel bioavailability after oral administration of paclitaxel or pro- drug to rats pretreated with quercetin, European Journal of Pharmaceutics and Biopharmaceutics, 2004, 57(2), 313–318.

[98] Choi JS, Han HK, The effect of quercetin on the pharmacokinetics of verapamil and its major metabolite, norverapamil, in rabbits, Journal of Pharmacy and Pharmacology, 2004, 56(12), 1537–1542.

[99] Kim KA, Park PW, Park JY, Short-term effect of quercetin on the pharmacokinetics of fexofenadine, a substrate of P-glycoprotein, in healthy volunteers, European Journal of Clinical Pharmacology, 2009, 65(6), 609–614.

[100] Dixon RA, Steele CL, Flavonoids and isoflavonoids a gold mine for metabolic engineering, Trends in Plant Science, 1999, 4(10), 394–400.

[101] Hodek P, Trefil P, Stiborova M, Flavonoids-potent and versatile biologically active compounds interacting with cytochromes P450, Chemico-Biological Interactions, 2002, 139 (1), 1–21.

[102] Dupuy J, Larrieu G, Sutra JF, Lespine A, Alvinerie M, Enhancement of moxidectin bioavailability in lamb by a natural flavonoid: Quercetin, Veterinary Parasitology, 2003, 112 (4), 337–347.

[103] Bardelmeijer HA, Beijnen JH, Brouwer KR, et al., Increased oral bioavailability of paclitaxel by GF120918 in mice through selective modulation of P-glycoprotein, Clinical Cancer Research, 2000, 6(11), 4416–4421.

[104] Choi JS, Han HK, Enhanced oral exposure of diltiazem by the concomitant use of naringin in rats, International Journal of Pharmaceutics, 2005, 305(1–2), 122–128.

[105] Lim SC, Choi JS, Effects of naringin on the pharmacokinetics of intravenous paclitaxel in rats, Biopharmaceutics & drug disposition, 2006, 27(9), 443–447.

[106] Yeum CH, Choi JS, Effect of naringin pre-treatment on bioavailability of verapamil in rabbits, Archives of Pharmacology Research, 2006, 29(1), 102–107.

[107] Lambert JD, Kwon SJ, Ju J, et al., Effect of genistein on the bioavailability and intestinal cancer chemopreventive activity of epigallocatechin-3-gallate, Carcinogenesis, 2008, 29(10), 2019–2024.

[108] Kekuda PT, Nishanth BC, Praveen Kumar SV, Kamal D, Sandeep M, Megharaj HK, Cow urine concentrate: A potent agent with antimicrobial and anthelminthic activity, Journal of Pharmacy Research, 2010, 3, 1025–1027.

[109] Khan A, Srivastava VK, Antitoxic and bioenhancing role of kamdhenu ark (cow urine distillate) on fertility rate of male mice (Mus musculus) affected by cadmium chloride toxicity, International Journal of Cow Science, 2005, 1, 43–46.

[110] Oladimeji FA, Adegbola AJ, Onyeji CO, Appraisal of Bioenhancers in Improving Oral Bioavailability: Applications to Herbal Medicinal Products, Journal of Pharmaceutical Research International, 2018, 1–23.

[111] Lee JA, Ha SK, Cho E, Choi I, Resveratrol as a bioenhancer to improve anti-inflammatory activities of apigenin, Nutrients, 2015, 7(11), 9650–9661.

[112] Doney JA, Carotenoids of enhanced bioavailability. US Patent, 2013, US 8613946 B2.

[113] Lee JA, Ha SK, Kim YC, Choi I, Effects of friedelin on the intestinal permeability and bioavailability of apigenin, Pharmacological reports: PR, 2017, 69(5), 1044–1048.

[114] Ajazuddin SS, Applications of novel drug delivery system for herbal formulations, Nanomedicine: Nanotechnology, Biology and Medicine, 2008, 4, 70–78.

[115] Samaligy MS, Afifi NN, Mahmoud EA, Evaluation of hybrid liposomes-encapsulated silymarin regarding physical stability and *in vivo* performance, International journal of pharmaceutics, 2006, 319(1–2), 121–129.

[116] Xiao L, Zhang YH, Xu JC, Jin XH, Preparation of floating rutinalginate – chitosan microcapsule, Chine Trad Herb Drugs, 2008, 2, 209–212.

[117] You J, Cui F, Han X, Wang Y, Yang L, et al., Study of the preparation of sustained-release microspheres containing zedoary turmeric oil by the emulsion solvent-diffusion method and evaluation of the self emulsification and bioavailability of the oil, Colloids and Surfaces B, 2004, 48(1), 35–41.

[118] Mei Z, Chen H, Weng T, Yang Y, Yang X, Solid lipid nanoparticle and microemulsion for topical triptolide, European Journal of Pharmaceutics and Biopharmaceutics, 2003, 56(2), 189–196.

[119] Su YL, Fu ZY, Zhang JY, Wang WM, Wang H, et al., Microencapsulation of Radix salvia miltiorrhiza nanoparticles by spray-drying, Powder Technology, 2008, 184, 114–121.

[120] Xiao YL, Luo JB, Yan ZH, Rong HS, Huang WM, Preparation and *in vitro* and *in vivo* evaluations of topically applied capsaicin transferosomes, Yao Xeu Xeu Bao, 2006, 41(5), 461–466.

[121] Singh HP, Utreja P, Tiwary AK, Jain S, Elastic liposomal formulation for sustained delivery of colchicine: *In vitro* characterization and *in vivo* evaluation of anti-gout activity, AAPSJ, 2009, 11(1), 54–64.

[122] Bhattacharya S, Ghosh A, Phytosomes: The emerging technology for enhancement of bioavailability of botanicals and nutraceuticals, International Journal of Aes Anti Med, 2009, 2(1), 225–229.

[123] Goyal A, Kumar S, Nagpal M, Singh I, Arora S, Potential of novel drug delivery systems for herbal drugs, Indian Journal of Pharmaceutical Education and Research, 2011, 45(3), 225–235.

[124] Grace XF, Seethalakshmi S, Manna KP, Chamundeeswari D Bioenhancers-a new approach in modern medicine, Indo American Journal of Pharmaceutical Research, 2013, 3(12), 1576–1580.

Deepak Kulkarni, Sachin Surwase, Shubham Musale,
Prabhanjan Giram

Chapter 10
Current trends on herbal bioenhancers

Abstract: Bioavailability is the key parameter for efficient execution of the therapeutic action of a drug. Multiple chemical entities with significant therapeutic potential struggle with inadequate pharmacokinetics, which is the result of its poor
solubility and permeability. Bioenhancer is the innovative concept to alter the pharmacokinetics of low bioavailability of drugs. Ayurvedic system of medicines reported
the first use of a herbal bioenhancer since time immemorial. Herbal bioenhancers are
superior in terms of safety, efficacy, and availability. Ginger, caraway, aloe, piperine,
quercetin, curcumin, etc. are some of the most widely used herbal bioenhancers.
Modern drug delivery science has shown growing interest in the use of herbal bioenhancers to improve the bioavailability, safety, and efficacy of active pharmaceutical
ingredients. Herbal bioenhancers are often used for bioavailability enhancement of
nutraceuticals, antibiotics, anticancer, antitubercular, and cardiovascular agents,
which result in their rapid onset of action. Bioenhancers acts by modulating the drug
membrane permeation and presystemic metabolism, which hamper the clinical success of multiple potent drugs. In this current scenario, the use of herbal bioenhancers
is not only limited to conventional dosage forms, but they are also used in multiple
novel drug delivery systems and nanotechnology. The modern applications employ
herbal bioenhancers to improve the bioavailability of drugs through various novel
drug delivery techniques, such as liposomes, transferosomes, ethosomes, nanoparticles, etc. by various routes of administration. This chapter provides some novel insights to the current trends in herbal bioenhancers, their biomedical applications,
and the current patent scenario.

Deepak Kulkarni, Srinath College of Pharmacy, Bajajnagar, Aurangabad 431136, Maharashtra,
India.
Sachin Surwase, BIO-IT Foundry Technology Institute, Pusan National University, Busan,
Republic of Korea.
Shubham Musale, Department of Pharmaceutics, Dr. D.Y. Patil Institute of Pharmaceutical
Science and Research, Pimpri-Pune 411018, Maharashtra, India
Prabhanjan Giram, Department of Pharmaceutics, Dr. D.Y. Patil Institute of Pharmaceutical Science
and Research, Pimpri-Pune 411018, Maharashtra, India, e-mail: prabhanjanpharma@gmail.com;
e-mail: prabhanjan.giram@dypvp.edu.in

https://doi.org/10.1515/9783110746808-010

10.1 Introduction

Bioavailability is the important parameter governing drug action. Drugs with better bioavailability provide efficient therapeutic action with minimum dose. Better bioavailability decreases the hepatic and nephrotic load as there is a decrease in the amount of dose required for therapeutic action [1]. Improvement in bioavailability can provide opportunity for multiple active pharmaceutical entities to execute promising therapeutic effects. Hence, bioavailability enhancement is a major area of research across the globe [2].

The factors that majorly affect the bioavailability are solubility, permeability, and first-pass metabolism. Herbal bioenhancers efficiently increase the absorption of drugs without a significant alteration and interference in the physiology of the body and action of drug [3]. The addition of bioenhancers provides additional advantage of easy availability, economy, and lesser side effects. In the current scenario, herbal bioenhancers are used to improve the absorption and, ultimately, the bioavailability of various drugs activities used to treat the diseases and disorders associated with the central nervous system (CNS), gastrointestinal tract, and the cardiovascular system (CVS) [4]. Nutraceuticals such as vitamins are also a class facing the challenge of inadequate bioavailability and, hence, herbal bioenhancers have great demand for bioavailability enhancement of nutraceuticals [5].

10.2 Classification and mechanism of action

Improvement in bioavailability is the ultimate purpose in the use of herbal bioenhancers. They are classified into three classes according to their mechanism of actions (Figure 10.1). Herbal bioenhancers improve the bioavailability of various drug molecules by inhibiting P-glycoprotein (P-gp) drug efflux, inhibition of cytochrome P-450 (CYP-450),, and enhancement of permeation [6].

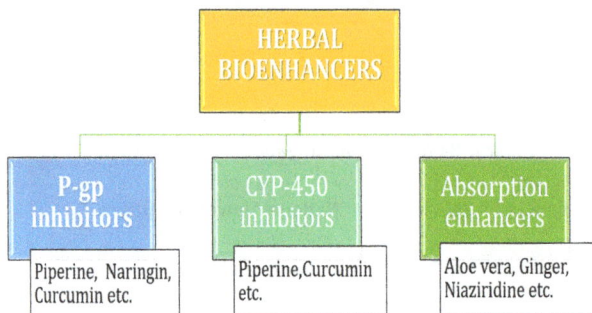

Figure 10.1: Classification of herbal bioenhancers.

10.2.1 Inhibition of P-glycoprotein drug efflux

P-gp is the efflux membrane transporter that regulates the intracellular uptake and distribution of various xenobiotics and toxins. This P-gp limits the permeability and absorption of various drugs due to the efflux mechanism, which ultimately results in low bioavailability. For efficient delivery and optimum bioavailability of drug, it is important to inhibit the P-gp efflux. The inhibition of P-gp efflux can be achieved by the blocking of the drug binding site of P-gp, alteration of the integrity of lipids in the cell membrane, and by disturbing the hydrolysis of adenosine triphosphate (Figure 10.2) [7]. Multiple bioenhancers, such as piperine, naringin, curcumin, etc. improve the bioavailability of drugs by inhibiting the P-gp drug efflux [8].

10.2.2 Inhibition of cytochrome P-450 (CYP-450) enzymes

The enzymes from CYP-450 family are largely responsible for the first-pass metabolism and elimination of multiple drugs. To improve the bioavailability of such drug molecules, it is important to inhibit these enzymes and to prevent the first-pass metabolism. Several herbal bioenhancers, such as piperine, curcumin, etc. inhibit the multiple enzymes CYPA1, CYP1B1, CYP1B2, CYP3A4, and CYP2E1, prevent the first-pass elimination of various drug molecules, and improves their bioavailability (Figure 10.3) [9, 10].

10.2.3 Absorption enhancers

Membrane permeability is the limiting factor for multiple drugs from the Biopharmaceutical Classification System (BCS) class III and BCS class IV. The inadequate permeation of these drugs reduces their absorption as well as therapeutic efficiency [11]. Improvisation in permeability can trigger the absorption of these lesser permeating drugs. This permeability-enhancement mechanism is executed by multiple herbal bioenhancers, such as aloe vera, ginger, niaziridine, and *Carum carvi*. These bioenhancers increase the permeation of drug molecules through biological membranes, resulting in better absorption and improved bioavailability (Figure 10.4) [12].

10.3 Current trends

Bioenhancers are mainly intended to be coadministered with active pharmaceutical ingredients to improve their bioavailability. Currently, herbal bioenhancers are the preferred choice of researchers working on bioavailability enhancement. Of late, herbal bioenhancers are used in bioavailability enhancement of drugs from multiple

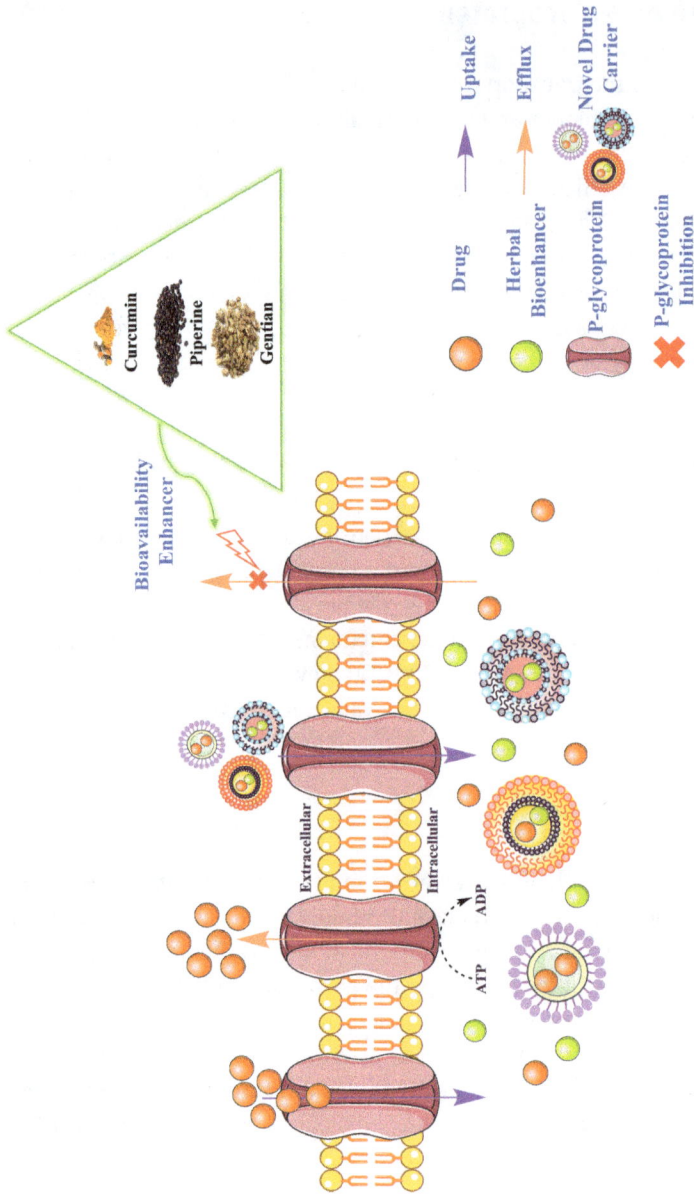

Figure 10.2: P-glycoprotein (P-gp)-mediated drug efflux inhibition mechanism of herbal bioenhancers delivered using novel drug delivery system.

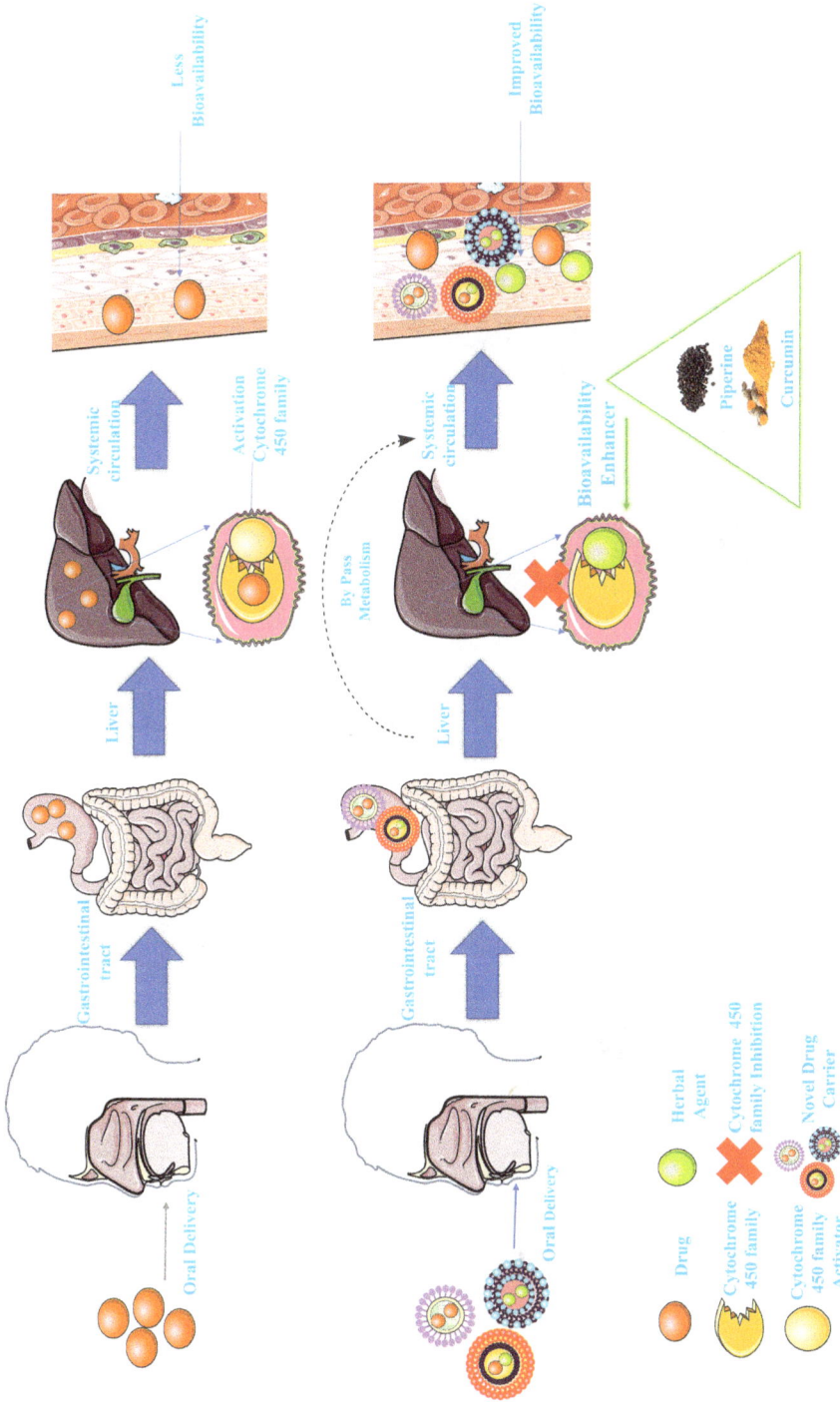

Figure 10.3: Cytochrome 450 inhibition mechanism of herbal bioenhancers delivered using novel drug delivery system.

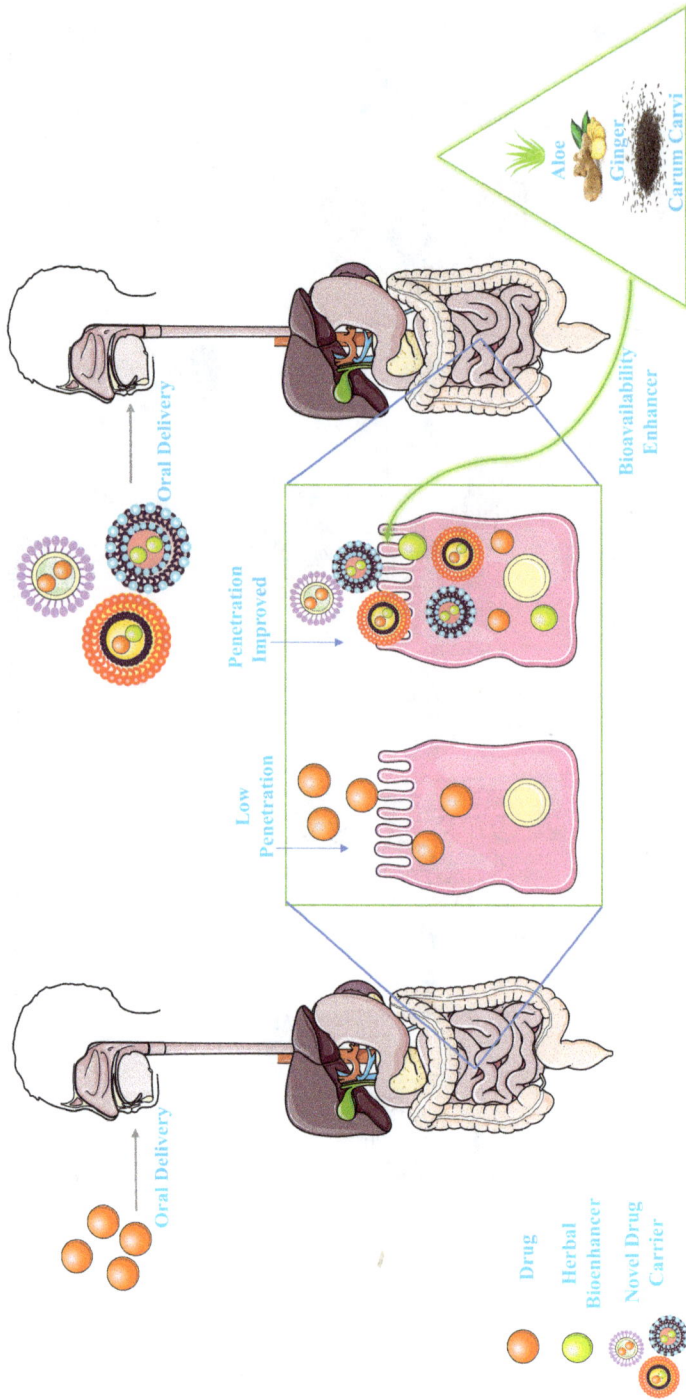

Figure 10.4: Absorption enhancement mechanism of herbal bioenhancers delivered using novel drug delivery system.

categories as well as neutraceuticals (Table 10.1). Piperine, curcumin, and ginger are the easily available substances across the world. These substances are milestone bio-enhancers and they have been proved through various research studies. In most of the recent researches it is found that these herbal bioenhancers are administered in the form of novel drug carriers, such as liposomes, microspheres, transferosomes, and ethosomes. The nanoformulations of herbal bioenhancers are also gaining popularity in biomedical applications [13, 14].

Table 10.1: Herbal bioenhancers for bioavailability improvement of various drugs.

Route of administration	Herbal bioenhancer	Drug candidate for bioavailability enhancement	Mechanism of action	References
Oral	Curcumin	Efflux transporter (P-gp) inhibition; metabolism (CYP3A4) inhibition	Midazolam: benzodiazepine	[15]
Oral	Emodin (anthraquinone derivative)	Efflux transporter (P-gp) inhibition	Digoxin: digitalis glycoside	[16]
Oral	Genistein (flavonoid)	Efflux transporter (MRP) inhibition	Epigallocatechin-3-gallate (EGCG): phenolic antioxidant	[17]
Oral	Gallic acid ester (organic acid)	Metabolism (CYP3A) inhibition	Nifedipine: calcium channel blocker	[18]
Oral	*Moringa oleifera* pods (traditional herbal medicine)	Metabolism (CYP450) inhibition	Rifampicin: semisynthetic rifamycin derivative	[19]
Oral	Naringin (flavonoid glycoside)	Metabolism (CYP3A4) inhibition	Tamoxifen: selective estrogen receptor modulator (SERM)	[20]
Oral	Peppermint oil (herbal)	Metabolism (CYP3A) inhibition	Cyclosporine: immunosuppressant	[21]
Oral	Piperine (alkaloid)	Metabolism (CYP450) inhibition	– Nimesulide: nonsteroidal anti-inflammatory – Carbamazepine: carboxamide derivative	[22, 23]
Oral	Quercetin (flavonoid)	Metabolism (CYP3A) inhibition	– Verapamil: Calcium channel blocker – Pioglitazone: thiazolidinedione	[24, 25]

Table 10.1 (continued)

Route of administration	Herbal bioenhancer	Drug candidate for bioavailability enhancement	Mechanism of action	References
Buccal	*Aloe vera* (gel, whole leaf)	Intercellular modulation	Didanosine: antiviral reverse transcriptase inhibitor	[26]
Pulmonary	HPBCD, CRYSMEB (cyclodextrin derivatives)	Tight junction modulation	– Mannitol: sugar alcohol	[27]

10.4 Herbal enhancers used in novel drug delivery and nanotechnology

Active pharmaceutical agents possess excellent in vitro therapeutics activity but poor in vivo activity due to poor membrane permeability and efflux by cell transporters such as p-gp, substrate for cytochrome P-450 metabolizer, which hamper target site entry and eventually decrease absorption, oral bioavailability, and therapeutic effect [5, 28]. However, when herbal enhancers are coadministered with active pharmaceutical ingredients, they play a very significant role in increasing the bioavailability or bio efficacy of various classes of drugs, such as anticancer, protein, peptide, antitubercular, antifungal, antiviral, nutraceuticals, and antihypertensives as reported in the literature [29, 30]. Development of several nanotechnology-based formulations offers an excellent platform for improvement in the bioactivity of drugs [31]. Liposomes, transferosomes, microspheres, nanoparticles, transdermal patches, nasal sprays, microcapsules, etc. are important novel drug delivery carriers and have been reported in literature along with herbal enhancer for improving therapeutics effects. Literature study report that different types of herbal enhancers have been used for enhancing the bioavailability of drugs. They are piperine, quercetin, naringin, glycyrrhizin, sinomenine, genistein, and nitrile glycoside [14]. In the literatures that are validated, piperine is used as the first oral bioavailability enhancers for nutrients and active pharmaceutical ingredients. This herbal enhancer improves the therapeutic effect of active pharmaceutical agents with poor solubility or permeability, when incorporated into novel drug delivery carries such as liposomes, transferosomes, microspheres, nanoparticles, transdermal patches, nasal sprays, microcapsules, and other novel drug delivery carriers, which is discussed in further section in detail [32, 33]. Herbal enhancer used for drug delivery and nanotechnology is as shown in (Figure 10.5).

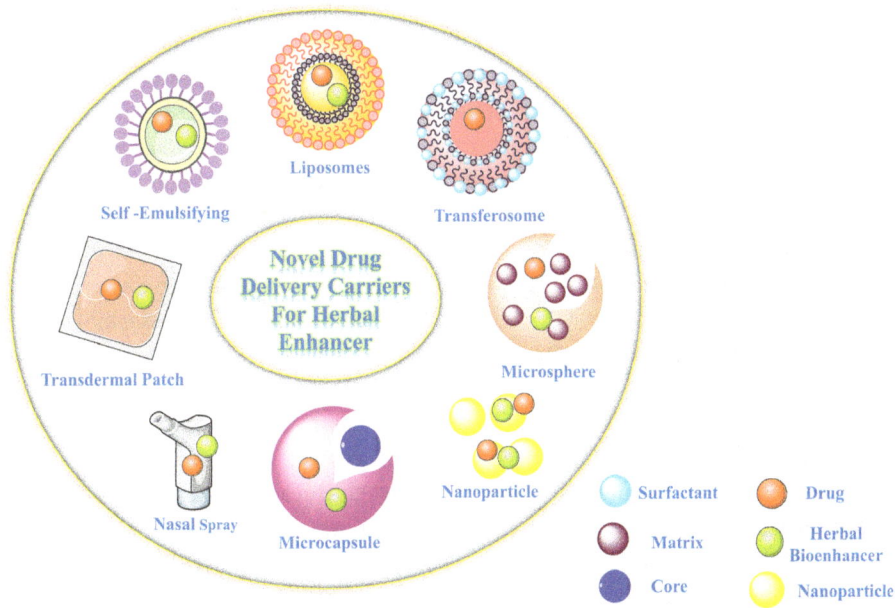

Figure 10.5: Novel drug delivery and nanotechnology-based carriers with herbal enhancers.

10.4.1 Liposome

Liposomes are phospholipid vesicular carriers made up of a lipid bilayer system composed of phospholipids and cholesterol. Phospholipids used for liposomes are DSPE, DEPE-PEG-2000, DOPC, DPPG, cholesterol, and triolein. Various conventional methods for formulation of liposomes are reported in literature reports. They include, thin-film hydration (Bangham method), reverse-phase evaporation technique, freeze drying of double emulsions, injection technique, detergent depletion, heating method, hydration methods, microfluidic method, membrane extrusion, electroformation methods, etc. These method have some disadvantages; they have poor stability, poor monodispersing of solvent, some organic solvents may cause environmental problem, etc. To overcome all these problems, alternative methods of preparation are developed based on supercritical fluids such as rapid expansion of supercritical solutions process, supercritical fluid extraction, super critical anti-solvent, depressurization of an expanded liquid organic solution-suspension, supercritical assisted liposome formation, and particles from gas saturated solutions [34]. Liposomes have some unique properties with potential to encapsulate hydrophilic or hydrophobic material at the interface of phospholipid double layer interface. They improve the aqueous solubility, intracellular uptake, as well as oral bioavailability of herbal substances or a combination with drug. Stability and biodistribution of liposomes increases the therapeutic activity of phytopharmaceuticals agents [35]. Liposomes have major limitation in the form of instability in plasma and premature leakage of drug content from its lipid matrix. Literature study shows different types of formulations are developed for encapsulation of herbal enhancers into liposomal system for enhancing the bioactivity of phytopharmaceuticals agents. Quercetin and curcumin have poor water solubility to overcome such formulation barrier, author loaded quercetin and curcumin in liposomal vesicular carrier for the anti-inflammatory activity via transdermal drug delivery. In this work anti-inflammatory activity performed in murine model, in this work liposomal vesicle carrier system enhances the penetration of quercetin and curcumin significantly in one platform. Quercetin and curcumin inhibit O-tetradecanoylphorbol-13-acetate induced at the wound site and also show better anti-inflammatory activity [36, 37]. In another literature study, Silymarin-loaded liposome was successfully evaluated for hepatoprotective activity. Results demonstrate significant increase in the bioavailability of Silymarin [38]. *Artemisia arborescens* L. essential oil was loaded in liposome and was evaluated for various physicochemical and biological characterization. In a literature study, *Artemisia arborescens* L. essential oil showed antiviral activity. This liposome carrier system was developed by the film hydration and simple sonication technique. This study reports that liposomal carrier system improves the penetration in cytoplasmic barrier, resulting in efficient targeting to cells [39]. In another literature study, a formulation of quercetin-loaded liposome carrier system was developed for antianxiety or cognitive improvement by the reverse evaporation method. In this study, the author and co-worker observed that quercetin-loaded liposome carrier system decreases the dose of quercetin and also improves the penetration of the bioactive component across the blood–brain barrier [40].

10.4.2 Transferosomes

Transferosomes are also known as transformable liposomes. They are similar to liposomes except in their structure. Due to the composition of the surfactant on the lipid bilayer, they form an elastic structure which significantly improves the cellular uptake of carrier-loaded drugs. Surfactant composition ranges between 10% and 25% and the ethanol composition ranges between 3% and 10%. Transferosomes have the potential to pass stratum corneum by virtue of osmotic pressure. Due to the easy penetration of the skin and the elasticity of the structure, they are widely used in transdermal drug delivery system for enhancement of the bioavailability of phytopharmaceuticals [41]. Various patented methods are reported in the preparation of transferosomes. Transferosomes are prepared on basis of stability, composition of vesicle, drug-carrying capacity, etc. Some conventional methods are reported in literature review for the formulation of transferosome, such as thin film hydration method, also called the rotary evaporation-sonication technique, as well as some novel techniques such as high-pressure homogenization technique, suspension homogenization, vortexing-sonication, centrifugation process, modified handshaking process, reverse-phase evaporation method, ethanol injection method, etc. [42]. In a literature, the author formulated vincristine-loaded transferosome to improve the antimitotic effect and enhance the lymph targeting of vincristine with minimal side effects [43]. In another literature study, capsaicin transferosome was developed by the high shear dispersing method. The evaluation of the formulation results indicates that capsaicin-loaded formulation enhances the penetration of the cell membrane [44]. In another study, the author formulated curcumin transferosome gel for antipsoriatic activity, which was developed by the de-solvation method. This study reported significant improvement in skin penetration [45]. Colchicine is a bioactive component used in the treatment of antigout therapy. Colchicine has poor penetration. To overcome this problem, colchicine transferosome has been successfully developed. The study results demonstrate that the transferosome carrier system improves the skin penetration of colchicine [46].

10.4.3 Microspheres

Microspheres are a spherical morphology of the biodegradable polymer matrix of polylactic acid or the copolymer of lactic acid as well as glycolic acid. Their radii range between 1 and 1,000 μm. They can be prepared by the spray-drying technique, extrusion technique, emulsification or gelation technique, etc. The phytopharmaceuticals-loaded microparticulate system is administered via I.V injection. The major advantage of microspheres is its site-specific drug delivery [47, 48]. In literature, several herbal enhancers are incorporated in the form of microspheres for enhancing the bioavailability. Isoniazid and rifampicin are widely used in the treatment of tuberculosis therapy. In a literature report, the author prepared isoniazid using the rifampicin drug-loaded microsphere carrier system by the complex co-acervation and modified emulsion technique simultaneously. In this system, hydro-alcoholic extracts of *Carum carvi* and *Ocimum sanctum* are encapsulated as a herbal bioavailability enhancer. The formulation results confirm increased encapsulation efficiency and sustained drug release due to the incorporation of herbal enhancer [49]. *Cynara scolymus* microspheres have been formulated by the spray drying technique for nutraceutical purposes. In this study, results show that *Cynara scolymus*-loaded microspheres demonstrate controlled release of the active component [50]. In another literature report, the authors formulated piperine and curcumin-loaded alginate microspheres by the emulsification and gelation method. Piperine was used as a herbal enhancer for improving the bioavailability of curcumin. This study demonstrates that encapsulation efficiency increased by the emulsification gelation method and increased the curcumin bioavailability by coadministration with a bioavailability enhancer [51]. Zedoary oil microspheres have been prepared by the emulsion solvent diffusion method for hepatoprotective activity. In this study report, the microsphere carrier system gave a controlled release of the active component and an increase in the bioavailability was observed [52].

10.4.4 Nanoparticles

Nanoparticles can be formulated with carbohydrate, lipid, proteins, natural, synthetic, and semi-synthetic polymers with size ranges between 10 and 1,000 nm. Polymer-based nanoparticles prepared by various methods are used in modern drug delivery applications. In the literature, the methods used for nanoparticle formulation have been divided into two groups: (1) polymerization of monomers, including emulsion method, miniemulsion, microemulsion, interfacial polymerization, and controlled living radical method; and (2) preformed monomers methods enlisted as emulsification solvent evaporation method, emulsification solvent diffusion, salting out, nanoprecipitation, dialysis, and supercritical fluid technology. They have several advantages such as increased therapeutic index, preventing degradation from various enzymes, reducing toxicity, etc. [53, 54]. In literature, herbal enhancers are incorporated in the nanoparticulate system for improving the bioavailability of the drug. Oleanolic acid is a pentacyclic triterpenoid, obtained from the Oleaceae family. It shows several pharmacological activities, such as immunomodulatory, cardiovascular, antidiabetic, hepatoprotective, anti-inflammatory, or anti-HIV action. In a literature study, the author developed oleanolic acid. The author incorporated chitosan-coated poly-(lactide)-co-glycolide nanoparticles and polyphenolic was used as bioenhancer. In this study, the author and coworkers observed that the codelivery of oleanolic acid and a bioenhancer, after administration, prevent fertility in a female patient. This administration gave a sustained release effect in the breast cancer cells treatment. The author concluded that this is the best approach for a safe chemotherapy in a female patient with various promising results [55]. Triptolide is an herbal constituent, traditionally well-known for its anti-inflammatory, antifertility, antineoplastic, or immunosuppressive activity. However, it shows poor water solubility as well as severe toxicity. In a literature report, the author formulated a triptolide nanoparticle for the anti-inflammatory activity via transdermal drug delivery. Results of this study confirmed that triptolide nanoparticle improve the penetration of the drug via stratum corneum by an increase in hydration [56]. In the literature report, green tea catechin (–)-epigallocatechin gallate (EGCG) was loaded in chitosan-tripolyphosphate nanoparticles. EGCG is a well-known anticancer, neuroprotective, and antioxidant agent. This developed formulation was analyzed in Swiss Outbred mice. In this study, result confirmed oral administration of EGCG chitosan-tripolyphosphate nanoparticles increased oral absorption [57]. In another literature report, radix salvia miltiorrhiza nanoparticles were formulated by the spray drying technique. Radix salvia is used in the treatment of various heart diseases. In this study, the author demonstrates that radix salvia miltiorrhiza-loaded nanoparticles carrier system improve the bioavailability of radix salvia [58].

10.4.5 Transdermal patches

Transdermal patches are one of the most widely used approaches for delivery of drugs across the skin by maintaining a controlled release profile of the drug to the systemic circulation, to avoid first-pass metabolism. Transdermal patches are divided into different types based on their therapeutic application, such as single-layer adhesive patch, multi-layer adhesive path, reservoir system, and microreservoir system. In the matrix system, transdermal patches are classified as drug-in-adhesive system, and matrix dispersion system. These transdermal drug delivery patches can be formulated by numerous methods, such as mercury substrate method [59], circular teflon mold method, glass substrate method, IPM membranes method, ethylene vinyl acetate copolymer membrane method, asymmetric TPX membrane method, aluminum-backed adhesive film method, etc. They have some key properties. For example, they avoid frequent dosing due to the continuous release of the drug, avoid first-past effect, improve patient compliance, etc. [60]. In literature, various herbal enhancers are incorporated in the transdermal patches to increase the bioavailability of the drug. Zidovudine (retrovir) is an antiviral agent, coadministered with antiretroviral agents used for the HIV-1 infection therapy. In a recent study, the author developed zidovudine transdermal patch using the solvent casting method, where T-anethole was incorporated as a herbal permeation enhancer. The author evaluated the physicochemical parameter as well as *ex-vivo* drug release in the study. Results indicate that an optimized concentration of transdermal patches enhance the drug release [61]. In another literature report, the olanzapine matrix-type of transdermal patch was formulated with natural oil, such as corn oil. A vegetable oil was used as a permeation enhancer. The author concluded that natural oil incorporated in transdermal patches significantly improves the olanzapine permeation across the skin [62]. Amlodipine besylate is a class of calcium channel blockers used in the treatment of hypertension therapy. The author prepared a transdermal patch by the solvent casting method. In this system, amlodipine besylate was used as an active ingredient and olive oil was used as an herbal enhancer [63]. In this study, the author observed a better release of the drug from the transdermal patch as compared to the control. In another literature study, indomethacin transdermal patch was developed for anti-inflammatory activity by the solvent evaporation method. Patchouli oil was used as a herbal enhancer. The study results indicate that increasing the concentration of patchouli increases the penetration of the drug [64].

10.4.6 Nasal sprays

Nasal spray is a novel drug delivery approach for the administration of a drug through the nasal cavity. Compared with the conventional drug delivery system, nasal spray has numerous advantages such as accurate dose of drug required to be administered to the patient by a metered dose valve. They delivered small as well as macromolecular drug for systemic or local therapy; prevent postnasal dripping and anterior leakage of the nasal droplets in the nasal cavity due to the dispersion of the liquid formulation; increase the dispersed area in the nasal cavity, etc. [65]. In a literature study, different types of herbal enhancers were incorporated in the nasal sprays for enhancing the bioavailability by improving the absorption in the nasal cavity. Raloxifene hydrochloride was used in the treatment of osteoporosis. For this, the author formulated raloxifene hydrochloride nasal spray. This study showed an improvement in the penetration of raloxifene hydrochloride via bovine nasal mucosa [66].

10.4.7 Microcapsule

Microcapsule can be developed by the encapsulation of the drug in polymeric microcapsules. The therapeutic drug molecule is coated by a thin film. Their size ranges between 1 and 100 µm. In a literature review, a microcapsule was developed by spray drying, sol–gel encapsulation method, supercritical CO_2, layer-by-layer, self-assembly, microfluidic techniques, in situ polymerization method, interfacial polymerization, etc. Natural, synthetic, and semisynthetic polymers can be used for the formulation of the shell material. Microcapsules can be used for targeted drug delivery with a sustained drug release of the drug. The core material used for the formulation also gives a unique feature to the microcapsule like solid or liquid particles can be encapsulated to gives protection from some reactive material, and safe or easy handling of toxic material [67, 68]. Galactagogue extract is used for nutraceutical purposes. In a literature report, the author developed an optimized chitosan/ triphosphate galactagogue-loaded microcapsule by ionotropic gelation and optimization by Box-Behnken design. In this study, the author observed that galactagogue encapsulated in a microcapsule carrier system increases its stability [69]. Thymoquinone chemically, known as monoterpenoid quinone, is a bioactive component obtained from the natural source of *Nigella sativa* L. seeds oil. Numerous therapeutical and biological activities reported in literature study include, gastroprotective, antioxidant, hepatoprotective, immunomodulator, anti-inflammatory, bronchodilator, anti-diabetic, anti-inflammatory purpose, etc. In a literature study, the author prepared thymoquinone-encapsulated microcapsule for functional yogurt preparation in nutraceutical application by the spray drying technique. The author concluded that this herbal active ingredient incorporated in the yoghurt production for functional food gave several health benefits. The thymoquinone microcapsule gave high stability to food preparation [70]. *Morinda citrifolia* L. is a class of Rubiaceae family used for various therapeutical purposes. In a literature study, the author formulated Morinda citrifolia L. microcapsule by the spray-drying method using different proportions of κ-carrageenan and maltodextrin, where the latter is a binding agent. In this study, the author and co-worker evaluated the prepared formulation. The result indicates that micro-encapsulated herbal food extract has high antioxidant activity. Spray drying temperature is good drying aid for κ-carrageenan or maltodextrin. The author suggested *Morinda citrifolia* L. fruit as a food additive for health benefit [71]. Propolis is a resin obtained from bees from the species *Apis mellifera*. It is used as a food additive. In another literature report, the author encapsulated propolis as a food additive by the complex co-acervation method. In this study, the author successfully developed formulation in powder form, free from alcohol, showing significantly release in food. The encapsulation process stored the antioxidant activity of the propolis-active component [72].

10.4.8 Self-emulsifying drug delivery (SEDDS)

Self-emulsifying drug delivery system (SEDDS) is a novel technological approach. It depends on the lipid, isotropic mixture of the solvent, the co-solvent, the concentration of the surfactant, the drug, the oil, and the oil-to-surfactant ratio. It forms a milky type of emulsion, with droplet size starting from sub-micrometric, followed by slight agitation in water or in the gastrointestinal fluid. This system differentiated the droplet size and the surfactant composition. The self-emulsifying drug delivery system droplet size is above 300 nm, while the self-nano emulsifying drug delivery system droplet size is below 100 nm and they contain less concentration of the surfactant. Numerous approaches are reported in the literature review for the delivery of herbal enhancers, such as self-emulsifying capsules, solid SEDDS, self-emulsifying controlled/sustained-release pellets, dry emulsion, self-emulsifying suppositories, self-emulsifying beads, self-emulsifying nanoparticles, etc. These drug delivery systems are formulated by different method, such as high-pressure homogenizer method, high energy approach, microfluidization, and sonication method. These systems also having some advantages over the conventional dosage form. They enhance the oral bioavailability of the drug, decrease the manufacturing cost, increase stability, and improve patient compliance in the administration of the drug delivery application [73, 74]. To improve the solubility of piperine and enhance its oral bioavailability, the author developed an optimized formulation of piperine SEDDS. It optimizes the formulation with ternary phase diagram and test of solubility. Intestinal permeability of formulation was analyzed by in vitro, in vivo, and in situ study. In this literature report, the author observed improvement in the dissolution rate, oral bioavailability, and the intestinal permeability of piperine [75].

In another literature study, the author developed an optimized formulation of Raloxifene drug-loaded in SEDDS and drug-loaded formulation with the incorporation of herbal bioavailability enhancers, piperine and quercetin. The results demonstrate that the drug-loaded SEDDS formulation increases the bioavailability of the drug, while, herbal enhancer-incorporated SEDDS showed no additionally increased oral bioavailability. From this result, the author concluded that the in vivo performance of susceptible active ingredient does not always improve the bioavailability by incorporating herbal enhancers in the drug delivery system [76]. Phyllanthin is a bioactive chemical component obtained from *Phyllanthus amarus*. It has shown numerous therapeutic effects. However, it exhibits poor oral bioavailability. The author formulated phyllanthin in SEDDS by the high-pressure homogenizer technique. In this study, the author concluded that SEDDS is the best approach for the oral delivery of phyllanthin. It may enhance the oral bioavailability as well as increase the therapeutic performance [77]. Isoliquiritigenin is a flavonoid component obtained from the root of Glycyrrhiza uralensis. It has various therapeutic properties reported, such as anti-asthmatic, anti-inflammatory, antianaphylaxis, antitumor, antioxidant, but shows poor bioavailability. An optimized formulation of isoliquiritigenin SEDDS was prepared by the high-pressure homogenizer technique for use in the treatment of ovalbumin-induced asthma. In this study, results demonstrate a significant increase in the bioavailability of Isoliquiritigenin compared with suspension, and used for asthma therapy [78].

Table 10.2: Herbal enhancers used in novel drug delivery and nanotechnology.

Formulation	Active ingredient	Category	Advantage	Development technique	References
Liposome Catechin	Catechin	Antioxidant, antiobesity, anti-inflammatory, antidiabetic	Improved bioavailability.	Reverse phase evaporation	[79, 80]
Liposome *Artemisia arborescens*	*Artemisia arborescens*	into cytoplasmic barrier, antiviral	Targeting of essential oils to cells, enhance penetration into the cytoplasmic barrier	Film method and sonication	[39]
Capsaicin transferosome	Capsaicin	Analgesic	Good topical absorption	High shear dispersion	[44]
Curcumin transferosome gel	Curcumin	Anti-psoriatic	Improvement in skin penetration	De-solvation method	[45]

Table 10.2 (continued)

Formulation	Active ingredient	Category	Advantage	Development technique	References
Cynara scolymus microspheres	*Cynara scolymus*	nutraceuticals	Controlled release of nutraceuticals	Spray drying technique	[50]
Zedoary oil microspheres	Zedoary	Hepatoprotective	Sustained release and higher bioavailability	emulsion solvent diffusion method	[52]
Taxol-loaded nanoparticles	Taxol	Anticancer	Enhances the bioavailability and sustained drug release	Emulsion solvent evaporation method	[81]
Radix *Salvia miltiorrhiza* nanoparticles	Radix *Salvia*	Coronary heart diseases, angina pectoris and myocardial infraction	Improves the bioavailability	Spray drying technique	[58]
Amlodipine besyalte with olive oil transdermal patch	Amlodipine besyalte	Calcium channel blocker	Better release of drug.	Solvent casting method	[63]
Indomethacin with patchouli oil transdermal patch	Indomethacin	Anti-inflammatory	Increase in the concentration of patchouli oil enhances the penetration.	Solvent evaporation technique	[64]
Opioid analgesic and aloe nasal spray	Opioid analgesic and aloe	Antihistamine	Increase in nasal decongestant activity		[82]
Nasal delivery of raloxifene hydrochloride	Raloxifene hydrochloride	Osteoporosis	Increased penetration in the nasal bovine mucosa	solvent evaporation technique	[66]
Galactagogue extract microcapsule	Galactagogue	Nutraceutical	Increase in stability	ionotropic gelation and Box-Behnken design	[69]

Table 10.2 (continued)

Formulation	Active ingredient	Category	Advantage	Development technique	References
Propolis microcapsule	Propolis	Nutraceutical	Increase in encapsulation, increase in antioxidant activity	complex co-acervation method.	[72]
Phyllanthin SEDDS	Phyllanthin	Hepatoprotective, anti-hyperuricemic activity in animal	Increased systemic availability	High pressure homogenizer method	[77]
Isoliquiritigenin SEDDS	Isoliquiritigenin	Anti-asthmatic	Improved bioavailability, improved asthmatic effect	High-pressure homogenizer method	[78]

10.4.9 Herbal enhancers for veterinary applications

Herbal medicine has the potential for bioavailability and is explored for veterinary application. It reduces dose and side effects of the coadministered active pharmaceutical agent. Herbal enhancers reported in the literature in veterinary applications for improving therapeutic efficiency decrease the dose of various categories of active pharmaceutical agents, such as antituberculosis, anticancer, antiviral, and antibiotics utilized mostly for the treatment of animals in veterinary practice. Herbal enhancers decrease the therapeutic concentration of the active pharmaceutical agents by improving permeability across the cellular transporter, influx, affect solubility, and metabolism [83]. In a literature, Dama et al. reported that the coadministration of pefloxacin with black pepper, long pepper, and ginger rhizome combination extract significantly improve the plasma concentration and plasma concentration of Gaddi goat's as compared with control pefloxacin. The results of the present study confirm significantly higher pharmacokinetic profile observed because of the herbal bioavailability enhancement effect. Overall, results confirmed bioenhancing effect of trikatu on the properties of pefloxacin used for the treatment of bacterial infection in animals [84]. Singh et al. reported in the literature that *Piper longum* significantly affects the pharmacokinetics of oxytetracycline used for mycoplasmal, bacterial, and various infections associated for veterinary use. This coadministration of piper longum herbal enhancer improves the therapeutic efficiency of oxytetracycline, with decreased quantity of administration [85][3]. Yan et al. studied curcumin herbal bioenhancer with the coadministration of docetaxel. Results show significant improvement in the bioavailability of SEDDS. Results of this study confirmed curcumin has inhibitory effect on p-gp and cytochrome P450 [CYP] 3A, responsible for significant improvement in the pharmacokinetics of docetaxel [86].

10.5 Recent case studies

Various nano/microcarriers have been used for the bioavailability enhancement of poorly absorbed water-soluble herbal constituents, such as flavonoids, tannins, glycosides, etc. (Table 10.3). Due to the large molecular size and poor lipid solubility, their ability to transport across lipid-rich cell membrane will be severely limited. Use of nano/microcarrier results in enhanced therapeutic action of poorly soluble plant extract.

Table 10.3: Recent case studies utilizing nano/microformulations for delivery of herbal bioenhancers/products.

Formulation	Herbal product	Method of preparation	Application	Route of administration	References
Phytosome	Bacopa	Bacopa-phyto-phospholipid complex	Activity enhancer	Oral	[87]
Phytosome	Rutin	Rutin-phospholipid complex	Solubility enhancement	Oral	[88]
Phytosome	Quercetin, kaempferol and isorhamnetin	*Ginkgo biloba* extract-phospholipid complexes	Bioavailability enhancement	Oral	[89]
Silybin Phytosome®	Silybin Flavonoids	Silybin-phospholipid complexation	Absorption enhancer	Oral	[90]
Ginseng Phytosome®	Ginsenosides	Phospholipid complexation	Absorption enhancer	Oral	[90]
Green tea Phytosome®	Epigallocatechin	Phospholipid complexation	Absorption enhancer	Oral	[90]
Hawthorn phytosome®	Flavonoids	Phospholipid complexation	Absorption enhancer	Oral	[90]
Curcumin phytosomes®	Curcumin	Curcumin–phospholipid complexation	Bioavailability enhancement	Oral	[91]

Table 10.3 (continued)

Formulation	Herbal product	Method of preparation	Application	Route of administration	References
Naringenin phytosomes®	Naringenin	Naringenin–phospholipid complex	Prolonged action of duration	Oral	[92]
Ethosomes	*Artemisia princeps* Pampanini	Rotary evaporation sonication method	Enhanced transdermal delivery	In vitro	[93]
Liposomes	Persicae Semen and Carthami Flos	Rotary evaporation	Bioavailability enhancement	In vitro	[94]
Liposomes	Propolis flavonoids	Ethanol injection method	Activity enhancement	In vitro	[95]
Liposomes	TOH, GTE, epicatechin (EC) and catechin (C)	Rotary evaporation	Activity enhancement	In vitro	[96]
Microspheres	Chelerythrine	Emulsion cross-linking method	Enhanced tumor delivery	In vitro	[97]
Microspheres	Zedoary oil	Emulsion solvent diffusion method	Bioavailability enhancement	Oral	[52]
Microspheres	Camptothecin	Solvent evaporation	Prolonged release	Intraperitoneally and intravenously	[97]
Microspheres	Quercetin	Solvent evaporation	Dose reduction	In vitro	[98]
Microspheres	*Cynara scolymus* extract	Spray-drying technique	Controlled release	Oral	[50]

10.6 Recently used in vitro, in vivo, and ex vivo models for bioavailability enhancement study

Multiple in vitro, in vivo and ex vivo models are in practice to study the enhancement in bioavailability using herbal bioenhancers. These study designs provide an idea about the therapeutic efficiency of herbal bioenhancers during clinical applications. Multiple animal models used to study the bioavailability enhancement are illustrated in Table 10.4.

Table 10.4: Recently used in vitro, in vivo and ex vivo models for bioavailability enhancement study.

Herbal bioenhancer	Drug candidate for bioavailability enhancement	Study design model	References
Menthol (alcohol)	Dideoxycytidine	Ex vivo (porcine buccal mucosa)	[99]
Aloe vera (gel and whole leaf)	Atenolol Insulin Indinavir	Ex vivo (rat intestinal tissue) In vitro (Caco-2 cells^2) In vivo (rat)	[100–102]
Curcumin (flavonoid)	Mycohenolic acid Norfloxacin Midazolam	In vitro (rat microsomes) In vivo (rabbit) In vitro (human liver microsomes)	[103]
Genistein (flavonoid)	Paclitaxel	In vivo (rat)	[104]
Gokhru extract	Metformin	In vitro (goat everted sac) In vitro (chicken everted intestine)	[105, 106]
Lysergol (alkaloid)	Berberine Curcumin	In vivo (rat) In vitro (rat liver microsomes)	[107, 108]
Naringin (flavonoid glycoside)	Diltiazem Clopidogrel Verapamil	In vivo (rat) Ex vivo (rat everted gut sac) In vivo (rabbit)	[109–111]
Piperine (alkaloid)	Phenol red Resveratrol Nevirapine	In vivo (rat, mice)	[112–114]
Quercetin (flavonoid)	Ranolazine Irinotecan Clopidogrel Doxorubicin	In vivo (rat), Ex vivo (rat and chick everted intestinal sac) In vitro (Caco-2 cells^2) In vivo (dog) In vitro (human MCF-7 ADRr cells 7)	[115–118]
Resveratrol	Methotrexate	In vitro (Caco-2 cells^2), ex vivo (rat everted intestine, rat kidney slices), in vivo (rat)	[119]

10.7 Current patents on herbal enhancers of pharmaceutical/nutraceutical agents

This section focuses on the current patents associated with herbal bioavailability enhancers used to enhance the bioavailability of pharmaceutical or nutraceuticals agents. Initially, several researchers filed US patent and European patent for the herbal ingredients associated with piperine as a herbal bioavailability enhancer. In these patents, they disclosed the invention associated with the improvement of gastrointestinal absorption of the nutritional compound containing pure alkaloid, Piperine. Piperine-based formulations have been delivered through several routes of administration, such as oral, topical, and parenteral. Patents also disclosed the method of extraction and purification of herbal bioenhancer alkaloid, piperine [120–122]. Another patent contains the use of lysergol as a bioactive enhancer and bioavailability facilitator for reducing the dose of broad-spectrum antibiotics as well as enhanced absorption of nutritional elements [123]. Several US patents have been filed related to the herbal extracts of *Carum*, *Zingiber*, and *Moringa* as bioavailability enhancers. These extracts are used in combination with each other as a bioenhancer [124–130]. Patents also disclosed a method for increasing the systemic exposure of cells selected from tumor cells and normal cells to the orally administered drug by administering the bioenhancer comprising an inhibitor of breast cancer resistance protein [13, 131]. More recently filed patents contain a herbal extract, in combination with several other bioenhancers, rather than single entity (Table 10.5) [132–134].

Table 10.5: Patent scenario of herbal bioenhancers.

Patent / Patent application no.	Herbal bioenhancer	Application	References
US005536506A	Piperine	Bioavailability enhancement of nutritional compounds	[120]
US5744161A	Piperine	Bioavailability enhancement of various drugs	[121]
US5439891A	Piperine	Bioavailability enhancement antitubercular and antileprotic drugs	[122]
EP0650728B1	Piperine	Bioavailability enhancement of antitubercular and antileprotic drugs	[123]
WO2003080059A1	Lysergol	Bioavailability enhancement of antibiotic	[135]
US20070020347A1	*Carum carvi* extracts extracts	Bioavailability enhancement of various drugs	[124]
US2003/0170326A1	*Zingiber officinale*	Bioavailability enhancement of various drugs	[125]
US6858588B2	*Moringa oleifera*	Bioavailability enhancement of drug and nutrients	[126]
EP1611148B1	*Moringa oleifera*	Bioavailability enhancement of drug and nutrients	[127]

Table 10.5 (continued)

Patent / Patent application no.	Herbal bioenhancer	Application	References
US20060057234A1	*Glycyrrhiza glabra*	Bioavailability enhancement of pharmaceutical/nutraceutical agent	[128]
US007514105B2	*Cuminum cyminum*	Bioavailability/bioefficacy enhancement of various drugs	[129]
US20100112101A1	*Stevia rebaudiana*	Bioavailability/bioefficacy enhancement of various drugs	[130]
US7030132B2	Camptothecin	Bioavailability enhancement	[131]
US7070814B2	*Cuminum cyminum*	Bioavailability/bioefficacy improvement of nutrients and vitamins	[132]
US20120058208A1	Plant extracts of curcumin, vanilla and ginger	Bioavailability enhancement	[133]
US9724311B2	Curcumin	Bioavailability enhancement	[134]

10.8 Future perspective

Drug delivery advancement is a never-ending area of research. Limitation of the absorption and bioavailability restrict the application of various drug molecules with significant therapeutic potential. This limited use of drugs with low bioavailability ultimately results in a rise in burden on healthcare for new drug development [136]. This problem of limited bioavailability of drugs provides great opportunity for researchers to work on various bioavailability enhancement mechanisms, including herbal bioenhancers. Bioenhancers from natural origin have various advantages of ease and economic availability, lesser side effects, maximum efficacy, and many more. Currently, bioenhancers are delivered in the form of various novel and nanocarriers [137]. According to the insight, SLICE news report, herbal medicine market is expected to reach 50 billion US dollars by 2030. The demand for herbal medication and ingredients is increasing across the globe, showing their importance [138]. The use of natural resources is due to their easy availability. Bioenhancers of natural origin will be the substances with incredible significance as they can bring multiple drugs back into the action by improving their pharmacokinetics [139].

List of abbreviations

BCS	Biopharmaceutical Classification System
CNS	Central nervous system
CVS	Cardiovascular system
CYP-450	Cytochrome P-450
EGCG	Epigallocatechin gallate
P-gp	P-glycoprotein
SEDDS	Self-emulsifying drug delivery system

References

[1]   Jhanwar B, Gupta S, Biopotentiation using herbs: Novel technique for poor bioavailable drugs, Int J PharmTech Res, 2014, 6(20), 443–454.
[2]   Verma CPS, Verma S, Ashawat MS, Pandit V, An Overview: Natural Bio-enhancer's in Formulation Development, Journal of Drug Delivery and Therapeutics, 2019, Nov 15, 9(6), 201–205.
[3]   Oladimeji FA, Adegbola AJ, Onyeji CO, Appraisal of Bioenhancers in Improving Oral Bioavailability: Applications to Herbal Medicinal Products, Journal of Pharmaceutical Research International, 2018, Nov 29, 1–23.
[4]   Peterson B, Weyers M, Steenekamp JH, Steyn JD, Gouws C, Hamman JH, Drug Bioavailability Enhancing Agents of Natural Origin [Bioenhancers] that Modulate Drug Membrane Permeation and Pre-Systemic Metabolism, Pharmaceutics, 2019, Jan, 11(1), 33.
[5]   Dudhatra GB, Mody SK, Awale MM, Patel HB, Modi CM, Kumar A, et al., A Comprehensive Review on Pharmacotherapeutics of Herbal Bioenhancers, The Scientific World Journal, 2012, Sep 17, 2012, e637953.
[6]   Chivte VK, Tiwari SV, Nikalge APG Bioenhancers: A brief review, Advanced Journal of Pharmacie and Life Science Research, 2017, 2, 1–18.
[7]   P-glycoprotein Inhibition for Optimal Drug Delivery - Md. Lutful Amin, 2013 [Internet]. [cited 2021 May 30]. Available from: https://journals.sagepub.com/doi/full/10.4137/DTI.S12519
[8]   Kumar-Sarangi M, Chandra-Joshi B, Ritchie B, Natural bioenhancers in drug delivery: An overview, Puerto Rico Health Sciences Journal, 2018, 37(1), 12–18.
[9]   Bibi Z, Role of cytochrome P450 in drug interactions, Nutrition & Metabolism, 2008, Oct 18, 5(1), 27.
[10] Tatiraju DV, Bagade VB, Karambelkar PJ, Jadhav VM, Kadam V, Natural bioenhancers: An overview, Journal of Pharmacognosy and Phytochemistry, 2013, 2(3), 55–60.
[11] Aungst BJ, Absorption Enhancers: Applications and Advances, The AAPS Journal, 2012, Mar 1, 14([1]), 10–18.
[12] Saxena V, Singh A, An update on bio-potentiation of drugs using natural options, Asian Journal of Pharmaceutical and Clinical Research, 2020, 13(11), 25–32.
[13] Chavhan SA, Shinde SA, Gupta INN Current trends on natural bio enhancers: A review, Internal Journal of Pharmacognosy and Chinese Medicine, 2018, 2, 2576–4772.
[14] Kesarwani K, Gupta R, Bioavailability enhancers of herbal origin: An overview, Asian Pacific Journal of Tropical Biomedicine, 2013, Apr 1, 3(4), 253–266.

[15]   Zhang W, Tan TMC, Lim L-Y, Impact of Curcumin-Induced Changes in P-Glycoprotein and CYP3A Expression on the Pharmacokinetics of Peroral Celiprolol and Midazolam in Rats, Drug Metabolism and Disposition: The Biological Fate of Chemicals, 2007, Jan 1, 35(1), 110–115.

[16]   Li X, Hu J, Wang B, Sheng L, Liu Z, Yang S, et al., Inhibitory effects of herbal constituents on P-glycoprotein in vitro and in vivo: Herb–drug interactions mediated via P-gp, Toxicology and Applied Pharmacology, 2014, 275(2), 163–175.

[17]   Lambert JD, Kwon S-J, Ju J, Bose M, Lee M-J, Hong J, et al., Effect of genistein on the bioavailability and intestinal cancer chemopreventive activity of (-)-epigallocatechin-3-gallate, Carcinogenesis, 2008, Oct 1, 29(10), 2019–2024.

[18]   Wacher VJ, Benet LZ Use of gallic acid esters to increase bioavailability of orally administered pharmaceutical compounds [Internet]. US6180666B1, 2001 [cited 2021 May 30]. Available from: https://patents.google.com/patent/US6180666B1/en

[19]   Pal A, Bawankule DU, Darokar MP, Gupta SC, Arya JS, Shanker K, et al., Influence of Moringa oleifera on pharmacokinetic disposition of rifampicin using HPLC-PDA method: A pre-clinical study, Biomedical Chromatography, 2011, 25(6), 641–645.

[20]   Choi J-S, Kang KW, Enhanced tamoxifen bioavailability after oral administration of tamoxifen in rats pretreated with naringin, Archives of Pharmacal Research, 2008, 31(12), 1631–1636.

[21]   Wacher VJ, Wong S, Wong HT, Peppermint Oil Enhances Cyclosporine Oral Bioavailability in Rats: Comparison with d-α-Tocopheryl Poly(ethylene glycol 1000) Succinate [TPGS] and Ketoconazole, Journal of Pharmaceutical Sciences, 2002, Jan 1, 91(1), 77–90.

[22]   Gupta SK, Bansal P, Bhardwaj RK, Velpandian T, COMPARATIVE ANTI-NOCICEPTIVE, ANTI-INFLAMMATORY AND TOXICITY PROFILE OF NIMESULIDE vs NIMESULIDE AND PIPERINE COMBINATION, Pharmacological Research, 2000, Jun 1, 41(6), 657–662.

[23]   Pattanaik S, Hota D, Prabhakar S, Kharbanda P, Pandhi P, Pharmacokinetic interaction of single dose of piperine with steady-state carbamazepine in epilepsy patients, Phytotherapy Research, 2009, 23(9), 1281–1286.

[24]   Choi J-S, Han H-K, The effect of quercetin on the pharmacokinetics of verapamil and its major metabolite, norverapamil, in rabbits, Journal of Pharmacy and Pharmacology, 2004, 56(12), 1537–1542.

[25]   Umathe SN, Dixit PV, Kumar V, Bansod KU, Wanjari MM, Quercetin pretreatment increases the bioavailability of pioglitazone in rats: Involvement of CYP3A inhibition. Biochemical Pharmacology, 2008, Apr 15, 75(8), 1670–1676.

[26]   Ojewole E, Mackraj I, Akhundov K, Hamman J, Viljoen A, Olivier E, et al., Investigating the Effect of Aloe vera Gel on the Buccal Permeability of Didanosine, Planta Medica, 2012, Mar 78, 4, 354–361.

[27]   Salem LB, Bosquillon C, Dailey LA, Delattre L, Martin GP, Evrard B, et al., Sparing methylation of β-cyclodextrin mitigates cytotoxicity and permeability induction in respiratory epithelial cell layers in vitro, Journal of Controlled Release, 2009, Jun 5, 136(2), 110–116.

[28]   Mukherjee PK, Harwansh RK, Bhattacharyya S, Chapter 10 - Bioavailability of Herbal Products: Approach Toward Improved Pharmacokinetics, PK Mukherjee, editor, Evidence-Based Validation of Herbal Medicine [Internet], Boston, Elsevier, 2015, cited 2021 May 24]. p. 217–45. Available from https://www.sciencedirect.com/science/article/pii/B9780128008744000106.

[29]   Ajazuddin AA, Qureshi A, Kumari L, Vaishnav P, Sharma M, et al., Role of herbal bioactives as a potential bioavailability enhancer for Active Pharmaceutical Ingredients, Fitoterapia, 2014, Sep 1, 97, 1–14.

[30]   Byeon JC, Ahn JB, Jang WS, Lee S-E, Choi J-S, Park J-S, Recent formulation approaches to oral delivery of herbal medicines, Journal of Pharmaceutical Investigation, 2019, Jan 1, 49(1), 17–26.

[31] Nalini: Novel nanosystems for herbal drug delivery - Google Scholar [Internet]. [cited 2021 May 24]. Available from: https://scholar.google.com/scholar_lookup?title=Novel%20Nano systems%20for%20Herbal%20Drug%20Delivery&author=T.%20Nalini&publication_ year=2017

[32] Pinisetti, D., J. Kakadiya and G.S. Chakraborthy, Recent trends on natural bioenhancers: An overview. Asian J. Res. Chem. Pharmaceut. Sci., 2019, 7: 708–725.

[33] Saraf S, Applications of novel drug delivery system for herbal formulations, Fitoterapia, 2010, 81(7), 680–689.

[34] Maja L, Željko K, Mateja P, Sustainable technologies for liposome preparation, The Journal of Supercritical Fluids, 2020, Nov, 1(165), 104984.

[35] Shashi K, Satinder K, Bharat P, A complete review on: Liposomes, International Research Journal of Pharmacy, 2012, 3(7), 10–16.

[36] Castangia I, Nácher A, Caddeo C, Valenti D, Fadda AM, Díez-Sales O, et al., Fabrication of quercetin and curcumin bionanovesicles for the prevention and rapid regeneration of full-thickness skin defects on mice, Acta Biomaterialia, 2014, Mar 1, 10(3), 1292–1300.

[37] Nanotherapeutics for anti-inflammatory delivery - ScienceDirect [Internet]. [cited 2021 May 24]. Available from: https://www.sciencedirect.com/science/article/abs/pii/ S1773224715300447

[38] El-Samaligy MS, Afifi NN, Mahmoud EA, Evaluation of hybrid liposomes-encapsulated silymarin regarding physical stability and in vivo performance, International Journal of Pharmaceutics, 2006, Aug 17, 319(1), 121–129.

[39] Sinico C, De Logu A, Lai F, Valenti D, Manconi M, Loy G, et al., Liposomal incorporation of Artemisia arborescens L, essential oil and in vitro antiviral activity. European Journal of Pharmaceutics and Biopharmaceutics., 2005, Jan 1, 59(1), 161–168.

[40] Priprem A, Watanatorn J, Sutthiparinyanont S, Phachonpai W, Muchimapura S, Anxiety and cognitive effects of quercetin liposomes in rats, Nanomedicine: Nanotechnology, Biology and Medicine, 2008, Mar 1, 4(1), 70–78.

[41] [PDF] "TRANSFEROSOMES-A REVIEW." ResearchGate [Internet]. [cited 2021 May 24]; Available from: https://www.researchgate.net/publication/337077339_TRANSFEROSOMES-A_REVIEW

[42] Opatha SAT, Titapiwatanakun V, Chutoprapat R, Transferosomes: A promising nanoencapsulation technique for transdermal drug delivery, Pharmaceutics, 2020, 12([9]), 855.

[43] Lu Y, Hqu S, Zhang L, Li Y, He J, Guo D. [Transdermal and lymph targeting transferosomes of vincristine]. Yao Xue Xue Bao. 2007 Oct 1;42(10):1097–1101.

[44] Xiao YL, Luo JB, Yan ZH, Rong HS, Huang WM, Preparation and in vitro and in vivo evaluations of topically applied capsaicin transferosomes, Yao Xeu Xeu Bao, 2006, 41(5), 461–466.

[45] Patel R, Singh SK, Singh S, Sheth NR, Gendle R, Development and characterization of curcumin loaded transferosome for transdermal delivery, Journal of Pharmaceutical Sciences and Research, 2009, 1(4), 71.

[46] Singh HP, Utreja P, Tiwary AK, Jain S, Elastic Liposomal Formulation for Sustained Delivery of Colchicine: In Vitro Characterization and In Vivo Evaluation of Anti-gout Activity, The AAPS Journal, 2009, Mar 1, 11(1), 54–64.

[47] Uyen NTT, Hamid ZAA, Tram NXT, Ahmad N Fabrication of alginate microspheres for drug delivery: A review, International Journal of Biological Macromolecules, 2020, 153, 1035–1046.

[48] Das MK, Ahmed AB, Saha D, MICROSPHERE A DRUG DELIVERY SYSTEM–A REVIEW, International Journal of Current Pharmaceutical Research, 2019, Jul 15, 34–41.

[49] Pingale PL, Pandharinath RR, Shrotriya PG, Study of herbal bioenhancers on various characteristics of isoniazid and rifampicin microspheres, International Journal of Infectious Diseases, 2014, Apr, 1(21), 235.

[50] Gavini E, Alamanni MC, Cossu M, Giunchedi P, Tabletted microspheres containing Cynara scolymus (var, Spinoso sardo) extract for the preparation of controlled release nutraceutical matrices. Journal of Microencapsulation., 2005, Aug 1, 22(5), 487–499.

[51] Di IŞILDAK. ENCAPSULATION OF CURCUMIN AND PIPERINE LOADED ALGINATE MICROSPHERES FOR DRUG DELIVERY.

[52] You J, Cui F, Han X, Wang Y, Yang L, Yu Y, et al., Study of the preparation of sustained-release microspheres containing zedoary turmeric oil by the emulsion–solvent-diffusion method and evaluation of the self-emulsification and bioavailability of the oil, Colloids and surfaces. B, Biointerfaces, 2006, Mar 1, 48(1), 35–41.

[53] Crucho CI, Barros MT Polymeric nanoparticles: A study on the preparation variables and characterization methods, Materials Science and Engineering: C, 2017, 80, 771–784.

[54] Ealia SAM, Saravanakumar MP, A review on the classification, characterisation, synthesis of nanoparticles and their application, IOP Conference Series: Materials Science and Engineering, 2017, Nov 263, 032019.

[55] Sharma M, Sharma S, Sharma V, Sharma K, Yadav SK, Dwivedi P, et al., Oleanolic–bioenhancer coloaded chitosan modified nanocarriers attenuate breast cancer cells by multimode mechanism and preserve female fertility, International Journal of Biological Macromolecules, 2017, Nov 1, 104, 1345–1358.

[56] Mei Z, Chen H, Weng T, Yang Y, Yang X, Solid lipid nanoparticle and microemulsion for topical delivery of triptolide, European Journal of Pharmaceutics and Biopharmaceutics, 2003, Sep 1, 56(2), 189–196.

[57] Dube A, Nicolazzo JA, Larson I, Chitosan nanoparticles enhance the plasma exposure of (-)-epigallocatechin gallate in mice through an enhancement in intestinal stability, European Journal of Pharmaceutical Sciences, 2011, 44(3), 422–426.

[58] Su YL, Fu ZY, Zhang JY, Wang WM, Wang H, Wang YC, et al., Microencapsulation of Radix salvia miltiorrhiza nanoparticles by spray-drying, Powder Technology, 2008, May 6, 184(1), 114–121.

[59] Gupta JRD, Irchhiaya R, Garud N, Tripathi P, Dubey P, Patel JR, Formulation and evaluation of matrix type transdermal patches of glibenclamide, International Journal of Pharmaceutical Sciences and Drug Research, 2009, 1(1), 46–50.

[60] Sonkar R, Prajapati SK, Chanchal DK, Bijauliya RK, Kumar S, A review on transdermal patches as a novel drug delivery system, International Journal of Life Sciences and Review, 2018, 4(4), 52–62.

[61] Mamatha J, Gadili S, Formulation PK, Evaluation of Zidovudine Transdermal Patch using Permeation Enhancers, Journal of Young Pharmacists, 2020, 12(2s), s45.

[62] Aggarwal G, Dhawan S, HariKumar SL, Natural Oils as Skin Permeation Enhancers for Transdermal Delivery of Olanzapine: In Vitro and In Vivo Evaluation, Current Drug Delivery, 2012, Mar 1, 9(2), 172–181.

[63] Mohanty D, Bakshi V, Singh MA, Aamiruddin M, Rashaid MA, Raj MP, et al., Formulation and Characterization of Transdermal Patches of Amlodipine Besylate Using Olive Oil as the Natural Permeation Enhancer, Indo American Journal of Pharmaceutical Research, 2016, 6, 6.

[64] Das A, Ahmed AB, Formulation and evaluation of transdermal patch of indomethacin containing patchouli oil as natural penetration enhancer, Asian Journal of Pharmaceutical and Clinical Research, 2017, 10(11), 320–325.

[65] Gao M, Shen X, Mao S, Factors influencing drug deposition in the nasal cavity upon delivery via nasal sprays, Journal of Pharmaceutical Investigation, 2020, 50(3), 251–259.

[66] Ahmed OA, Badr-Eldin SM, In situ misemgel as a multifunctional dual-absorption platform for nasal delivery of raloxifene hydrochloride: Formulation, characterization, and in vivo performance, International journal of nanomedicine, 2018, Oct, 11(13), 6325–6335.

[67] Arefin P, Habib MS, Chakraborty D, Bhattacharjee SC, Das S, Karmakar D, et al. An overview of microcapsule dosage form.

[68] Veiga RDSD, Rad S-B, Corso MP, Canan C, Essential oils microencapsulated obtained by spray drying: A review, Journal of Essential Oil Research, 2019, Nov 2, 31(6), 457–473.

[69] Yousefi M, Khorshidian N, Mortazavian AM, Khosravi-Darani K, Preparation optimization and characterization of chitosan-tripolyphosphate microcapsules for the encapsulation of herbal galactagogue extract, International Journal of Biological Macromolecules, 2019, Nov, 1(140), 920–928.

[70] Abedi A-S, Rismanchi M, Shahdoostkhany M, Mohammadi A, Hosseini H, Microencapsulation of Nigella sativa seeds oil containing thymoquinone by spray-drying for functional yogurt production, International Journal of Food Science & Technology, 2016, 51(10), 2280–2289.

[71] Krishnaiah D, Sarbatly R, Nithyanandam R, Microencapsulation of Morinda citrifolia L. extract by spray-drying, Chemical Engineering Research & Design, 2012, May 1, 90(5), 622–632.

[72] Nori MP, Favaro-Trindade CS, Matias de Alencar S, Thomazini M, de Camargo Balieiro JC, Contreras Castillo CJ, Microencapsulation of propolis extract by complex coacervation. LWT - Food Science and Technology., 2011, Mar 1, 44(2), 429–435.

[73] Mishra V, Nayak P, Yadav N, Singh M, Tambuwala MM, Aljabali AA, Orally administered selfemulsifying drug delivery system in disease management: Advancement and patents, Expert Opinion on Drug Delivery, 2021, 18(3), 315–322.

[74] Park H, Ha E-S, Kim M-S, Current Status of Supersaturable Self-Emulsifying Drug Delivery Systems, Pharmaceutics, 2020, Apr, 12(4), 365.

[75] Shao B, Cui C, Ji H, Tang J, Wang Z, Liu H, et al., Enhanced oral bioavailability of piperine by self-emulsifying drug delivery systems: In vitro, in vivo and in situ intestinal permeability studies, Drug Delivery, 2015, Aug 18, 22(6), 740–747.

[76] Thakur PS, Singh N, Sangamwar AT, Bansal AK, Investigation of Need of Natural Bioenhancer for a Metabolism Susceptible Drug – Raloxifene, in a Designed Self-Emulsifying Drug Delivery System, AAPS PharmSciTech, 2017, Oct 1, 18(7), 2529–2540.

[77] Hanh ND, Mitrevej A, Sathirakul K, Peungvicha P, Sinchaipanid N, Development of phyllanthin-loaded self-microemulsifying drug delivery system for oral bioavailability enhancement, Drug Development and Industrial Pharmacy, 2015, Feb 1, 41(2), 207–217.

[78] Cao M, Zhan M, Wang Z, Wang Z, Li X-M MM, Development of an Orally Bioavailable Isoliquiritigenin Self-Nanoemulsifying Drug Delivery System to Effectively Treat Ovalbumin-Induced Asthma, International journal of nanomedicine, 2020, Nov, 13(15), 8945–8961.

[79] Fang J-Y, Hwang T-L, Huang Y-L, Fang C-L, Enhancement of the transdermal delivery of catechins by liposomes incorporating anionic surfactants and ethanol, International Journal of Pharmaceutics, 2006, Mar 9, 310(1), 131–138.

[80] Chen G, Li D, Jin Y, Zhang W, Teng L, Bunt C, et al., Deformable liposomes by reverse-phase evaporation method for an enhanced skin delivery of (+)-catechin, Drug Development and Industrial Pharmacy, 2014, Feb 1, 40(2), 260–265.

[81] Preparation of paclitaxel-loaded poly(D,L-lactic acid) nanoparticles– 《Acta Academiae Medicinae Militaris Tertiae》 2006年15期 [Internet]. [cited 2021 May 24]. Available from: https://en.cnki.com.cn/Article_en/CJFDTotal-DSDX200615010.htm

[82] Wiersma JG Herbal based nasal spray [Internet]. US5948414A, 1999 [cited 2021 May 24]. Available from: https://patents.google.com/patent/US5948414A/en?oq=US+5948414

[83] Yurdakok-Dikmen B, Turgut Y, Filazi A, Herbal Bioenhancers in Veterinary Phytomedicine, Frontiers in Veterinary Science, Internet]. 2018 [cited 2021 May 30];5. Available from https://www.frontiersin.org/articles/10.3389/fvets.2018.00249/full.

[84] Dama MS, Varshneya C, Dardi MS, Katoch VC, Effect of trikatu pretreatment on the pharmacokinetics of pefloxacin administered orally in mountain Gaddi goats, Journal of Veterinary Science, 2008, 9(1), 25.

[85] Singh M, Varshneya C, Telang RS, Srivastava AK Alteration of pharmacokinetics of oxytetracycline following oral administration of Piper longum in hens, Journal of Veterinary Science, 2005, 6, 3.

[86] Yan Y-D, Marasini N, Choi YK, Kim JO, Woo JS, Yong CS, et al., Effect of dose and dosage interval on the oral bioavailability of docetaxel in combination with a curcumin self-emulsifying drug delivery system (SEDDS), European journal of drug metabolism and pharmacokinetics, 2012, Sep 1, 37(3), 217–224.

[87] Habbu P, Madagundi S, Kulkarni R, Jadav S, Vanakudri R, Kulkarni V, Preparation and evaluation of Bacopa–phospholipid complex for antiamnesic activity in rodents, Drug Invention Today, 2013, Mar 1, 5(1), 13–21.

[88] Zhang J, Tang Q, Xu X, Li N, Development and evaluation of a novel phytosome-loaded chitosan microsphere system for curcumin delivery, International Journal of Pharmaceutics, 2013, May 1, 448(1), 168–174.

[89] Chen Z, Sun J, Chen H, Xiao Y, Liu D, Chen J, et al., Comparative pharmacokinetics and bioavailability studies of quercetin, kaempferol and isorhamnetin after oral administration of Ginkgo biloba extracts, Ginkgo biloba extract phospholipid complexes and Ginkgo biloba extract solid dispersions in rats, Fitoterapia, 2010, Dec 1, 81(8), 1045–1052.

[90] Yanyu X, Yunmei S, Zhipeng C, Qineng P, The preparation of silybin–phospholipid complex and the study on its pharmacokinetics in rats, International Journal of Pharmaceutics, 2006, Jan 3, 307(1), 77–82.

[91] Maiti K, Mukherjee K, Gantait A, Saha BP, Mukherjee PK, Curcumin–phospholipid complex: Preparation, therapeutic evaluation and pharmacokinetic study in rats, International Journal of Pharmaceutics, 2007, Feb 7, 330(1), 155–163.

[92] Maiti K, Mukherjee K, Gantait A, Saha BP, Mukherjee PK, Enhanced therapeutic potential of naringenin-phospholipid complex in rats, Journal of Pharmacy and Pharmacology, 2006, 58(9), 1227–1233.

[93] Yang HG, Kim HJ, Kim HS, Park SN, Ethosome formulation for enhanced transdermal delivery of Artemisia princeps Pampanini extracts, Applied Chemistry for Engineering, 2013, 24(2), 190–195.

[94] Mignet N, Seguin J, Ramos Romano M, Brullé L, Touil YS, Scherman D, et al., Development of a liposomal formulation of the natural flavonoid fisetin, International Journal of Pharmaceutics, 2012, Feb 14, 423(1), 69–76.

[95] Yuan J, Liu J, Hu Y, Fan Y, Wang D, Guo L, et al., The immunological activity of propolis flavonoids liposome on the immune response against ND vaccine, International Journal of Biological Macromolecules, 2012, Nov 1, 51(4), 400–405.

[96] Yin J, Becker EM, Andersen ML, Skibsted LH, Green tea extract as food antioxidant. Synergism and antagonism with α-tocopherol in vegetable oils and their colloidal systems, Food Chemistry, 2012, Dec 15, 135(4), 2195–2202.

[97] Machida Y, Onishi H, Kurita A, Hata H, Morikawa A, Machida Y, Pharmacokinetics of prolonged-release CPT-11-loaded microspheres in rats, Journal of Controlled Release, 2000, May 15, 66(2), 159–175.

[98] Chao P, Deshmukh M, Kutscher HL, Gao D, Sundara Rajan S, Hu P, et al., Pulmonary targeting microparticulate camptothecin delivery system: Anti-cancer evaluation in a rat orthotopic lung cancer model, Anti-cancer Drugs, Internet]. 2010 Jan [cited 2021 May 30];21(1). Available from, https://www.ncbi.nlm.nih.gov/pmc/articles/PMC3859198/.

[99] Shojaei AH, Khan M, Lim G, Khosravan R, Transbuccal permeation of a nucleoside analog, dideoxycytidine: Effects of menthol as a permeation enhancer, International Journal of Pharmaceutics, 1999, Dec 10, 192(2), 139–146.

[100] Beneke C, Viljoen A, Hamman J, In Vitro Drug Absorption Enhancement Effects of Aloe vera and Aloe ferox, Scientia Pharmaceutica, 2012, Jun, 80(2), 475–486.

[101] Chen W, Lu Z, Viljoen A, Hamman J, Intestinal Drug Transport Enhancement by Aloe vera, Planta medica, 2009, May, 75(6), 587–595.

[102] Wallis L, Malan M, Gouws C, Steyn D, Ellis S, Abay E, et al., Evaluation of Isolated Fractions of Aloe vera Gel Materials on Indinavir Pharmacokinetics: In vitro and in vivo Studies, Current Drug Delivery, 2016, May 1, 13(3), 471–480.

[103] Basu NK, Kole L, Kubota S, Owens IS, Human Udp-Glucuronosyltransferases Show Atypical Metabolism of Mycophenolic Acid and Inhibition by Curcumin, Drug metabolism and disposition: the biological fate of chemicals, 2004, Jul 1, 32(7), 768–773.

[104] Li X, Choi J-S, Effect of genistein on the pharmacokinetics of paclitaxel administered orally or intravenously in rats, International Journal of Pharmaceutics, 2007, Jun 7, 337(1), 188–193.

[105] Ayyanna C, Mohan Rao G, Sasikala M, Somasekhar P, Arun Kumar N, Pradeep Kumar MVS, Absorption enhancement studies of metformin hydrochloride by using tribulus terrestris plant extract, International Journal of Pharmacy & Technology, 2012, 4(1), 4118–4125.

[106] Kumar A, Bansal M Formulation and evaluation of antidiabetic tablets: Effect of absorption enhancer, World Journal of Pharmaceutical Research, 2014, 3, 1426–1445.

[107] Patil S, Dash RP, Anandjiwala S, Nivsarkar M, Simultaneous quantification of berberine and lysergol by HPLC-UV: Evidence that lysergol enhances the oral bioavailability of berberine in rats, Biomedical Chromatography, 2012, 26(10), 1170–1175.

[108] Shukla M, Malik MY, Jaiswal S, Sharma A, Tanpula DK, Goyani R, et al., A mechanistic investigation of the bioavailability enhancing potential of lysergol, a novel bioenhancer, using curcumin, RSC Advances, 2016, Jun 21 6(64), 58933–58942.

[109] Choi J-S, Han H-K, Enhanced oral exposure of diltiazem by the concomitant use of naringin in rats, International Journal of Pharmaceutics, 2005, Nov 23, 305(1), 122–128.

[110] Lassoued MA, Sfar S, Bouraoui A, Khemiss F, Absorption enhancement studies of clopidogrel hydrogen sulphate in rat everted gut sacs, Journal of Pharmacy and Pharmacology, 2012, 64(4), 541–552.

[111] Yeum C-H, Choi J-S, Effect of naringin pretreatment on bioavailability of verapamil in rabbits, Archives of Pharmacal Research, 2006, Jan 1, 29(1), 102.

[112] Bajad S, Bedi KL, Singla AK, Johri RK, Piperine Inhibits Gastric Emptying and Gastrointestinal Transit in Rats and Mice, Planta Medica, 2001, 67(2), 176–179.

[113] Johnson JJ, Nihal M, Siddiqui IA, Scarlett CO, Bailey HH, Mukhtar H, et al., Enhancing the bioavailability of resveratrol by combining it with piperine, Molecular Nutrition & Food Research, 2011, 55(8), 1169–1176.

[114] Kasibhatta R, Naidu MUR, Influence of Piperine on the Pharmacokinetics of Nevirapine under Fasting Conditions, Drugs in R&D, 2007, Nov 1, 8(6), 383–391.

[115] Babu PR, Babu KN, Peter PLH, Rajesh K, Babu PJ, Influence of quercetin on the pharmacokinetics of ranolazine in rats and in vitro models, Drug Development and Industrial Pharmacy, 2013, Jun 1, 39(6), 873–879.

[116] Bansal T, Awasthi A, Jaggi M, Khar RK, Talegaonkar S, Pre-clinical evidence for altered absorption and biliary excretion of irinotecan (CPT-11) in combination with quercetin: Possible contribution of P-glycoprotein, Life Sciences, 2008, Aug 15, 83(7), 250–259.

[117] Lee JH, Shin Y-J, Oh J-H, Lee Y-J, Pharmacokinetic interactions of clopidogrel with quercetin, telmisartan, and cyclosporine A in rats and dogs, Archives of Pharmacal Research, 2012, Oct 1, 35(10), 1831–1837.

[118] Scambia G, Ranelletti FO, Panici PB, De Vincenzo R, Bonanno G, Ferrandina G, et al., Quercetin potentiates the effect of adriamycin in a multidrug-resistant MCF-7 human breast-cancer cell line: P-glycoprotein as a possible target, Cancer Chemotherapy and Pharmacology, 1994, Nov 1, 34(6), 459–464.

[119] Jia Y, Liu Z, Wang C, Meng Q, Huo X, Liu Q, et al., P-gp, MRP2 and OAT1/OAT3 mediate the drug-drug interaction between resveratrol and methotrexate, Toxicology and Applied Pharmacology, 2016, Sep 1, 306, 27–35.

[120] Majeed M, Badmaev V, Rajendran R Use of piperine to increase the bioavailability of nutritional compounds [Internet]. US5536506A, 1996 [cited 2021 May 30]. Available from: https://patents.google.com/patent/US5536506A/en

[121] Majeed M, Badmaev V, Rajendran R Use of piperine as a bioavailability enhancer [Internet]. US5744161A, 1998 [cited 2021 May 30]. Available from: https://patents.google.com/patent/US5744161A/en?oq=US5744161A

[122] Kapil RS, Zutshi U, Bedi KL, Singh G, Johri RK, Dhar SK, et al. Process for preparation of pharmaceutical composition with enhanced activity for treatment of tuberculosis and leprosy [Internet]. US5439891A, 1995 [cited 2021 May 30]. Available from: https://patents.google.com/patent/US5439891A/en

[123] Kapil RS, Zutshi U, Bedi KL, Singh G, Johri RK, Dhar SK, et al. Pharmaceutical compositions containing piperine and an antituberculosis or antileprosy drug. European Patent Number, EP0650728B1. 2002;

[124] Qazi G, Bedi K, Johri R, Tikoo M, Tikoo A, Sharma S, et al. Bioavailability enhancing activity of Carum carvi extracts and fractions thereof [Internet]. US20070020347A1, 2007 [cited 2021 May 30]. Available from: https://patents.google.com/patent/US20070020347A1/en?oq=US20070020347A1%2c

[125] Qazi G, Bedi K, Johri R, Tikoo M, Tikoo A, Sharma S, et al. Bioavailability enhancing activity of Zingiber officinale Linn and its extracts/fractions thereof [Internet]. US20030170326A1, 2003 [cited 2021 May 30]. Available from: https://patents.google.com/patent/US20030170326A1/en

[126] Khanuja SPS, Arya JS, Tiruppadiripuliyur RSK, Saikia D, Kaur H, Singh M, et al. Nitrile glycoside useful as a bioenhancer of drugs and nutrients, process of its isolation from moringa oleifera [Internet]. US6858588B2, 2005 [cited 2021 May 30]. Available from: https://patents.google.com/patent/US6858588B2/en

[127] Khanuja SPS, Arya JS, Tiruppadiripuliyur RSK, Saikia D, Kaur H, Singh M, et al. Novel nitrile glycoside useful as a bio-enhancer of drugs and nutrients, process of its isolation from moringa oleifera [Internet]. EP1611148B1, 2008 [cited 2021 May 30]. Available from: https://patents.google.com/patent/EP1611148B1/en?oq=EP1611148B1

[128] Khanuja SPS, Kumar S, Arya JS, Shasany AK, Singh M, Awasthi S, et al. Composition comprising pharmaceutical/nutraceutical agent and a bio-enhancer obtained from Glycyrrhiza glabra [Internet]. US20060057234A1, 2006 [cited 2021 May 30]. Available from: https://patents.google.com/patent/US20060057234A1/en?oq=US20060057234A1

[129] Qazi GN, Bedi KL, Johri RK, Tikoo MK, Tikoo AK, Sharma SC, et al. Bioavailability/bioefficacy enhancing activity of Cuminum cyminum and extracts and fractions thereof [Internet]. US7514105B2, 2009 [cited 2021 May 30]. Available from: https://patents.google.com/patent/US7514105B2/en

[130] Gokaraju GR, Gokaraju RR, D'Souza C, Frank E Bio-availability/bio-efficacy enhancing activity of stevia rebaudiana and extracts and fractions and compounds thereof [Internet]. US20100112101A1, 2010 [cited 2021 May 30]. Available from: https://patents.google.com/patent/US20100112101A1/en

[131] Schellens JHM, Schinkel AH Method of improving bioavailability of orally administered drugs, a method of screening for enhancers of such bioavailability and novel pharmaceutical

compositions for oral delivery of drugs [Internet]. US7030132B2, 2006 [cited 2021 May 30]. Available from: https://patents.google.com/patent/US7030132B2/en?oq=US7030132B2

[132] Qazi GN, Bedi KL, Johri RK, Tikoo MK, Tikoo AK, Sharma SC, et al. Bioavailability / bioefficacy enhancing activity of Cuminum cyminum and extracts and fractions thereof [Internet]. US7070814B2, 2006 [cited 2021 May 30]. Available from: https://patents.google.com/patent/US7070814B2/en

[133] Jacob CV Synergistic Composition for Enhancing Bioavailability of Curcumin [Internet]. US20120058208A1, 2012 [cited 2021 May 30]. Available from: https://patents.google.com/patent/US20120058208A1/en?oq=US2012%2f0058208A1

[134] Deshpande J, Juturu V Curcumin compositions and uses thereof [Internet]. US20150297536A1, 2015 [cited 2021 May 30]. Available from: https://patents.google.com/patent/US20150297536A1/en?oq=US2015%2f0297536A1

[135] Khanuja S, Arya J, Srivastava S, Shasany A, Kumar TS, Darokar M, et al. Antibiotic pharmaceutical composition with lysergol as bio-enhancer and method of treatment [Internet]. US20070060604A1, 2007 [cited 2021 May 30]. Available from: https://patents.google.com/patent/US20070060604A1/en

[136] Verma S, Rai S Scholars Academic Journal of Pharmacy (SAJP) ISSN 2347-9531 (Print).

[137] Randhawa GK, Kullar JS, Rajkumar, Bioenhancers from mother nature and their applicability in modern medicine, International Journal of Applied and Basic Medical Research, 2011, 1(1), 5–10.

[138] Herbal Medicine Market Share Global forecast to 2030 | insight slice [Internet]. [cited 2021 May 30]. Available from: https://www.insightslice.com/herbal-medicine-market

[139] Zafar N, Pharm M, Herbal Bioenhancers: A Revolutionary Concept in Modern Medicine, Zafar World Journal of Pharmaceutical Research [Internet], 2017, 6(16), 381–397. Available from: jpr.net - Google Search [Internet]. [cited 2021 May 30].

Swarnali Das Paul, Harish Sharma, Gyanesh Sahu,
Chanchal Deep Kaur, Sanjoy Kumar Pal

Chapter 11
Future perspectives of herbal bioenhancer

Abstract: Bioenhancers or drug facilitators are excipients that when added with therapeutically active agents improve their absorption, permeation, solubility, and bioavailability. Nowadays, bio-enhancers have secured a significant position in drug delivery research and applications as they help in the reduction of dose, toxicity, and drug resistance, as also improve potential use of drugs. Bioenhancers of herbal origin have gained special attention as they are ecologically safe, inexpensive, easily solicited, non-obsessive, pharmacologically dormant, and nonallergenic nature. In the last five years, many reports have been published claiming cheaper, safe, and good bioavailability of chemotherapeutic drugs, when used in combination with natural bioenhancers. At the same time, a few reports have also shown no change in pharmacokinetic parameters, which might be due to the rapid metabolism of bioenhancer or different target sites of active pharmaceutical ingredients. However, almost 60–65% of findings showed improvement in pharmacokinetic/pharmacodynamic parameters, and some of them are in clinical trial. This chapter aims to compile literature on the future perspective of herbal bioenhancers. For this purpose, information related to their novel delivery approaches such as liposomes, transfersomes, ethosomes, and nanoparticles have been described. The ecological benefits of these bioenhancers are discussed, along with several examples. Further, their applications in a different category of diseases including viral diseases, cancer, tuberculosis, ocular diseases, and gastrointestinal problems are compiled. Advances in bioenhancers have also raised challenges in regulatory control. Therefore, a brief of regulatory aspects of using herbal bioenhancers is also discussed in this chapter.

Swarnali Das Paul, Professor in Pharmaceutics, Shri Shankaracharya College of Pharmaceutical
Sciences, Bhilai, Chhattisgarh 490020, India, e-mail: swarnali34@gmail.com
Chanchal Deep Kaur, Shri RawatpuraSarkar Institute of Pharmacy, Kumhari, Chhattisgarh, India
Harish Sharma, Shri RawatpuraSarkar Institute of Pharmacy, Kumhari, Chhattisgarh, India;
Rungta Institute of Pharmaceutical Sciences, Kohka, Bhilai, Chhattisgarh, India
Gyanesh Sahu, Rungta Institute of Pharmaceutical Sciences, Kohka, Bhilai, Chhattisgarh, India;
Rungta College of Pharmaceutical Sciences and Research, Kohka, Bhilai, Chhattisgarh, India
Sanjoy Kumar Pal, Department of Biological Sciences, SSIT, Skyline University, Kano, Nigeria

https://doi.org/10.1515/9783110746808-011

11.1 Introduction

The concept of bioenhancer was derived from Trikatu, which means three acrids, mentioned in traditional Ayurveda . These three agents were long pepper, black pepper, and ginger, which were used in a combination for treating different ailments. The role of bioavailability enhancers was first identified by Bose in 1929. He reported the role of long pepper when administered with adhatodavasaka leaves in increasing its activity [1, 2]. However, in 1979, the world's first bioenhancer was discovered by an Indian scientist in Jammu at the Indian Institute of Integrative Medicine, formerly Regional Research Lab. He discovered the role of piperine as a bioavailability enhancer, at that time. After getting the international patent and completion of phase IIIb clinical trials, the Drug Control General of India (DCGI) granted the license to the market drug "Risorine" in India, to be used for antituberculosis treatment. The formulation contains 300 mg of isoniazid (INH), 200 mg of rifampicin, and 10 mg of piperine [3]. Similarly, the herbal *Carum carvi* L. was found to be a good bioenhancer and modifies the kinetics of antitubercular treatment, favorably [4]. Piperine was also found to boost the curcumin serum levels, absorption, and bioavailability in humans and rodents with minimal toxicity [5]. Curcumin's pleiotropic properties are due to its capacity to affect a variety of signaling molecules. Curcumin's safety, tolerability, and nontoxicity at large doses have all been proven in human clinical trials. Curcumin has been shown to have therapeutic potential against a wide range of human ailments when administered alone or in combination with other medicines [6].

After this progress, the Ayurveda concept of bioenhancers has merged with synthetic medicines for many other benefits besides bioavailability enhancing. To date, there are many unexplored areas of bioenhancers. These include interaction with active molecules, possible mode of actions, possible combinations with other therapeutic agents, their clinical outcomes, and evaluation of toxicity. Hence, exploring novel bioenhancers with versatile mechanisms and fewer side effects is the need of the hour [7].

11.2 Application of different herbal bioenhancer

11.2.1 Bioavailability/bioefficacy-enhancing activity

To gain the maximum therapeutic efficacy of any drug, bioavailability should be maximized, because the extent of bioavailability immediately impacts plasma concentrations. Therefore, for years, the development of a product with maximum bioavailability has been of utmost interest, as most of them have unwanted poisonous or aspect results, are expensive, and require frequent administration as well as extensive management.

However, bioavailability enhancement by replacing the principal therapeutic agent with a secondary agent has gained extensive recognition. However, as per data from Ayurvedic literature, the usage of herbal bioenhancers as a secondary agent is a very good option for increasing the bioavailability of poorly soluble drugs [8].

By enhancing bioavailability, highly priced tablets may become less expensive and decrease in the toxic outcomes by reducing the specified dose of medicine may be possible. Poorly bioavailable capsules remain subtherapeutic because a first-rate part of a dose never reaches the plasma or exerts its pharmacological impact, except when very huge doses are given, which can also lead to severe side effects. Any big development in bioavailability will result in reducing the dose or the dose frequency of that unique drug. Intersubject variability is a special concern for a drug with a narrow safety margin. Incomplete oral bioavailability includes bad dissolution or low aqueous solubility, terrible intestinal membrane permeation, degradation of the drug in gastric or intestinal fluids, and presystemic intestinal or hepatic metabolism. Many healing treatments are also accompanied by a lack of essential nutraceuticals within the direction of therapy. Bioenhancers improve nutritional status through growing bioavailability/bioefficacy of diverse nutraceuticals, which includes metals and nutrients [9].

Bioavailability enhancement may be accomplished by many mechanisms, together with promoting the absorption of the medication, inhibiting or decreasing the charge of biotransformation of medication in the liver or intestines, modifying the signaling method among host and pathogen, making sure elevated accessibility of the medication to the pathogens, while adjusting the immune response in a manner that the general requirement of the drug is decreased considerably. Besides the above mechanisms, bioenhancers are also useful for promoting the transport of nutrients and medication across the blood–brain barrier, which is essential in many cases including different central nervous system (CNS) disorders, cerebral infections, and epilepsy.

Fundamental classes of medicine that have shown increased bioenhancement encompass respiratory, cardiovascular, gastrointestinal tract, CNS, antibiotics, and anticancer. Some examples include sulfadiazine, tetracyclines, phenobarbitone, rifampicin, vasicine, ethambutol, pyrazinamide, nimesulide, phenytoin, dapsone, carbamazepine, coenzyme Q 10, indomethacin, β-carotene, amino acids, ciprofloxacin glucose, curcumin, and numerous other medicines [10].

11.2.2 Antitubercular and antileprotic drugs

As mentioned earlier, the first reported bioenhancer, piperine, was employed in the treatment of tuberculosis in humans. Rifampicin is one of the first-line drugs for the treatment of both tuberculosis and leprosy. But this drug was effective in a much higher dose due to low bioavailability and, therefore, exhibited toxicity. After combining piperine with this drug, the dose profile (from 450 to 200 mg) was reduced along with the treatment period. Further, the bioavailability of rifampicin was increased by 60%. Rifampicin inhibits the transcription of the polymerase by acting on RNA polymerase in human cells, which is being catalyzed by *Mycobacterium smegmatis*. Piperine enhanced this inhibition activity of RNA polymerase, several folds. It also arouses the binding capacity of rifampicin in resistant strains of bacteria [11, 12].

11.2.3 Medical adjuvants for antibiotics/chemotherapy

Nowadays, the use of antibiotics and antimicrobials has highly increased; therefore, the problem of drug resistance and addiction has increased. Hence, a high dose of such drugs is required for exerting the same therapeutic effect due to reduced drug absorption and resisting efflux pumps.

Despite advances in the field of pharmacology and traditional chemistry, the production of novel synthetic antibiotics, modification of antimicrobial compounds, and identification of appropriate enzyme targets for inhibitor development, current worldwide medicate advancement endeavors may not be sufficient to supply spearheading antimicrobials for the coming decade [13]. Given the rise in acquired resistance to traditional antibiotics, it makes sense to attempt mixing traditional antibiotics with bioenhancing plant extracts to achieve antimicrobial synergism [14]. The use of such a conventional and herbal combination therapy against hard-to-eradicate bacteria may open up newer avenues for infectious disease treatment. Combination therapy may be used to broaden the antimicrobial spectrum, prevent resistant mutants from emerging, and reduce side effects [15].

The use of bioenhancer along with the main drug has led to increased bioavailability and minimized drug dosage. Piperine, the main alkaloid found in the plant's black piper (*Piper nigrum* Linn) and long pepper (*Piper longum* Linn) are well known for increasing bioavailability and, hence, improving medication and nutraceutical efficacy, in addition to being an efflux pump inhibitor [16–19]. Piperine used in the antituberculosis medicine "Risorine," has shown evidence of increasing bioavailability [20]. The studies explored the anticancer and cancer-protective activity of a piperine-free *P. nigrum* extract against breast cancer cells. The resistance of cancer cells to multiple chemotherapeutic agents and the side effects of some agents pose a problem for the successful treatment of breast cancer. Currently, the progress of multidrug resistance (MDR) is a major problem to chemotherapy. Over a long-term treatment, several patients suffer from MDR, which can decrease therapeutic efficiency and lead to treatment failure and a decrease chance of survival. Therefore, the search for new potent chemotherapeutic agents from natural compounds is one way to detect new compounds for cancer treatment [21]. Piperine also reduced the cytotoxic aflatoxins by inhibiting CYP-P450 enzyme, which activates mycotoxins into harmful products [22].

Numerous medicinal plants have served as anticancer resources, and over 60% of current anticancer drugs, such as *topotecan, vinblastin, paclitaxel,* and *etotecan* are plant-derived compounds [23–25].

Antibiotics and an ethanolic extract of *Ficus exasperata* leaf have been shown to have synergistic efficacy against *E. coli* and *Staphylococcus albus* [26]. The harmful effects of an increasing number of mutagenic and environmental carcinogens can be prevented or minimized by using herbal bioenhancers. The antimutagenic properties of tulsi (*Ocimum tenuiflorum*) [27]; Haldi (*Curcuma longa*) [28]; amla (*Phyllamentus embelica*) [29]; and neem (*Azadirachta indica*) [30] have been scientifically established. These promising antimutagenic herbals could be applied to bacteria to prevent spontaneous mutations, thereby reducing bacterial antibiotic resistance. Certain Ayurvedic preparations such as Brahma Rasayana and Amalaki Rasayana have been shown to enhance the DNA repair mechanism [31, 32]. These preparations could be applied to counteract the spontaneous or induced mutations in bacteria. Another report revealed enhanced bioavailability for Nevirapine drug was found when combined with piperine. Nevirapine, a nucleoside inhibitor, is used with other antiretroviral agents for the treatment of HIV-1 [33].

11.2.4 Cardiovascular disease

Breviscapine, a familiar bioactive flavonoid extracted from traditional medicine, has been widely used in ischemic cerebrovascular and cardiovascular diseases, to prolong the duration of the drug in the circulation, reduce the frequency of injection administration, and subsequently afford patient compliance [34].

Ginkgo biloba phytosomes (GBP): It exhibits significant cardioprotective activity by lowering the levels of serum marker enzymes and lipid peroxidation and elevating the levels of catalase, glutathione, glutathione peroxidase, superoxide dismutase, and glutathione reductase [35].

11.2.5 Anti-inflammatory action

Triptolide (TP): It has been shown to have anti-inflammatory, antineoplastic, antifertility, and immunosuppressive activity. However, its clinical use was limited due to some serious toxicity. The mechanism for triptolide-induced hepatotoxicity was related to reactive oxygen species (ROS)-inducing lipid peroxidation and DNA damage.

Glycyrrhizic acid: It is a triterpene glycoside that possesses a wide range of biological and pharmacological activities. When extracted from the plant, it can be obtained in the form of mono-ammonium glycyrrhizin and ammonium glycyrrhizin. Glycyrrhizic acid has been used in China and Japan as a hepatoprotective drug in cases of chronic hepatitis and it shows anti-inflammatory action [36].

11.2.6 Nutraceuticals

Bioenhancer has wide applications in the nutrition field in enhancing the absorption and bioavailability of foods or nutrients by acting on the gastrointestinal tract. As per a reported clinical study, the herbal bioenhancer, piperine, can increase the bioavailability of vitamins against placebo by 50–60%. Data suggested the reported mechanism is owing to the thermogenic properties of piperine [3, 37].

11.3 Recent advances of bioenhancers

11.3.1 Bioenhancer: piperine

Drug: 18β-glycyrrhizic acid

Delivery system: transdermal patches
Alsaad et al. prepared and evaluated patches of glycyrrhizic acid with a synthetic polymer, carbopol 934. It was a reservoir-type patch in which they used piperine as a bioenhancer. The result showed that the patch containing herbal bioenhancer had tremendous potential as compared to those without bioenhancer [38].

11.3.2 Bioenhancer: piperine

Drug: celecoxib

Delivery system: oral delivery
Srivastava et al. worked on the colon cancer cells to study the synergistic antiproliferative effect of piperine with celecoxib. They reported that, by using bioenhancer, oral bioavailability increased to 129%. Further, this formulation was expressively cytotoxic to HT-29 cells. They suggested it is a novel approach for the treatment of colon cancer [39].

11.3.3 Bioenhancer: *Artemisia annua* L

Drug: vancomycin, erythromycin, chloramphenicol and kanamycin

Delivery system: oral delivery
Rolta et al. worked on *A. annua* as bioenhancer with different antibacterial and antifungal agents to overcome the drug resistance. They found a methanolic extract of artemisia had a greater antioxidant effect as compared to petroleum ether extract. In an antimicrobial study against *Candida* strains, both the extracts produced potent inhibitory action as compared to plain drugs [40].

11.3.4 Bioenhancer: *Dunaliella salina (D. salina)*

Drug: β-carotene

Delivery system: oral delivery
El-Baz et al. developed oral tablets of *D. salina* powder by direct compression technique by using a novel solubilizer, Sepitrap™ 80, and crospovidone. Both these solubilizers and crospovidone helped in reducing the disintegration time and enhanced the dissolution rate of β-carotene. The authors found that the tablets of *D. salina* powder had a promising antifibrotic potential in rats [41].

11.3.5 Bioenhancer: naringin

Drug: resveratrol

Delivery system: oral delivery
Chakraborty et al. worked on a combination of resveratrol and naringin, which exhibited intense protection against ischemia injury-induced myocardial toxicity, as compared to resveratrol alone. Both the results of pharmacokinetic and pharmacodynamics were satisfactory [42, 43].

11.3.6 Bioenhancer: piperine

Drug: silybin

Delivery system: oral delivery
Bi et al. demonstrated the potential of piperine as a bioenhancer with silybin. This product boosted the therapeutic effect in a liver-injured rat model. Piperine enhanced

the absorption of silybin and inhibited the biliary excretion in sandwich–cultured rat hepatocytes. Further, they mentioned piperine did not affect the phase-2 metabolism of silybin [44].

11.4 The future of bioenhancers

11.4.1 Bioenhancing activity through different routes of drug administration

The bioavailability can be increased by different mechanisms including increasing the polarity of the drug through chemical modification, prodrug formation, film coating, targeted delivery, encapsulation in suitable delivery systems, micro or nanosization, etc. The mechanism of drug and bioenhancer, when administered through different routes, is summarized in Figure 11.1.

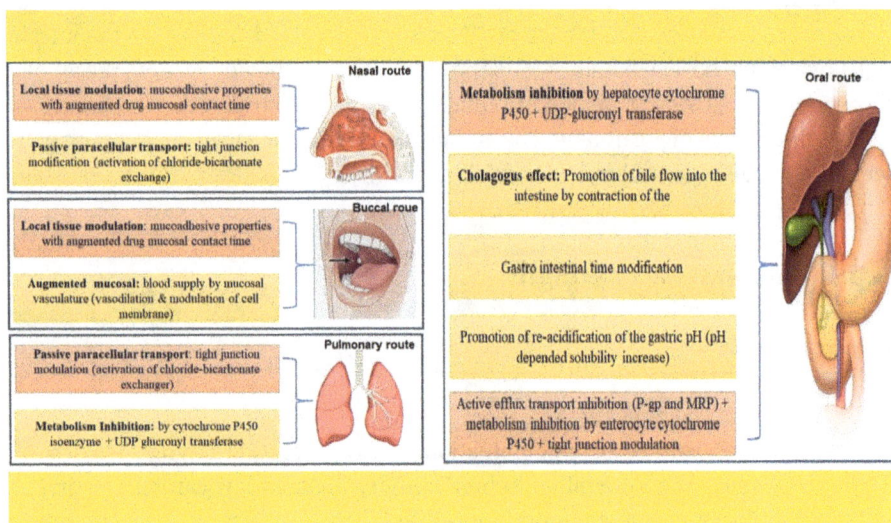

Figure 11.1: Illustration of the main mechanisms of action of bioenhancers for different routes of drug delivery.

11.4.2 New drug delivery systems for traditional bioenhancer

Table 11.1 depicts numerous examples of drugs and bioenhancers in different dosage forms for delivering through different routes [45–62].

Many researchers have worked on new delivery systems for entrapping bioavailability enhancers of herbal origin from a different route. New drug delivery systems like liposomes, nanoparticle, transferosomes, and many others are reported to increase the bioavailability of traditional bioenhancer. A few important examples are summarized here. Catechins liposome when delivered through the transdermal route increased permeation through the skin. It was prepared by the rotary evaporation sonication method with an encapsulation efficiency 93.0 ± 0.1%. When delivered through oral route, the water solubility of flavonoids and lignans nanoparticles, was increased. These nanoparticles were prepared by the nanosuspension method with 90% encapsulation. Flavonoid was entrapped in phytosomes for antioxidant activity by phospholipids complexation and also stabilized the ROS, when given by subcutaneous route. Similarly, when curcumin was entrapped in phytosomes, it increased antioxidant activity and bioavailability through the oral route. However, to date, many other novel delivery systems are unexplored for their efficiency in delivering bioenhancers. More researches should be focused on this area as it has a wide scope of applications [63].

Table 11.1: Delivery of drugs by using natural bioenhancers through different routes.

S. no.	Bioenhancer (class)	Research compound	Mode of action	Study design model	References
		Intranasal route for delivery of drugs with bioenhancer			
1.	Aloe vera (plant)	Didanosine	Intercellular modulation	In vitro (Franz diffusion cells)	[45]
2.	Chitosan (deacetylated chitin)	FITC–dextran	Mucoadhesion; changes in lipid organization	In vitro (T146 cells 1)	[46]
3.	Chitosan (deacetylated chitin)	Corticosteroid	Mucoadhesion	In vivo (pig), ex vivo (porcine buccal mucosa)	[47]
4.	Cod-liver oil extract (cod fish)	Ergotamine tartrate	No mechanism specified	Ex vivo (hamster cheek pouch)	[48]
5.	Oleic acid (cod fish)	Insulin	No mechanism specified	In vitro (dissolution test), in vivo (rat)	[49]
		Oral route for delivering drugs with bioenhancer			
1.	Aloe vera	Atenolol	Tight junction modulation	Ex vivo (rat intestinal tissue)	[50]

Table 11.1 (continued)

S. no.	Bioenhancer (class)	Research compound	Mode of action	Study design model	References
			Oral route for delivering drugs with bioenhancer		
2.	Caraway	Rifampicin, isoniazid, pyrazinamide	Local mucosal tissue modulation	*In vivo* (human)	[51]
3.	Curcumin	Midazolam	Efflux transporter inhibition; CYP3A4 inhibition	*In vivo* (rat)	[52]
4.	Emodin	Digoxin	Efflux transporter inhibition	*In vitro* (MDR1-MDCKII cells 6, Caco-2 cells2	[53]
5.	Gallic acid ester	Nifedipine	CYP3A metabolism inhibition	*In vitro* (human liver microsomes)	[54]
			Buccal route for delivering drugs with bioenhancer		
1.	Chitosan	Endogenous polypeptide hormone	Tight junction modulation	*In vivo* (sheep)	[55]
2.	Chitosan	Morphine	Increased mucoadhesion	*In vivo* (sheep, human)	[56]
3.	Chitosan–TBA (thiolated polymer	Insulin	Increased mucoadhesion	*In vivo* (rat)	[57]
4.	TMC (chemically modified chitosan)	Mannitol, a sugar alcohol	Tight junction modulation, increased mucoadhesion	*In vivo* (rat)	[58]
			Pulmonary route for delivering drugs with bioenhancer		
1.	Aprotinin, bestatin (protease inhibitors)	Granulocyte-colony stimulating factor	Metabolism inhibition	*In vivo* (rat)	[59]
2.	Chitosan (chemically modified biopolymer)	Somatostatin analog	Tight junction modulation	*In vitro* (Calu-3 cells 5); in vivo (rat)	[60]

Table 11.1 (continued)

S. no.	Bioenhancer (class)	Research compound	Mode of action	Study design model	References
			Pulmonary route for delivering drugs with bioenhancer		
3.	Citric acid (chelating agents)	Insulin	metabolism inhibition	*In vivo* (rat)	[61]
4.	Sodium taurocholate (bile salt)	Insulin	Metabolism enhancement, enzymatic degradation inhibition	*In vitro* (Caco-2 cells 2), in vivo (dog)	[62]
5.	TMC (cationic polymers)	Octreotide	Tight junction modulation	*In vitro* (Calu-3 cells 5); in vivo (rat)	[60]

11.4.3 Reduce the cost of drug development

The high cost of drug development, as well as the imminent patent expiration of many bestselling drugs are major roadblocks to long-term commercial viability. In addition, intellectual property (IP) specifications that are related to trade, such as patenting of products have been extended to include a large number of countries, where generics have previously dominated [37]. Bioenhancers is a novel phenomenon discovered using Ayurveda, a traditional Indian medical method (as mentioned by Charaka, Sushruta, and other apothecaries in the traditional system of medicine). The idea may be useful not only in lowering the toxicity side effects but also, importantly, the drug development costs, which could have a very positive impact on the country's economy, as desired by the WHO. Ayurveda can greatly aid the drug discovery process by the use of reverse pharmacology. This may provide new ways of detecting active compounds and reduced drug development costs. A reduction in the cost of medication could make treatment accessible to a more extensive segment of society, including financially challenged patients [17]. When coadministered or pre-treated with a variety of medications and nutraceuticals, available scientific research studies on bioenhancers have shown to have a significant improving effect on bioavailability. Many natural agents such as piperine, curcumin, *Zingiber offcinale*, glycyrrhizin, niaziridin, *Aloe vera*, *Cuminum cyminum*, allicin, *Carum carvi*, lysergol, sinomenine, *Stevia rebaudiana*, genistein, *Ammanniamultiflora*, capmul, capsaicin, quercetin, and naringin [15] are potential bioenhancing agents and, thus, require urgent scientific attention.

11.4.4 Ecological aspect

The dosage of anticancer chemotherapeutic drugs such as taxol can be decreased, with the help of bioenhancers. This will also decrease the drug toxicity because of the lower dose of taxol given for treatment. The ecological implication of this is huge, as taxol is extracted from the bark of the Pacific yew tree, which is one of the world's slowest growing trees. Currently, several trees must be cut down for the treatment of one patient. With the integration of bioenhancers, fewer trees need to be sacrificed [3, 17].

11.4.5 Regulatory guidelines

The hurdles in the use of bioenhancers are that large-scale production of bioenhancers is not an easy task. Phytomolecules as bioenhancers are extracted in meager amounts from natural sources, and this imposes a big hurdle in their use. From the commercialization point of view, large-scale use of bioenhancers is needed, rather than laboratory scale. Secondly, it is important to get regulatory approval for them. Detailed study of their physicochemical and pharmacokinetic properties is required to ensure their safety profile. And lastly, without sufficient clinical studies, they cannot be incorporated into formulations and marketed directly for public use. The bioenhancing effect of phytochemicals as natural bioenhancers of the wide variety of drugs and nutraceuticals in animals and humans needs a lot of experimentation on animals. Lack of information on the mechanism of action, adverse effects, and evaluation of toxicity indices of the extracts have to be taken into account. Research should be focused on all these parameters of safety, compatibility with drugs and nutraceuticals, toxicity, efficacy, and mechanism of action of these bioenhancers. Finally, optimization of pharmacokinetics of these bioenhancers is required to establish them as effective bioenhancers.

Regulation is a difficult task when it comes to traditional drug products. The Quality Council of India and the Department of AYUSH together have developed two brands for traditional medicines: Premium mark and AYUSH mark [37]. This enactment of a product certification program initiated by the Department of AYUSH for several AYUSH goods is to gain consumer trust. The program is based on protocol implementation. The two levels of this program are a) AYUSH Standard Mark, based on compliance with domestic regulatory requirements; and b) AYUSH Premium Check, which is based on GMP prerequisites that comply with WHO rules and having stricter guidelines. In the case of products containing bioenhancers, the US-FDA and EMEA have set a few standards for physicochemical and pharmacological properties. But no comprehensive standards have been set for medicines containing herbal bioenhancers. The regulatory authorities should take the initiative to set standards for such products.

11.4.6 Industry and future of bioenhancers

In the last two decades, big pharmaceutical companies have become increasingly interested in therapeutic herbs. This is due to the increased awareness and interest in medicinal plants [64] among the public and scientific community. By 2050, global trade in medicinal plants and their products is estimated to be worth $5 trillion, with both China and India emerging as key players [65]. Looking to this projection, many biopharmaceutical giants have begun including herbal divisions in their drug development efforts. On the other hand, many mini- and micro-sized Ayurvedic firms that lack good laboratories and skilled personnel are not indulging in patents or new drug discovery research. It is also true that many Ayurvedic drug companies have avoided entering molecular research because they desire to remain firm to the traditional path, brand loyalty, and specialized consumer base. The majority of Ayurvedic enterprises focus on producing traditional Ayurvedic products (based on the classical Ayurvedic text). A novel proprietary product developed by a firm gives them exclusive marketing rights in an unchallenged competitive environment, and they might even have a commodity monopoly, in some cases. Without the required push from the big pharmaceutical companies, no medicine, be it traditional or herbal, can be successful in the market. Increased research work is, thus, needed to convince the international scientific community on the efficacy of herbal bioenhancers. Though many herbal bioenhancers do show tremendous promise in the lab and preclinical studies, the real impetus comes after the success of a properly designed clinical trial. The knowledge of Ayurvedic remedies has already resulted in many drugs or standardized extracts with identified active compounds, viz., *gum guggulu*, *brahmi*, *reserpine*, *flavopiridol*, etc. [66]. Proper coordination of the institution–industry-regulators may see many more herbal products getting international recognition in the future.

11.5 Conclusion

By using bioenhancers, the dosage of the drug is reduced and the hazards of drug resistance are curtailed. This concept is especially applicable for drugs like anticancer, antimicrobials, and other potent drugs. Herbal bioenhancer is less toxic, easily available, and has a wide mechanism of action. However, development of herbal bioenhancers has also created new challenges for regulatory control. Moreover, nanodrug products that use bioenhancers need to have separate regulations as they are different from traditional products. In this issue, the European Medicines Evaluation Agency and United States Food and Drug Administration have taken the initiative to recognize possible regulatory and scientific challenges. Another point to be considered is the evaluation of ecological aspects of using bioenhancer. This is important because herbal enhancer is one of the essential parts of the ecological system, and therefore, may have a profound effect on it. Several researches are being conducted with bioenhancers, but very few of them are entering into clinical studies. To establish these products' potential and use, more clinical trials should be conducted. The reported data on bioenhancer research and clinical trial can further help commercialize these products. To date, commercialization of these products is limited due to lack of proper data, insufficient clinical trial results, and lower research interest among researchers and others. Proper coordination of the institution–industry-regulators may see many more herbal products getting international recognition in the future.

List of abbreviations

CNS	Central nervous system
MDR	Multidrug resistance
D. salina	*Dunaliella salina*

References

[1] Patwardhan B, Vaidya AD, Natural products discovery: Accelerating the clinical candidate development using reverse pharmacology approaches, Indian Journal of Experimental Biology, 2010, 48, 220–227.

[2] Iswariya T, Pradesh A, Gupta S, Bioavailability enhancers: An overview, IJARIIT, 2019, 5, 825–829.

[3] Atal N, Bedi KL, Bioenhancers: Revolutionary concept to market, Journal of Ayurveda and Integrative Medicine, 2010, 1, 96–3.

[4] Choudhary N, Khajuria V, Gillani ZH, Tandon VR, Arora E, Effect of Carum carvi, a herbal bioenhancer on pharmacokinetics of antitubercular drugs: A study in healthy human volunteers, Perspectives in Clinical Research, 2014, 5, 80–84.

[5] Shoba G, Joy D, Joseph T, Majeed M, Rajendran R, Srinivas PS, Influence of piperine on the pharmacokinetics of curcumin in animals and human volunteers, Planta Medica, 1998, 4, 353–356.

[6] Gupta SC, Patchva S, Aggarwal BB, Therapeutic roles of curcumin: Lessons learned from clinical trials, The AAPS Journal, 2013, 15, 195–23.

[7] Jhanwar B, Gupta S, Biopotentiation using herbs: Novel technique for poor bioavailable drugs, International Journal of PharmTech Research, 2014, 6, 443–454.

[8] Oladimeji FA, Adegbola AJ, Onyeji CO, Appraisal of Bioenhancers in Improving Oral Bioavailability: Applications to Herbal Medicinal Products, Journal of Pharmaceutical Research International, 2018, 29, 1–23.

[9] Breedveld P, Beijnen JH, Schellens JH, Use of P-glycoprotein and BCRP inhibitors to improve oral bioavailability and CNS penetration of anticancer drugs, Trends in Pharmacological Sciences, 2006, 27, 17–7.

[10]   Kang MJ, Cho JY, Shim BH, Kim DK, Lee J, Bioavailability enhancing activities of natural compounds from medicinal plants, Journal of Medicinal Plants Research, 2009, 3, 1204–1211.
[11]   Balakrishnan V, Varma S, Chatterji D, Piperine augments transcription inhibitory activity of rifampicin by several fold in Mycobacterium smegmatis, Current Science, 2001, 80, 1302–1305.
[12]   Kapil RS, Zutshi U, Bedi KL, Singh G, Johri RK, Dhar SK, Kaul JL, Sharma SC, Pahwa GS, Kapoor N, Tickoo AK, inventors. Process for preparation of pharmaceutical composition with enhanced activity for treatment of tuberculosis and leprosy. United States patent US 5,439,891. 1995 Aug 8.
[13]   Gupta PD, Birdi TJ, Development of botanicals to combat antibiotic resistance, Journal of Ayurveda and Integrative Medicine, 2017, 8, 266–275.
[14]   Chanda S, Rakholiya K, Combination therapy: Synergism between natural plant extracts and antibiotics against infectious diseases, Microbiol Book Series, 2011, 1, 520–529.
[15]   Dudhatra GB, Mody SK, Awale MM, Patel HB, Modi CM, Kumar A, Kamani DR, Chauhan BN, A comprehensive review on pharmacotherapeutics of herbal bioenhancers, The Scientific World Journal, 2012, 2012, 637.
[16]   Badmaev V, Majeed M, Prakash L, Piperine derived from black pepper increases the plasma levels of coenzyme Q10 following oral supplementation, The Journal of Nutritional Biochemistry, 2000, 11, 109–113.
[17]   Kesarwani K, Gupta R, Bioavailability enhancers of herbal origin: An overview, Asian Pacific Journal of Tropical Biomedicine, 2013, 3, 253–266.
[18]   Majumdar SH, Kulkarni AS, Kumbhar SM, "Yogvahi (Bioenhancer)": An Ayurvedic concept used in modern medicines, International Research Journal of Pharmacy and Medical Sciences, 2018, 1, 20–25.
[19]   Tatiraju DV, Bagade VB, Karambelkar PJ, Jadhav VM, Kadam V, Natural bioenhancers: An overview, Journal of Pharmacognosy and Phytochemistry, 2013, 2, 55–5.
[20]   Drabu S, Khatri S, Babu S, Lohani P, Use of herbal bioenhancers to increase the bioavailability of drugs, Research Journal of Pharmaceutical, Biological and Chemical Sciences, 2011, 2, 7–9.
[21]   Szakács G, Paterson JK, Ludwig JA, Booth-Genthe C, Gottesman MM, Targeting multidrug resistance in cancer, Nature Reviews. Drug Discovery, 2006, 5, 219–234.
[22]   Raguz S, Yagüe E, Resistance to chemotherapy: New treatments and novel insights into an old problem, British Journal of Cancer, 2008, 99, 387–391.
[23]   Reen RK, Wiebel FJ, Singh J, Piperine inhibits aflatoxin B1-induced cytotoxicity and genotoxicity in V79 Chinese hamster cells genetically engineered to express rat cytochrome P4502B1, Journal of Ethnopharmacology, 1997, 58, 165–173.
[24]   Newman DJ, Cragg GM, Snader KM, Natural products as sources of new drugs over the period 1981–2002, Journal of Natural Products, 2003, 66, 1022–1037.
[25]   Cragg GM, Newman DJ, Plants as a source of anti-cancer agents, Journal of Ethnopharmacology, 2005, 100, 72–79.
[26]   Odunbaku OA, Ilusanya OA, Akasoro KS, Antibacterial activity of ethanolic leaf extract of Ficusexasperata on Escherichia coli and Staphylococcus albus, Scientific Research and Essays, 2008, 3, 562–564.
[27]   Stajkovi O, Beric-Bjedov T, Mitic-]ulafic D, Stankovic S, Vukovic-Gacic B, Simic D, Knezevic-Vukcevi J, Antimutagenic properties of Basil (OcimumbasilicumL.) in Salmonella typhimurium TA100, Food Technology and Bioprocess Technology, 2007, 45, 213–214.
[28]   Ragunathan I, Panneerselvam N, Antimutagenic potential of curcumin on chromosomal aberrations in Allium cepa, Journal of Zhejiang University. Science. B, 2007, 8, 470–475.
[29]   Torabizadeh M, Saeiadpour S, Lakshmi P, Antimutagencity of Phyllamentus Embelica, IOSR Journal of Dental and Medical Sciences, 2012, 1, 13–16.

[30] Alabi OA, Anokwuru CP, Ezekiel CN, Ajibaye O, Nwadike U, Fasasi O, Abu M, Anti-mutagenic and anti-genotoxic effect of ethanolic extract of neem on dietary aflatoxin induced genotoxicity in mice, The Journal of Biological Sciences, 2011, 11, 307–317.

[31] Swain U, Sindhu KK, Boda U, Pothani S, Giridharan NV, Raghunath M, et al., Studies on the molecular correlates of genomic stability in rat brain cells following Amalakirasayana therapy, Mechanisms of Ageing and Development, 2012, 133, 112–117.

[32] Guruprasad K, Subramanian A, Singh VJ, Sharma RSK, Gopinath PM, Sewram V, et al., Brahma rasayana protects against ethyl methane sulfonate or methyl methane sulfonate induced chromosomal aberrations in mouse bone marrow cells, BMC Complementary and Alternative Medicine, 2012, 12, 113–119.

[33] Kasibhatta R, Naidu MU, Influence of piperine on the pharmacokinetics of nevirapine under fasting conditions: A randomised, crossover, placebo-controlled study, Drugs in R&D, 2007, 8, 383–391.

[34] Zhong H, Deng Y, Wang X, Yang B, Multivesicular liposome formulation for the sustained delivery of breviscapine, International Journal of Pharmaceutics, 2005, 301, 15–24.

[35] Vandana SP, Suresh RN, Cardioprotective activity of Ginkgo biloba Phytosomes in isoproterenol-induced myocardial necrosis in rats: A biochemical and histoarchitectural evaluation, Exp.Toxicol.Pathol., 2008, 60, 397–7.

[36] Hou J, Zhou SW, Formulation and preparation of glycyrrhizic acid solid lipid nanoparticles, ACTA Academiae Medic Inaemilitaristertiae, 2008, 30, 1043–1045.

[37] Madhavan H, Innovation system and increasing reformulation practices in the ayurvedic pharmaceutical sector of south India, Asian Medicine, 2014, 9, 236–271.

[38] Alsaad AA, Formulation & evaluation of β-glycyrrhetinic acid patches with natural bioenhancer, Materials Today: Proceedings, 2021, May, 3.

[39] Srivastava S, Dewangan J, Mishra S, Divakar A, Chaturvedi S, Wahajuddin M, Kumar S, Rath SK, Piperine and Celecoxib synergistically inhibit colon cancer cell proliferation via modulating Wnt/β-catenin signaling pathway, Phytomedicine, 2021, 84, 153484.

[40] Rolta R, Sharma A, Sourirajan A, Mallikarjunan PK, Dev K, Combination between antibacterial and antifungal antibiotics with phytocompounds of Artemisia annua L: A strategy to control drug resistance pathogens, Journal of Ethnopharmacology, 2021, 266, 113420.

[41] El-Baz FK, Ali SI, Basha M, Kassem AA, Shamma RN, Elgohary R, Salama A, Design and evaluation of bioenhanced oral tablets of Dunaliellasalina microalgae for treatment of liver fibrosis, Journal of Drug Delivery Science and Technology, 2020, 59, 101845.

[42] Chakraborty M, Bhattacharjee A, Ahmed MG, Sindhi Priya ES, Shahin H, Taj T, Effect of Naringin on myocardial potency of Resveratrol against ischemia reperfusion induced myocardial toxicity in rat, Synergy, 2020, 10, 100062.

[43] Hernández-Aquino E, Zarco N, Casas-Grajales S, et al., Naringenin prevents experimental liver fibrosis by blocking TGFβ-Smad3 and JNK-Smad3 pathways, World Journal of Gastroenterology. 2017, 23, 4354–14.

[44] Bi X, Yuan Z, Qu B, Zhou H, Liu Z, Xie Y, Piperine enhances the bioavailability of silybin via inhibition of efflux transporters BCRP and MRP2, Phytomedicine, 2019, 54, 98–10.

[45] Ojewole E, Mackraj I, Akhundov K, Hamman J, Viljoen A, Olivier E, Wesley-Smith J, Govender T, Investigating the effect of Aloe vera gel on the buccal permeability of didanosine, Planta Medica, 2012, 78, 354–361.

[46] Portero A, Remuñán-López C, Nielsen HM, The potential of chitosan in enhancing peptide and protein absorption across the TR146 cell culture model – an in vitro model of the buccal epithelium, Pharmaceutical Research, 2002, 19, 169–174.

[47] Şenel S, Kremer MJ, Kaş S, Wertz PW, Hıncal AA, Squier CA, Enhancing effect of chitosan on peptide drug delivery across buccal mucosa, Biomaterials, 2000, 21, 2067–2071.

[48] Tsutsumi K, Obata Y, Takayama K, Loftsson T, Nagai T, Effect of cod-liver oil extract on the buccal permeation of ergotamine tartrate, Drug Development and Industrial Pharmacy, 1998, 24, 757–762.
[49] Morishita M, Barichello JM, Takayama K, Chiba Y, Tokiwa S, Nagai T, Pluronic® F-127 gels incorporating highly purified unsaturated fatty acids for buccal delivery of insulin, International Journal of Pharmaceutics, 2001, 212, 289–293.
[50] Beneke C, Viljoen A, Hamman J, In vitro drug absorption enhancement effects of Aloe vera and Aloe ferox, Scientiapharmaceutica, 2012, 80, 475–486.
[51] Choudhary N, Khajuria V, Gillani ZH, Tandon VR, Arora E, Effect of Carum carvi, a herbal bioenhancer on pharmacokinetics of antitubercular drugs: A study in healthy human volunteers, Perspectives in Clinical Research, 2014, 5, 80–84.
[52] Zhang W, Tan TM, Lim LY, Impact of curcumin-induced changes in P-glycoprotein and CYP3A expression on the pharmacokinetics of peroral celiprolol and midazolam in rats, Drug Metabolism and Disposition, 2007, 35, 110–115.
[53] Li X, Hu J, Wang B, Sheng L, Liu Z, Yang S, Li Y, Inhibitory effects of herbal constituents on P-glycoprotein in vitro and in vivo: Herb–drug interactions mediated via P-gp, Toxicology and Applied Pharmacology, 2014, 275, 163–175.
[54] Wacher VJ, Benet LZ, inventors; AvMaxInc, assignee. Use of gallic acid esters to increase bioavailability of orally administered pharmaceutical compounds. United States patent US 6,180,666. 2001 Jan 30.
[55] Hinchcliffe M, Jabbal-Gill I, Smith A, Effect of chitosan on the intranasal absorption of salmon calcitonin in sheep, Journal of Pharmacy and Pharmacology, 2005, 57, 681–687.
[56] Illum L, Watts P, Fisher AN, Hinchcliffe M, Norbury H, Jabbal-Gill I, Nankervis R, Davis SS, Intranasal delivery of morphine, Journal of Pharmacology and Experimental Therapeutics, 2002, 301, 391–399.
[57] Krauland AH, Guggi D, Bernkop-Schnürch A, Thiolated chitosan microparticles: A vehicle for nasal peptide drug delivery, International Journal of Pharmaceutics, 2006, 307, 270–277.
[58] Hamman JH, Stander M, Kotze AF, Effect of the degree of quaternisation of N-trimethyl chitosan chloride on absorption enhancement: In vivo evaluation in rat nasal epithelia, International Journal of Pharmaceutics, 2002, 232, 235–242.
[59] Machida M, Hayashi M, Awazu S, The effects of absorption enhancers on the pulmonary absorption of recombinant human granulocyte colony-stimulating factor (rhG-CSF) in rats, Biological & Pharmaceutical Bulletin, 2000, 23, 84–86.
[60] Florea BI, Thanou M, Junginger HE, Borchard G, Enhancement of bronchial octreotide absorption by chitosan and N-trimethyl chitosan shows linear in vitro/in vivo correlation, Journal of Controlled Release, 2006, 110, 353–361.
[61] Todo H, Okamoto H, Iida K, Danjo K, Effect of additives on insulin absorption from intratracheally administered dry powders in rats, International Journal of Pharmaceutics, 2001, 220, 101–110.
[62] Johansson F, Hjertberg E, Eirefelt S, Tronde A, Bengtsson UH, Mechanisms for absorption enhancement of inhaled insulin by sodium taurocholate, European Journal of Pharmaceutical Sciences, 2002, 17, 63–68.
[63] Saraf S, Applications of novel drug delivery system for herbal formulations, Fitoterapia, 2010, 81, 680–689.
[64] Arora P, Healthcare biotechnology firms in India: Evolution, structure and growth, Current Science, 2005, 89, 458–463.
[65] Nirmal SA, Pal SC, Otimenyin SO, Aye T, MostafaElachouri M, Kundu SK, Thandavarayan RA, Mandal SC, Contribution of herbal products in global market, Pharma Review, 2013, 95–104.
[66] Differding E, The Drug Discovery and Development Industry in India – Two Decades of Proprietary Small-Molecule R&D, ChemMedChem, 2017, 12, 786–818.

Rupali Patil, Pratiksha Aher, Punam Bagad, Shraddha Ekhande

Chapter 12
Herbal bioenhancers in veterinary phytomedicine

Abstract: "Yogvahi" is a Sanskrit term for the novel concept of Bioenhancers, available to enhance the efficacy of drugs, when used in combination. Bose, in 1929, first used the term Bioenhancer for the increased antihistaminic activity of Vasaka using long pepper. Some distinguished discoveries of Ayurvedic Yogvahi are Trikatu, bhasmas, and cow urine distillate. WHO also acknowledged the use of piperine (PIP) as a bioenhancer and in modern medicines; it is used in tubercular therapy, in combination with rifampicin and isoniazid. These beneficial compounds in veterinary medicine reduce dosage, ensure better treatment rate, shorten treatment time, and eliminate drug resistance or adverse effects, all of which have financial consequences. There have been a few failures as a consequence of these developments, since the fundamental mechanisms of active substance intake (influence on solubility, active substance efflux, and transport proteins, enhance gastrointestinal permeability) and active substance metabolism (inhibition/induction of drug-metabolizing enzymes, thermogenic result) are still unknown. Data from laboratory animals could not be attributed due to species-specific variations in these processes. Plant products in veterinary patients have a mixed record of effectiveness, ranging from good and stable to unsuccessful and risky. Most trials, on the other hand, lack consistent endpoints, and observation periods are typically brief; the clinical relevance of reported findings is not always apparent. Also, data comparing herbal therapies to well-known prescription medications are often unavailable. While the database on herbs is growing, veterinarians interested in prescribing safe, effective, natural, plant-based compounds should look up information on the compound or product in question in the most recent scientific literature. Bioenhancers used in veterinary phytomedicines include long pepper, black pepper, turmeric, ginger, caraway, black cumin, liquorice, aloe, and others. The importance of bridging the gap between the ancient concept of Ayurveda and the current use of herbal bioenhancers in veterinary medicine is highlighted in this chapter.

Rupali Patil, GES's Sir Dr. M. S. Gosavi College of Pharmaceutical Education and Research, Nashik 425005, Maharashtra, India, e-mail: ruupalipatil@gmail.com
Pratiksha Aher, Punam Bagad, Shraddha Ekhande, GES's Sir Dr. M. S. Gosavi College of Pharmaceutical Education and Research, Nashik 425005, Maharashtra, India

https://doi.org/10.1515/9783110746808-012

12.1 Introduction

12.1.1 Bioenhancers/bioavailability enhancers/biopotentiers

Bioenhancers are compounds that enhance the bioavailability and efficacy of active ingredients, when coupled with them, without having any activity of their own at the quantity given [1]. Apart from antibiotics, anticancer medicines, cardiovascular drugs, anti-inflammatory drugs, central nervous system medicines, and so on, they also improve vitamin and nutrient absorption [2].

Increased bioavailability refers to the amount of medicine in the bloodstream that is available for action. Increased Bio efficacy refers to a drug's increased effectiveness as a result of increased bioavailability as well as by other processes [3].

12.1.2 History

Bioenhancers or bioavailability enhancers did not exist as a term or a chapter in any current scientific literature prior to 1979. The phrase "bioavailability enhancers" was coined by Indian scientists in 1979 at the Indian Institute of Integrative Medicine, Jammu, previously RRL (Regional Research Laboratory), Jammu [4]. Following that, Dr. C. K. Atal, the Director of the Institute and his research team at RRL Jammu systematically explored and established the notion of bioavailability enhancers [1]. PIP, the world's first bioenhancer, was discovered and scientifically confirmed by the institute, which used sparteine and vasicine to create the world's first experimentally bioenhanced medication [5].

Using sparteine and vasicine, the institution found and systematically approved PIP as the globe's first bioenhancer, resulting in the world's first experimentally bioenhanced medicines [5]. Dr. Atal led a rifampicin-based bioenhanced antitubercular (anti-TB) medicine investigation, leading to the production of the world's first bioenhanced anti-TB drug formulation. On World Tuberculosis Day 2011, the Indian government officially issued this Drugs Controller General of India-approved formulation at the Anusandhan Bhawan in Delhi, and it was also delivered to Mr. Bill Gates, Chairman of Microsoft, during a program at Le Meridian in Delhi on the same day [6].

PIP, a bioenhancer discovered in 1979, added a new section in medical science. Since then it has sparked worldwide interest and research into the topic, resulting in the discovery of a slew of novel bioenhancers. PIP is still the most powerful and well-studied bioenhancer in the market. It is safe, effective, cost-efficient, and simple to synthesize for commercial use. It is also a broad-spectrum bioenhancer affecting a variety of current drug classes (Figure 12.1) [7].

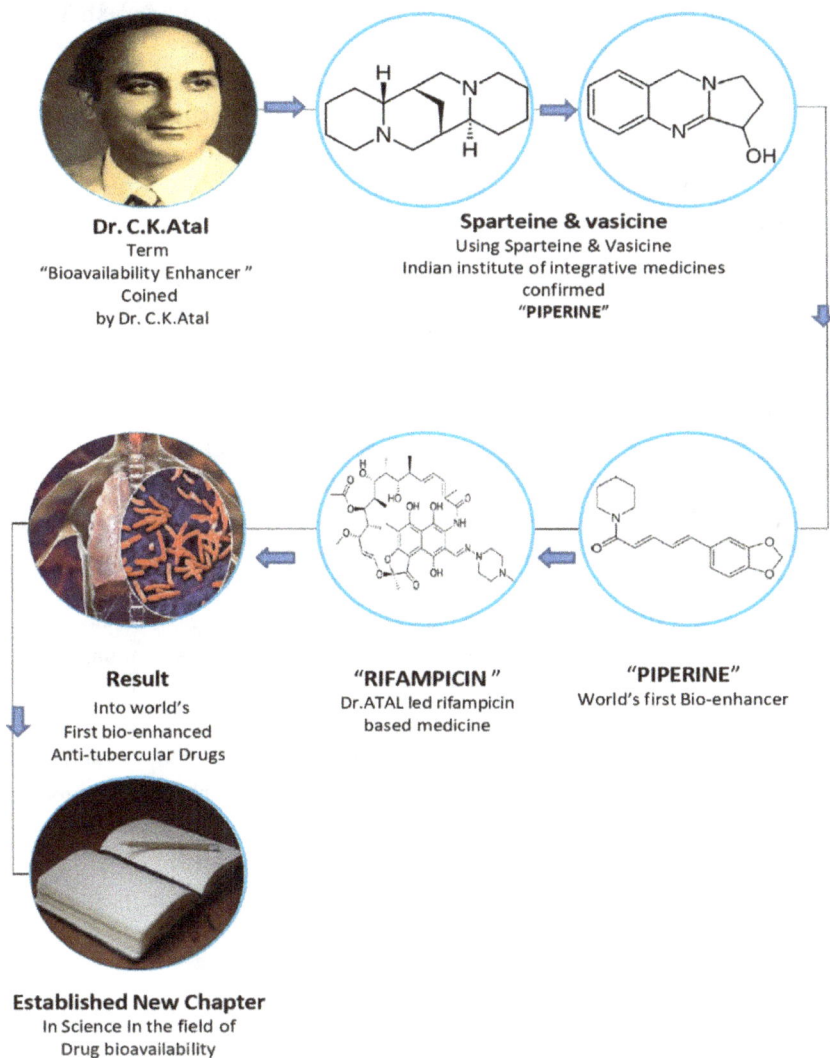

Dr. C.K.Atal
Term
"Bioavailability Enhancer"
Coined
by Dr. C.K.Atal

Sparteine & vasicine
Using Sparteine & Vasicine
Indian institute of integrative medicines
confirmed
"PIPERINE"

Result
Into world's
First bio-enhanced
Anti-tubercular Drugs

"RIFAMPICIN"
Dr.ATAL led rifampicin
based medicine

"PIPERINE"
World's first Bio-enhancer

Established New Chapter
In Science In the field of
Drug bioavailability

Figure 12.1: Pictorial representation of history of bioenhancer.

12.2 Need of bioenhancers in veterinary phytomedicines

Lipid solubility and molecular size are the crucial elements of substances passing through the cellular membrane and getting absorbed systemically after oral or topical administration [8]. Despite having high bioactivity in vitro, several plant extracts and phytoconstituents show low or no bioactivity in vivo with poor absorption and bioavailability due to inappropriate molecular size or insufficient lipid solubility, or both [9].

After oral consumption, several elements of the plant extract containing multiple constituents are sometimes degraded in the gastric environment. They lower the dose, decrease the treatment time, and thereby lessen the risk of drug resistance. They make the treatment cost-effective and reduce toxicity of the drug side effects due to dose economy [10].

12.3 Difficulties related to herbal bioenhancer

Although bioenhancers have been successful in drug delivery, not all approaches have been equally successful. New bioenhancers provide a number of key issues that are identified. Nanomaterials' physicochemical properties have been modified for the improvement of qualities such as lengthy blood circulation, improved useful surface area, safeguarding integrated medication from degradation, passing biological obstacles, and site-specific targeting [11].

Large-scale production is another problem in herbal bioenhancer research and development. Scaling up laboratory or pilot technology for ultimate commercialization is always necessary [9]. Low concentrations of nanomaterials, agglomeration, and the chemical process are all obstacles in scaling up. For better performance, it is always easy to change nanomaterials at the laboratory scale than at the larger scale. It is also a challenge to maintain the size and composition of nanomaterials at a large level while increasing bioavailability [12].

New regulatory issues have arisen as a result of advancements in herbal bioenhancers. Regulations that take into account the physicochemical and pharmacokinetic

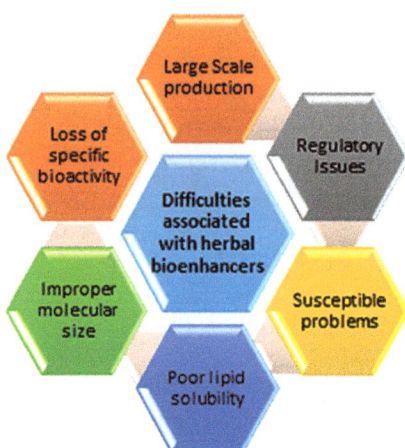

Figure 12.2: Pictorial representation of difficulties associated with herbal bioenhancer.

differences between nano drug products and conventional medicinal products are becoming increasingly relevant [10].

The FDA of the United States and the EMEA of European Medicines Evaluation Agency have started to lead in recognizing some potential technical and monitoring concerns [13]. Difficulties associated with herbal bioenhancers are represented in Figure 12.2.

12.4 Natural compounds as bioenhancer in veterinary care

12.4.1 Quercetin

12.4.1.1 Biological source

Quercetin, 3,3′,4′,5,7-pentahydroxyflavone, is the utmost plentiful nutritive flavonoid, present in fruits (especially citrus), apples, barks, buckwheat, broccoli, dark cherries, and berries, such as blueberries and cranberries, green leafy vegetables, green tea, seeds, nuts, flowers, olive oil, onions, red grapes, and red wine.

It cannot be generated by the human body. The word "Quercetum" means "Oak Forest" in Latin. It has a yellowish tinge [14].

12.4.1.2 Uses

Quercetin has antioxidant, radical scavenging, anti-inflammatory, anti-atherosclerotic, anticancer, and antiviral properties [15].

12.4.1.3 Mechanism of action

It works as a cytochrome P 3A4 inhibitor as well as a modulator of P-glycoprotein (P-gp), increasing the active pharmaceutical ingredient (API) bioavailability [16].

12.4.1.4 Combination of quercetin with other drugs

It is given in combination with drugs such as paclitaxel, verapamil, diltiazem, tamoxifen, fexofenadine, etoposide, and epigallocatechin gallate.

12.4.1.5 Examples

- **Pioglitazone**

The effect of quercetins on the bioavailability of frequently used antidiabetic medication, pioglitazone (Pioglit), was studied in female rats. Quercetin was given pretreatment. Pioglit was given by oral (10 mg/kg) and IV (5 mg/kg) routes at different timeslots. Bioavailability of Pioglit was increased by 75% after oral and 25% after IV pretreatment by quercetin [17].

- **Paclitaxel**

Quercetin pretreatment (2, 10, 20 mg/kg) exhibited increased paclitaxel (Pacli) bioavailability in rats after paclitaxel and prodrug as paclitaxel [18].

- **Verapamil**

Pharmacokinetics after verapamil and nor-verapamil was evaluated in rabbits with and without quercetin (5.0 and 15 mg/kg). Quercetin pretreatment, 30 min before verapamil dose had a significant impact on verapamil pharmacokinetics, whereas simultaneous quercetin administration had little influence on the oral verapamil exposure.

Twofold rise in the C_{max} and AUC (area under the curve) of verapamil, without major variations in T_{max} (time taken to reach the maximum concentration) and $t_{1/2}$ of verapamil, was experiential in rats treated with 15 mg/kg quercetin compared to the control group. Significant rise in the absolute and relative bioavailability of verapamil when equated to the control group was experiential in quercetin pretreated rabbits [19].

- **Doxorubicin**

Presence or absence of quercetin significantly affects the pharmacokinetic parameters of doxorubicin. Significant rise in peak plasma concentration of doxorubicin was observed in the presence of quercetin with AUC (31.2–136.0%) of oral doxorubicin (p 0.05 for 0.6 mg/kg, p 0.01 for 3 and 15 mg/kg) when equated to the control group [20].

- **Etoposide**

Rats received either etoposide per se orally or in combination with quercetin. High-performance liquid chromatography (HPLC) with a fluorescence detector was used to measure the plasma levels of etoposide. Pharmacokinetic characteristics of etoposide were considerably affected in the presence of quercetin. The inclusion of quercetin resulted in a substantial (P 0.055; 5 mg/kg; P 0.01;15 mg/kg,) absolute bioavailability between 12.7% and 13.6% [21] (Table 12.1).

Table 12.1: Bioenhancement of drugs by quercetin [4].

Category	Drug	Route of administration	Dose of quercetin
Anticancer	Doxorubicin	Oral	0.6, 3, or 15 mg/kg
Antihypertensive	Valsartan	Oral	10 mg/kg
Antiplatelet	Clopidogrel	Oral	250 mg/kg
Anticancer	Etoposide	Oral	5 or 15 mg/kg
Cardiac glycoside	Digoxin	Oral	40 mg/kg

12.4.2 *Stevia*

12.4.2.1 Biological source

It consists of leaves of *Stevia rebaudiana* (family: Asteraceae), commonly known as honey leaf [22].

12.4.2.2 Uses

Leaves are used to increase the bioavailability and efficacy of various medications, including antibiotics, antiobese, antidiabetic, antifungal, antiulcer, antiviral, anti-cancer, cardiovascular, anti-inflammatory, antiarthritic, anti-TB or anti-leprosy, anti-histaminic, C-reactive protein inhibitors, cholesterol-lowering, immune-suppressants, and immune-modulators [23].

12.4.2.3 Dose

The dose range of the extract of stevia as a bioenhancer is 0.01–50 mg/kg. The dose range of bioenhancers prepared from fractions or pure components of stevia ranges from 0.01 to 40 mg/kg, with a preference for 30 mg/kg, regardless of the amount of medicine in the composition. When used as a bioenhancer, stevia leaf dosage ranges from 0.01 to 250 mg/dose of medicinal, nutraceutical, or herbal extract.

12.4.2.4 Examples

Bioenhancing activity of *S. rebaudiana* extract containing 3-O-acetyl-11-keto-beta-boswellic acid (AKBA), an active substance in 5-Loxin R (US Patent 2004/0073060A1), was studied in Sprague Dawley (SD) rats.

5-Loxin R shows selectively enhanced bioavailability by 30% AKBA. It is now being used as a treatment for osteoarthritis. It shows superior efficacy over standard *B. serrata* extract. SD rats of either sex received either 100 mg of *B. serrata* extract standardized to 30% 5-Loxin or a mixture of 100 mg of 5-Loxin and 10 mg of *S. rubuadiana* extract standardized to 33% steviosides. An addition of 10% bioenhancer increased the serum AKBA concentrations substantially [24].

12.4.3 Piperine

12.4.3.1 Biological sources

PIP, a yellow crystalline alkaloid, is available in *Piper nigrum* Linn and *Piper longum* Linn [25].

12.4.3.2 Uses

It increases the bioavailability of certain dietary substances and medications [1]. It is used as a condiment and flavoring agent in a variety of savory dishes for a long time. *Piper* species have been used to cure a range of diseases in folk medicine, including seizure disorders activity. PIP is a frequently used bioenhancer that has been found to improve the bioactivity of antibiotics, antivirals, antifungals, cardiovascular medications, anti-TB medications, anti-inflammatory medications, and nutraceuticals. PIP increases the bioavailability of acyclovir by 70% [26].

12.4.3.3 Mechanism of action

Role of PIP in drug bioavailability can be explained in two ways:
1 Specific mechanism: Increase in the blood flow to the gastrointestinal tract and decrease in the formation of hydrochloric acid inhibits the breakdown of certain drugs. Increase in gut emulsifying content and increase in the levels of enzymes such as- glutamyl transpeptidase have a role in both active as well as passive transport of nutrients to the intestinal cell.
2 Nonspecific mechanisms that block enzymes involved in drug biotransformation to avoid drug inactivation and clearance [27].

12.4.3.4 Dose

According to various rat studies, PIP's bioenhancing dose is about 15 mg day^{-1}, not more than 20 mg per day in distributed doses, which is 10,000 to 40,000 times lower than PIP's lethal dose 50. Though variation in effective bioenhancing dose of PIP is observed based on the pharmacological substance, around 10% (w/w) of the active substance is acceptable for most drugs [28].

12.4.3.5 Drugs in combination with piperine

PIP augments bioavailability of various compounds used in veterinary practice, including amoxycillin, curcumin, ciprofloxacin, nevirapine, oxytetracycline, pyrazinamide, phenytoin, propranolol, rifampicin, and theophylline [26].

12.4.3.6 Examples

Wistar rats received diet, regular, a high-fat, a high fat plus black pepper, and a high-fat with PIP (the active ingredient in black pepper), for 10 weeks. Liver, kidney, heart, intestine and aorta of rats with high-fat intake showed elevated levels of thiobarbituric acid-reactive substances (TBARS) and conjugated dienes (CD), but superoxide dismutase (SOD), catalase (CAT), glutathione peroxidase (GPx), glutathione-S-transferase (GST), and reduced glutathione (GSH) levels were significantly decreased. Concurrent administration of PIP or black pepper decreased TBARS and CD levels while keeping the levels of SOD, CAT, GPx, GST, and GSH near normal. Consuming black pepper or PIP can assist cell recovery from oxidative damage caused by a high-fat diet [29].

PIP is an antibiotic bioenhancer that improves the pharmacological bioavailability and efficacy by modifying the drug metabolism. It is a nutritional bioenhancer and stimulates the gastrointestinal tract by improving the nutrient bioavailability and absorption. Inhibition of a variety of cytochrome p450 (CYP) (particularly CYP3A4) and hepatic or duodenal UDP (uridine 5'-diphosphate) glucuronosyl transferases depends on the route of administration, dose, and period of exposure. PIP has a potentiating effect on drug bioavailability and bioefficacy in laboratory animals as well as human volunteers [26].

- **Pefloxacin** (fluoroquinolone antibiotic) plasma levels and bioavailability in Gaddi goats was increased after oral treatment of an Ayurvedic formulation, Trikatu, a blend of black pepper, long pepper, and ginger rhizomes. In oxytetracycline-treated chickens that are given oral *Piper longum* 7 days before therapy, a similar increase in bioavailability and pharmacokinetic changes were found [30].

- **Beta-carotene:** Effectiveness of black pepper fruit extract containing at least 98% pure PIP was studied for its potential to expand blood reactivity to β-carotene through oral administration. Subjects received either a daily dose of 15 mg beta-carotene and PIP (5 mg) or a placebo for 16 days. Supplementing with beta-carotene plus PIP led to a significant increase in serum beta-carotene. After 1 day of beta-carotene plus PIP supplementation, AUC was 60% higher than after one day of beta-carotene plus placebo treatment [31].
- **Rifampicin,** combined with PIP, decreases the dose of antiTB medicine by roughly half while maintaining therapeutic efficacy with the regular dose (450 mg) [32].
- **Piperine, cefotaxime, and amoxicillin trihydrate:** PIP shows significant increase in the bioavailability of β-lactam antibiotics such as amoxicillin and cefotaxime. Increased bioavailability of these antibiotics may be due to PIP's effect on microsomal metabolizing enzymes or the enzyme system [33].

12.4.4 Allicin

12.4.4.1 Biological source

Allium sativum L. (family: Amaryllidaceae), usually recognized as Garlic, is an aromatic herbaceous spice used in traditional medicine since ancient times. *A. sativum* bulbs (diallyl disulfide) shows the presence of sulfur-containing substances, including ajoenes (*E*-ajoene, *Z*-ajoene), thiosulfate (allicin), vinyl dithiins (2-vinyl-(4 H)-1,3-dithiin, 3-vinyl-(4H)-1,2-dithiin), and sulfides. After that, slice off the garlic and break down the parenchyma, the alliinase enzyme transforms allicin, the primary cysteine sulfoxide to allicin [34].

12.4.4.2 Uses

Allicin shows antiplatelet, antioxidant, antibacterial, anticancer, immune-modulating effect, antidiabetic, antiparasitic, antifungal, antioxidant and anti-inflammatory, and antiviral properties [35].

12.4.4.3 Mechanism of action

Bioenhancing properties of garlic are mostly due to its inhibitory action on a number of CYP enzymes as well as physiological alterations in circulation that slow drug clearance and elimination. Garlic extracts have been shown to inhibit CYP enzymes in vitro (CYP2C9, CYP2C19, CYP3A4, and CYP3A5). Garlic powder was found to have no effect on CYP3A4 in a clinical investigation. Garlic oil, in particular, has been proven to reduce CYP2E1. In the presence of garlic extract, in an in vivo research, stimulatory effects were seen on both efflux and uptake transporters [36].

12.4.4.4 Combination with other drugs

It is used in combination with amphotericin B, captopril, and propranolol.

12.4.4.5 Examples

– **Cu^{2+}**
Cu^{2+} and allicin-Cu^{2+} show dose-dependent fungicidal action against *Saccharomyces cerevisiae*, and the presence of allicin greatly amplifies its lethal effects [37].

– **Amphotericin B**
Fungicidal action of amphotericin B (AmB) increases in the presence of allicin [3]. By decreasing the ergosterol transit from plasma membrane to the vacuole membrane, allicin facilitates AmB-induced vacuole membrane degradation. AmB's fungicidal impact, when combined with allicin, may be due to the induction of vacuole disruption and not related to potassium ion efflux. AmB's allicin-mediated action is only manifested when ergosterol is present in the plasma membrane [38].

– **Propranolol**
Homogenate of garlic at moderate dosages in propranolol treatment may have a favorable effect in rats with cardiac arrhythmia during hypertension treatment by reducing cholesterol, triglycerides, glucose levels, systolic blood pressure, cholesterol, triglycerides, and sugar levels. It increases bioavailability and half-life, and lowers propranolol clearance and elimination rate constant [39].

– **Captopril**
In rats, given captopril or hydrochlorothiazide after a myocardial infarction, garlic increased survival and cardiac function [40].

12.4.5 Cyminum

12.4.5.1 Biological source

Cumin (*Cuminum cyminum* L.) is a spice made from the dried seeds of an aromatic herb in the Apicaceae family [41].

12.4.5.2 Uses

C. cyminum exhibits estrogenic, hypolipidemic, anti-nociceptive and anti-inflammatory, anti-convulsant, anticancer, antibacterial, anti-tussive, antioxidant, and antifungal activities. Oil shows the presence of *p*-mentha-1,4-dien-7-al, cumin aldehyde, -terpinene, and -pineneas [42].

12.4.5.3 Dose

Dose range of *Cyminum* is from 2 to 20 mg/kg, and dose range of extract is from 10 to 30 mg/kg. Bioavailability increases by 25–335% after using Cyminum extract or fractions [26].

12.4.5.4 Combination with other drugs

Cumin has a bioenhancing impact when combined with antibiotics, antivirals, antifungals, cardiovascular medications, anti-TB medications, anti-inflammatory medications, and nutraceuticals [36].

12.4.5.5 Examples

Cuminum and cyminum extracts and fractions as well as medications from the categories of antibiotics, antivirals, antifungals, cardiovascular medications, anti-TB medications, and anti-inflammatory medications, were examined for their bioavailability/bioefficacy increasing function. Cumin was given orally, parenterally, nasally, inhaled (including nebulizers), rectally, vaginally, and transdermally to improve bioavailability [42]. Percent increase in bioavailability of some medicines due to bioactive fractions from *C. cyminum* are enlisted in Table 12.2.

Table 12.2: Percent increase in bioavailability of some medicines in the presence of bioactive fractions from *C. cyminum* [26].

Therapeutic category/class	Drugs	% Enhancement in bioavailability
1. Antibiotics a. Fluroquinolones	Ciprofloxacin	52
	P-floxacin	47
	O-floxacin	61

Table 12.2 (continued)

Therapeutic category/class	Drugs	% Enhancement in bioavailability
b. Macrolides	Erythromycin	75
	Roxithromycin	67
	Azithromycin	83
c. Cephalosporins	Cefalexin	60
	Cefadroxin	90
d. Penicillins	Amoxycillin	75
	Cloxacillin	94
e. Aminoglycosides	Kanamycin	95
2. Antifungal	Fluconazole	170
	Ketoconazole	136
3. Antiviral	Acyclovir	110
	Zidovudine	330
4. CNS drugs	Alprazolam	60
5. Anticancer	Methotrexate	125
	5-Fluorouracil	335
	Doxorubicin	85
	Cisplatin	70
6. Cardiovascular medications	Amlodipine	55
	Lisinopril	83
	Propranolol	135
7. Anti-inflammatory/antiarthritic	Diclofenac	65
	Piroxican	70
	Nimesulide	168
8. Anti-TB/anti-leprosy medications	Rifampicin	250
	Dapsone	60
	Ethionamide	78
	Cycloserine	89

Table 12.2 (continued)

Therapeutic category/class	Drugs	% Enhancement in bioavailability
9. Anti-histamines/respiratory disorders	Salbutamol	110
	Theophylline	87
	Bromhexine	50
10. Corticosteroids	Dexamethasone	85
	Betamethasone	95
11. Immunosuppressants	Cyclosporin A	156
	Tacrolimus	75
12. Anti-ulcer	Ranitidine	117
	Cimetidine	123

12.4.6 Caraway

12.4.6.1 Biological source

Caraway, dried and crushed *Carum carvi* (Caraway) seeds (family: Umbelliferae) contains caraway oil [43].

12.4.6.2 Uses

It improves the bioavailability of antibiotic, antifungal, antiviral, and anticancer medications [44].

12.4.6.3 Dose

The dose range of the bioenhancer extract is 5–100 mg/kg whereas the dose range of bioactive fraction is 1–55 mg/kg [26].

12.4.6.4 Combination with other drugs

Antimicrobial, antifungal, antiviral, anti-TB, antileprosy, anti-inflammatory/anti-arthritic, cardiovascular, anti-histaminic, respiratory distress relieving drugs,

Table 12.3: Percentage bioavailability augmentation of some medicines in the presence of *C. Carvi* [42].

Category of drugs		Compound	Dose (mg/kg)	% Enhancement by Carum carvi
Antibiotics	Macrolides	Azithromycin	25	55
		Erythromycin	45	70
		Roxithromycin	15	65
	Cephalosporins	Cefixime	40	80
		Cefdinir	40	89
	Penicillins	Amoxicillin	45	75
		Cloxacillin	25	110
	Aminoglycosides	Amikacin	50	85
		Ofloxacin	20	65
		Norfloxacin	40	55
Antifungal		Fluconazole	65	65
		Amphotericin B	78	78
		Ketoconazole	55	55
Antiviral		Acyclovir	40	78
		Zidovudine	10	92
CNS drugs		Haloperidol	0.5	95
Anti-cancer		5-Fluorouracil	25	90
Cardiovascular		Atenolol	5	100
		Propranolol	8	68
Anti-inflammatory/anti-arthritic		Nimesulide	10	100
		Rofecoxib	2.5	75
Anti-tuberculosis/anti-leprosy		Rifampicin	40	110
		Pyrazinamide	12.5	45
		Dapsone	10	56
		Ethionamide	25	68
Antihistamines		Salbutamol	0.8	75
		Theophylline	30	70
		Loratadine	1	76

Table 12.3 (continued)

Category of drugs	Compound	Dose (mg/kg)	% Enhancement by Carum carvi
Corticosteroids	Prednisolone	4	65
	Dexamethasone	0.05	72
	Betamethasone	0.1	80
Immunosuppressants	Cyclosporine A	10	100
	Tacrolimus	5	90
Antiulcer	Ranitidine	30	67
	Cimetidine	40	72
	Omeprazole	2	76
Vitamins	Vitamin A	1	19
	Vitamin B1	10	42
Antioxidants	B-carotene	15	55
	Silymarin	5	38
Herbal products	Curcumin	50	48
	Rutin	40	45
Amino acids	Methionine	20	28
	Lysine	40	29
	Valine	25	19
	Isoleucine	25	34

immunosuppressants, antiulcers, and nutraceuticals are some of the drugs that are combined with caraway for bioenhancement (Table 12.3). Compositions can be administered orally/parenterally, topically, through inhalation (including nebulizers), rectally, and vaginally in veterinary medicine [45].

12.4.6.5 Examples

The extract or fractions of *C. carvi* were found to be 20–110% more active, and in the range of 10–150 mg/kg, it has been shown to be highly efficient. In the presence of PIP, *C. carvi* showed strong activity, ranging from 25% to 95%, with PIP dosages in preparations ranging from 3 to 15 mg/kg. Bioactive fractions from *C. carvi* were supplemented [26].

12.4.7 Black cumin

12.4.7.1 Biological source

It is dried *Nigella sativa* Linn. seed (family: Ranunculaceae). *N. sativa* is a South and Southwest Asian annual blooming plant. Seeds are sometimes known as black seeds or black cumin [46].

12.4.7.2 Uses

They are used in folk medicine throughout the world to cure and prevent bronchial asthma, cough, diarrhoea, abdominal pain, and dyslipidaemia at a local level [47].

12.4.7.3 Mechanism of action

It acts by increasing the absorption of the drug by enhancing permeability of the intestine. P-gp action in in silico tests in contrast to the primary amino acid sequence of P-gp from rats was reduced by fatty acids in black cumin (predominantly eicosadeinoic acid [36].

12.4.7.4 Example

The transport of amoxicillin across the gut was examined in vitro using everted rat intestinal sacs. Separately, 3 and 6 mg methanol and hexane extracts of *Nigella* seeds were co-infused with amoxicillin (6 mg/mL). When compared to controls, in-vitro tests utilizing *Nigella* extracts in methanol and hexane greatly increased amoxicillin penetration. At the same dose levels, significant increase in permeation was observed for hexane extract compared to methanol extract. In vivo studies of coadministration of amoxicillin (25 mg/kg) and *Nigella* seed hexane extract (25 mg/kg) in Wistar albino rats shows that with oral amoxicillin, alone or in combination with other antibiotics, the C_{max} of amoxicillin in rat plasma was higher [48].

12.4.8 Curcumin

12.4.8.1 Biological source

Curcumin is the main active constituent present in *Curcuma longa* (family: Zingiberaceae) [4, 49]. Curcumin, a polyphenol, is the major component of turmeric, an herbal medicine and nutritive spice. This bright yellow spice, which is made from the plant's rhizome, has a lengthy history of usage in Chinese and Indian traditional medicine [50].

12.4.8.2 Uses

Turmeric is one of the most popularly used spices and condiments in Indian cuisine. It is also used in ointments and creams as a coloring agent. Curcumin acts an anti-inflammatory agent along with its conventional uses. Curcumin is used in antimicrobial and anticancer medicines as a bioenhancer [4]. Curcumin is a polyphenol with strong antioxidant properties. It protects the myocardium from free radical damage by scavenging free radicals. Curcumin has cardiovascular, anti-ischemic, anti-inflammatory, hepatoprotective, and antidiabetic properties [51].

12.4.8.3 Mechanism of action

Curcumin inhibits drug metabolizing enzymes such as CYP3A4 in the liver and likewise works by modification of the drug transporter, P-gp. Curcumin works by suppressing API-metabolizing enzymes, which results in an increase in celiprolol, midazolam C_{max}, and AUC [4, 26]. Curcumin's reduction of nonspecific drug metabolizing enzyme could be another method that improves medication bioavailability [4]. As a result, enhancement of overall concentration (C_{max}) and absolute bioavailability of concomitant drugs (AUC) is observed [52]. The bioenhancing property of curcuma longa is identical to that of PIP. The quantity of UDP glucuronyl transferase in the intestinal and liver tissues is reduced by *Curcuma longa*. Moreover, it alters the physiological function of the gastro-intestinal tract, resulting in improved drug absorption [53].

12.4.8.4 Examples

Curcumin enhances the pharmacokinetics of P-gp substrate drugs such as celiprolol, midazolam, and paclitaxel in rats [54–56], and marbofloxacin in boiler chickens [52, 55].

– **Docetaxel**
Curcumin potentiates the anti-tumor activity of docetaxel in rats. Curcumin (100 mg/kg) administration to a group of rats 30 min before docetaxel increases AUC 8-fold and C_{max} tenfold as compared to the control group, from 282.6 and 18.4 to 2,244.1 and 68.0 ng/mL and 102.5 11.5 to 1024.2 121.7 ng/mL, respectively. Docetaxel bioavailability increased eightfold in the treatment group, with 5.5% and 43.7% bioavailability in the control and treatment groups, respectively. Bioavailability of docetaxel was improved

by alterations in the role and expression of proteins involved in the transport and metabolism [4].

– **Norfloxacin**
Norfloxacin is a broad-spectrum antibacterial often used to treat urinary tract infections, infections of the lungs, and gastrointestinal system. The absorption profile of norfloxacin is very poor. Norfloxacin (100 mg/kg) was given to group 1, and group 2 received norfloxacin, following a curcumin (60 mg/kg) pretreatment. The concentrations of norfloxacin in the blood were measured at regular intervals. The AUC of norfloxacin in Group-2 animals increased from 2.67-0.42 to 4.06-1.24 (g/mL h). Curcumin doubles the activity of norfloxacin. Curcumin has been discovered to have anticancer properties, removing carcinogens in the body. Combination of curcumin with an anticancer drug like docetaxel boosts the drug's anticancer effect [4, 26, 57, 58].

– **Cyclophosphamide**
A combination of curcumin with a bioenhancer such as PIP improved the therapeutic effectiveness of curcumin against cyclophosphamide (CP)-induced cardiotoxicity in rats. They tested curcumin alone and found that different doses of curcumin, combined with PIP, have a strong protective effect against CP-induced cardiotoxicity. With comparison to the curcumin alone-treated group, the combination groups show a significant amount of synergistic activity. Curcumin (50 mg/kg, p.o.), combined with PIP (20 mg/kg, p.o.), showed the most cardioprotection among the combination classes and they concluded that curcumin and its combination with PIP showed effective protection against CP-induced myocardial toxicity [51].

– **Other examples/combination of curcumin and PIP**
Curcumin, alone and in conjunction with PIP, showed only a mild antimalarial effect and was unable to reverse the **artemisinin**-resistant phenotype or significantly impact the growth of the tested clone, when administered in combination with artemisinin [59].

Curcumin, in a mixture with PIP, suppresses **diethylnitrosamine** (DENA)-induced hepatocellular carcinoma (HCC) in rats better than curcumin alone [60].

Curcumin, in combination with PIP, provided neuroprotection against olfactory bulbectomy-induced depression in rats, oxidative-nitrosative stress-induced modulating neuro-inflammation and apoptosis [61].

– **Piperine**
Medicinal capabilities of C. longa are unavailable due to metabolism in the intestinal wall and liver, as well as due its limited bioavailability. Hepatic and intestinal glucuronidation is inhibited by PIP. Curcumin alone given to rats (2 g/kg) for 4 h, results in modest serum concentrations. After 1–2 h of the drug, a concomitant dose of PIP enhanced the serum levels of curcumin along with increased T_{max} (154%), decreased $t_{1/2}$ and Cl_B (154%), and increased bioavailability (154%).

Blood concentrations were either untraceable or very less in individuals after a single dose (2 g) of curcumin. After administration of the drug, PIP exhibited considerably greater levels, from 0.25 to 1 h (P 0.01 at 0.25 and 0.5 h; P 0.001 at 1 h); bioavailability was increased by 200%. PIP improves the serum levels, bioavailability, and absorption of curcumin in both humans and rats, with no significant side effects [26, 62].

12.4.9 Ginger

12.4.9.1 Biological source

Ginger consists of dried rhizome of *Zingiber officinale* Roscoe, which belongs to the Zingiberaceae family [63–65]. Active components transformed from gingerols are shogaols, zingerone, and paradol. Odor of ginger is primarily due to its volatile oil components, ranging from 1% to 3% yield. 6-Gingerol is the main pungent component of ginger [4, 26, 65].

12.4.9.2 Uses

Ginger works as a bioenhancer by boosting drug absorption via modulating the intestinal function [4]. Antiulcer action, antithrombotic action, antimicrobial action, antifungal action, anti-inflammatory action, antidiabetic action, antiemetic action, anthelmintic action, antipyretic and analgesic, antiapoptotic action, antioxidant, anticancer action are some of the properties of ginger. The mucosal membrane of the gastrointestinal tract responds strongly to ginger [4, 26, 63].

12.4.9.3 Mechanism of action

It impedes the metabolic related reactions, CYP2C9, CYP2C19, CYP2D6, and CYP3A4, in humans [36].

12.4.9.4 Dose

Z. officinale extract and bioactive fractions improve the bioefficacy of a various medicines, natural products, and vital nutraceuticals. In most cases, an adequate amount of PIP or *P. nigrum* or *P. longum* extract is coupled with one or more additives from medications, nutrients, antioxidant, or herbal remedies [66]. Bioavailability of *Z. officinale* alone is 30–75%; however, PIP increases the bioavailability by 10–85% when combined with *Z. officinale*. PIP is dosed at 4–12 mg/kg as a bioenhancer, whereas *Z. officinale* extract is dosed at 10–30 mg/kg. PIP doses vary from 6 to 10 mg/ kg, while bioactive fraction dosages from *Z. officinale* range from 4 to 12 mg/kg.

By the inclusion or exclusion of PIP, the extracts or bioactive fractions of Z. officinale maintain their bioenhancing effect [4, 26].

12.4.9.5 Examples

The antitumor medicines, methotrexate and 5-fluorouracil, had their bioavailability increased by 87% and 110%, respectively, due to ginger's bioactive component. It was also revealed that it enhanced the bioavailability of acyclovir, an antiviral medicine, by 82%. The bioactive fraction improves bioavailability of herbal products and nutraceuticals. Mixing of vitamin A with ginger extract, increases the bioavailability of the vitamin by 30%. As a result, ginger, alone or in conjunction with PIP, boosts the bioavailability of a wide range of pharmaceuticals, including antibiotics and nutraceuticals [4]. When ginger was coadministered with pefloxacin in rabbits, the pharmacokinetic profile of the antibiotic changed dramatically, with a rise in the maximum concentration, AUC, and half-life. It has also been proven to improve methotrexate, 5-fluorouracil, and acyclovir bioavailability (36). It boosts rifampicin and ethionamide bioavailability by 65% and 56%, respectively (4). It also improves the bioavailability of antibiotics, including cefadroxil (65%), cephalexin (85%), and amoxicillin (85%). Cloxacillin (90%), azithromycin (90%), and erythromycin (105%) are the antibiotics with the highest percentage of bioenhancement due to the herb [62, 67].

By a combination of PIP with rifampicin, the therapeutic efficacy of this anti-TB medicine can be lowered by nearly half while maintaining therapeutic efficacy at levels comparable to the usual dose (450 mg) [66]. Coadministration of ginger extracts maximizes the bioavailability of APIs, vitamins, and amino acids [68].

12.4.10 Glycyrrhizin

12.4.10.1 Biological source

Dried, unpeeled or peeled root, and stolon of liquorice, *Glycyrrhiza glabra* (family: Leguminosae) contain saponin glycoside, Glycyrrhizin [(3, 18)-30-hydroxy-11,30-dioxilean -12-en-3-yl-2-o-glucopyranouronosyl-D-glucopyranosiduronic acid] [4, 53, 62, 69].

12.4.10.2 Uses

Glycyrrhizin, due to its expectorant effect, is commonly used in cough treatments for bronchitis and sore throat conditions. It also has anti-inflammatory, antiallergic, anti-asthmatic, antiulcer, antirheumatoid activity. It is a diuretic and laxative that strengthens the immune system and stimulates the adrenal gland. It is used to treat liver illness and aids in the detoxification of drugs by the liver. Glycyrrhizin has 50 times more sweetness than sugar. It is predominantly used in the treatment of peptic ulcers and stomach illnesses, as well as the treatment of respiratory and intestinal channels. Glycyrrhizin shows anti-hepatotoxic, anti-inflammatory, anticancer, and antiviral activities. Antimicrobial compounds and vitamins such as B1 and B12 are transported more efficiently across the intestinal membrane [4, 26, 36, 53, 65].

12.4.10.3 Mechanism of action

Glycyrrhizin may work by the efflux transporter P-gp inhibitor, which is found in the intestine. The ability of glycyrrhizin to improve absorption is dependent on its conversion to glycyrrhetic acid by the intestine bacterial enzyme- glucuronidase [4].

Glycyrrhizin had no effect on CYP3A4, but some compounds isolated from liquorice extract, viz. (3R)-vestitol, 4-hydroxyguaiacol apioglucoside and liquiritigenin 7, 4′ diglucoside, possess significant CYP3A4 action. Glycyrrhizic acid and other liquorice metabolites may not decrease the in vitro activity of the intestinal efflux transporter P-gp. Glycyrrhizin and glycyrrhetinic acid have activator effects on the P-gp function. Glabridin, a liquorice metabolite, inhibits the enzymes CYP3A4, 2B6 and 2C9, according to a study. Another metabolite of liquorice, glycyrrhetinic acid, inhibits CYP3A, 2C9, and 2C19 in vitro and in vivo, as well as activating UDP glucuronosyl transferases in the rat liver. Glycyrrhizin and glycyrrhetinic acid inhibited nucleoside transporters, resulting in lower ribavirin bioavailability due to decreased ribavirin transit in the digestive tract [36].

12.4.10.4 Dose

The effective concentrations of this bioenhancer range from 0.05% to 50%, 0.1% to 10%, and 0.25% to 20% of the total bulk of antifungal, antibacterial, and nutraceutical chemicals, correspondingly [4, 70].

12.4.10.5 Examples

Diammonium glycyrrhizinate given orally in rats with **aconitine** shows increase in C_{max} (1.64 times), increased AUC (1.63 times), and improved absolute bioavailability of aconitine (1.85 times), without any significant change in the half-life and clearance (Cl). Glycyrrhizinate administration via tail vein and hepatic vein does not alter pharmacokinetic parameters of aconitine significantly. It may suppress P-gp in the intestine, resulting in an increase in absorption [4, 71].

Significant improvement in the efficacy and bioavailability of anti-infectives, antibiotics, and anticancer drugs was observed with plant-derived glycoside 'glycyrrhizin'.

Glycyrrhizin increases the intake of **antibiotics** and the added drugs crosses the biological membranes. This increases their plasma concentrations and, consequently, bioavailability. It makes antibiotics like rifampicin, ampicillin, tetracycline, and nalidixic acid more effective against Gram -ve bacteria, *E. coli*. Antibiotics like rifampicin, tetracycline, and nalidixic acid, as well as antifungal medicines like clotrimazole, are more effective against Gram +ve bacteria like *Bacillus subtilis* and *Mycobacterium smegmatis*.

Taxol works by preventing the Michigan Cancer Foundation-74 cancer cells from proliferating and multiplying. Glycyrrhizin increases the action of taxol (Paclitaxel) by a factor of five. Taxol, combined with glycyrrhizin, has more activity than taxol given alone [4, 26]. Because of its nontoxicity and usefulness at low concentration, glycyrrhizin has a lot of potential as a bioavailability enhancer for antitumor drugs, antifungal drugs, antibacterial drugs, and nutraceuticals. It also aids in reducing the dose-dependent contrary properties of chemotherapeutic drugs, as well as in the development of antimicrobial resistance [4].

12.4.11 *Aloe vera*

12.4.11.1 Biological source

It is a dried juice made by cutting the bases of *Aloe perryi* leaves, *Aloe vera* or *Aloe barbadensis*, and *Aloe ferox*, all of which belong to the Liliaceae family [72].

12.4.11.2 Uses

Aloe vera, a perennial and succulent xerophyte with immunomodulatory, wound and burn healing, hypoglycaemia, antitumor, gastroprotective, antifungal, and anti-inflammatory properties, is widely utilized in both human and veterinary medicine [36]. Aloe vera is high in phytochemicals and aids in the absorption of vitamins C and E [53, 73]. It is popularly referred to as Indian Aloe because it includes Aloein and Emodin, which help to increase vitamin C and E intake. With aloes, absorption is gentler and vitamins last long in the bloodstream, increasing vitamin C and E bioavailability in humans. Because of its future nutritional and therapeutic significance as an herbal bioenhancer, Aloe vera is a very promising plant [62, 74].

12.4.11.3 Mechanism of action

The bioavailability of vitamin C and E is increased in humans as they stay longer in the plasma. It can also prevent activated human neutrophils from releasing reactive oxygen free radicals.

12.4.11.4 Examples

Vitamin E and vitamin C were benefited by aloe vera. Gel of aloe vera and extracts of the whole-leaf elevated vitamin C and E plasma concentrations and enhanced absorption. The gel extract was mainly helpful in delaying and enhancing ascorbate absorption. Its plasma concentration was greatly prolonged, even after an overnight fast of 24 h. Vitamin E absorption and plasma levels were enhanced by both the gel and whole-leaf extracts, notably after 8 h [26, 65].

Ethanolic extract of aloe vera improves hypoglycaemic result of glipizide in streptozotocin-induced diabetic rats. Coadministration of aloe vera and pantoprazole in patients of mustard gas sufferers having symptoms of gastroesophageal reflux is better than individual therapy. This may be associated with the activation of endogenous prostagland in the synthesis responsible for cytoprotective effects on the gastric mucosa. In a rat model, aloe consumption was shown to increase P-gp and CYP3A activities, lowering cyclosporine bioavailability; consequently, decreased bioavailability of associated absorbed/metabolized medications could be expected [36, 75, 76].

12.4.12 Lysergol

12.4.12.1 Biological source

It is an alkaloid found in certain species of fungi like ergot fungus and in the morning glory plants (*Ipomoea sp.*) (Family: Convolvulaceae), which include the hallucinogenic seeds of *Rivea corymbosa*, *Argyreia nervosa*, and *Ipomoea violacea*. It is present in the Claviceps and Rhizopus genera of lower fungus (4). It's also known as clavine because it is a dimethylergoline derivative [65, 77].

12.4.12.2 Uses

Lysergol (9,10-Dihydro-6-methylergoline-8-methanol), a 5-hydroxytryptamine agonist, is a vasoactive medication that maintains normal blood flow and effective plasma concentrations (4). It is an excellent herbal bioenhancer because it improves the antibiotics' ability to fight bacteria [9, 77]. Many antibiotics' bactericidal activities are improved by it [62].

12.4.12.3 Mechanism of action

It encourages the killing effects of certain antibiotics on bacteria. They promote drug uptake from gastrointestinal membranes and cell membranes, which increases the biological action of medicines at the target site [4, 69, 78].

12.4.12.4 Dose

The effective dosage level of lysergol as a bioenhancer and bioavailability enhancer is 1–10 g/mL, although the optimum dose level is 10 g/mL [26].

12.4.12.5 Examples

Lysergol increases the bioavailability of antibiotics like rifampicin, tetracycline, and ampicillin in Gram + ve and Gram −ve bacteria like *B. subtilis*, *M. smegmatis*, and *E. coli*. Rifampicin activity against *E. coli* (Gram −ve bacteria) was increased 6–12 times after the addition of lysergol. Similarly, against *B. subtilis* (Gram + ve bacteria) and *M. smegmatis*, lysergol shows increased rifampicin bioavailability 3- to 4.6-fold and 4.5–6 times, correspondingly.

Use of lysergol as a bioenhancer in antibiotic preparations improves the antibiotic absorption in the body and cells, and also helps to reduce the overall antibiotic dosage through a synergistic action that reduces antibiotic resistance development [4, 26, 53, 79].

Lysergol increased berberine bioavailability in SD rats after oral therapy [26, 80, 81]. It also improved the bioavailability of curcumin. Curcumin's effect on the major efflux transporters, BCRP (breast cancer resistance protein) and P-gp, in humans was investigated by in situ permeation and in vitro pharmacokinetic studies with particular substrates (digoxin for P-gp probe, sulfasalazine for BCRP probe) and inhibitors to determine the mechanism of increased bioavailability (verapamil for P-gp and pantoprazole for BCRP). Improvement in the bioavailability of lysergol is due to the BCRP efflux transport mechanism [36, 82].

12.4.13 Capsaicin

12.4.13.1 Biological source

Capsaicin is the active constituent in chilli peppers, namely *Capsicum annum* L. (family: Solanaceae) [9, 65].

12.4.13.2 Uses

Capsaicin improves the bioavailability of a variety of medicines [83].

12.4.13.3 Mechanism of action

Absorption of capsicum enhances AUC of the drugs [69, 78].

12.4.13.4 Examples

It improved theophylline absorption and bioavailability from sustained-release gelatin capsules [53, 65, 84].

Capsaicin boosts theophylline and ciprofloxacin absorption [62]. In a rabbit study, theophylline was given orally with or without capsaicin, and the second maintenance dose of bioenhancer was administered after 11 h, increasing the theophylline plasma levels [85, 86].

Penetration of naproxen through the skin was 4-fold increased after the application of capsaicin–ethanol mixture [87].

No significant differences in the AUC, clearance, or distribution volume were observed after theophylline administration, with and without oral ground *Capsicum* fruit suspension. Single dose of capsicum shows significant variation in the elimination rate constant (kel) and it does not change after repeated dose.

Single oral dose of suspension of *Capsicum* fruit does not affect theophylline and its metabolites 1,3-dimethyluric acid (1,3 DMU) and I-methyluric acid (I-MU). The amount of I-MU, conversely, was considerably abridged following a 7-day repeated intake of *Capsicum* fruit. Finally, it was discovered that a single intake of *Capsicum* fruit could have an effect on the theophylline kinetics, whereas a recurrent dose altered the xanthine oxidase metabolic route [84].

PIP and capsaicin were evaluated as modulators for drug transport over the nasal epithelium. To test the special properties of the selected pepper constituents on drug penetration, a nasal epithelial cell line (Roswell Park Memorial Institute 2650) and excised sheep nasal tissue were utilized as models. Fluorescein isothiocyanate-dextran 4,400 (MW 4,400 Da) was employed as a huge molecular weight marker molecule for paracellular transport, whereas rhodamine 123 was used as a substrate for P-gp-mediated efflux. Capsaicin reduced the P-gp efflux more than PIP, although PIP increased drug penetration via other channels. Capsaicin was non-cytotoxic up to 200 M, while PIP was noncytotoxic up to 500 M, as evidenced by cell survival of more than 80% in cell cytotoxicity assays [88].

12.4.14 Sinomenine

12.4.14.1 Biological source

Sinomenine, an alkaloid derived from the root of the climbing plant *Sinomenium acutum Thunb* (family: Menispermaceae), is commonly found in China and Japan [65, 69, 78].

12.4.14.2 Uses

Sinomenine is commonly used to treat rheumatic and arthritic conditions. It is used to increase bioavailability of paeoniflorin [4, 26].

12.4.14.3 Mechanism of action

The rise in bioavailability of paeoniflorin may be correlated with inhibition of paeoniflorin efflux transport via P-gp in the small intestine by sinomenine. This combination has the potential to help with inflammation and arthritic pain [69].

12.4.14.4 Examples

According to co-incubative tissue culture assays of paeoniflorin and sinomenine in an everted rat gut model, paeoniflorin per se shows linear association between intake and incubation time but the association became non-linear when combined with sinomenine. After 45 min of incubation, paeoniflorin absorption increased significantly. Coadministration of paeoniflorin with sinomenine at 16 and 136 M increases absorption 1.5-fold and 2.5-fold, respectively [89].

In comparison to paeoniflorin alone, oral coadministration of sinomenine and paeoniflorin raises sinomenine plasma concentration with a C_{max} of 13.7 g/mL. Oral bioavailability of paeoniflorin was increased 12-fold in rats given sinomenine concurrently. Gastric gavage was used to administer a solitary dosage of paeoniflorin and with sinomenine hydrochloride to unrestrained conscious male SD rats. Simultaneous injection of sinomenine increases paeoniflorin C_{max} and AUC, delays t_{max}, and decreases CL3 and volume of distribution.

Combination of verapamil and quinidine, with paeoniflorin, resulted in increased concentrations 2.1- and 1.5-fold, respectively, demonstrating that they function in the identical way as sinomenine.

The absorption profile of a P-gp substrate, showed digoxin increase 2.5-fold when it was treated with sinomenine, presenting that paeoniflorin is probably a P-gp substrate like digoxin. As a result, the mechanism causal to the improvement of paeoniflorin bioavailability by sinomenine could be attributable to the suppression of the efflux transporter, P-gp. As a result, sinomenine can be added to any P-gp substrate to increase its bioactivity. Sinomenine has anti-inflammatory properties; thus, when combined with a medicine like paeoniflorin, it has a higher anti-rheumatoid arthritis effect.

12.4.15 Ammaniol

12.4.15.1 Biological source

Ammannia multiflora Roxb. (family: Lythraceae) contains ammaniol [26, 53, 65, 69, 78].

12.4.15.2 Mechanism of action

Ammaniol has the ability to boost glucose absorption. It also shows potent anti-hyperglycemic properties [69, 78].

12.4.15.3 Examples

Combination of nalidixic acid with methanolic extract of *A. multiflora* exhibits substantial bioenhancing effect against two *E. coli* strains, CA8000 and DH5. Methanolic extract of *A. multiflora* had strong bioenhancing effect and four times lower dose of nalidixic acid. The antimycobacterial activity of the methanolic extract of *A. multiflora* against *Mycobacterium tuberculosis* strain H37Rv was examined, and it shows modest action against this microorganism [69, 78].

12.4.16 Peppermint oil

12.4.16.1 Biological source

It is obtained from *Mentha piperita* L., (family: Lamiaceae).

12.4.16.2 Uses

The main component in peppermint is menthol and essential oil, and is mainly accountable for the anti-spasmodic properties. Other constituents present in peppermint oil are limonene, cineole, menthone, menthofuran, isomenthone, menthyl acetate, isopulegol, menthol, and pulegone.

Peppermint leaf and oil are used all over the world in traditional medicine, flavoring, and cosmetic and pharmaceutical goods. Of all the volatile oils, peppermint oil is the most often utilized. Peppermint is an antiseptic, astringent, carminative anti-spasmodic, antiemetic, diaphoretic, mild bitter, analgesic, anticatarrhal, antimicrobial, rubefacient, stimulant, and emmenagogue. For respiratory congestion, peppermint oil vapor was utilized as an inhalant. Bronchitis, coughs, and irritation of the throat and oral mucosa were all treated with peppermint oil-infused tea.

Peppermint oil also plays important role in the treatment of irritable bowel syndrome, Crohn's disease, ulcerative colitis, biliary tract diseases, and gallbladder and liver complications. Menstrual cramps can be relieved with peppermint oil. Externally, peppermint oil helps with neuralgia, myalgia, headaches, migraines, and chicken pox [90].

12.4.16.3 Mechanism of action

It acts mainly by CYP3A inhibition.

12.4.16.4 Examples

It boosts cyclosporine's oral bioavailability. The C_{max} and AUC of cyclosporine were nearly tripled by the coadministration of 100 mg/kg peppermint oil [53, 55, 65, 91–94]. In clinical investigations, caraway oil was used with peppermint oil or menthol to treat functional dyspepsia [95].

12.4.17 5′ Methoxy-hydnocarpin (5′- MHC)

12.4.17.1 Biological source

5′- MHC is obtained from leaves of *Barberis fremontii* Torr. (Family: Berberidaceae). It has no antimicrobial activities, but it blocks multidrug resistance -dependent berberine efflux from *S. aureus* cells, essentially disabling the bacterial resistance mechanism to the antimicrobial action of berberine [69, 78, 96–98].

12.4.17.2 Dose

The dose used is 100 µg/mL.

12.4.17.3 Example

Berberin is an example.

12.4.18 Resveratrol (RSV)

12.4.18.1 Biological source

Resveratrol (RSV) is a stilbenoid generated by plants in response to injury or disease attack by bacteria or fungi. Blueberries, grapes, raspberries, and mulberries have it in their skin.

12.4.18.2 Uses

RSV aids apigenin bypassing hepatic metabolism. Apigenin (a natural anti-inflammatory agent) and RSV combination increases plasma apigenin levels 2.39 times more than apigenin alone [65, 99, 100].

12.4.18.3 Examples

Plasma diclofenac concentrations were shown to be higher after RSV therapy compared to the control phase. RSV treatment increased maximum plasma concentration, C_{max}, AUC, $t_{1/2}$, and considerably reduced the elimination rate constant (kel), apparent oral clearance (CL/F) compared to the control. The geometric mean ratios for diclofenac's C_{max}, AUC, $t_{1/2}$, kel, and CL/F show that diclofenac and RSV have a clinically significant interaction. RSV-mediated suppression of the CYP2C9 enzyme may be responsible for diclofenac's altered pharmacokinetics. Combination of diclofenac with RSV may be a method to lower dosage and also reduce gastrointestinal adverse effects of diclofenac [99].

RSV substantially enhanced rat intestinal absorption of methotrexate (MTX). RSV also blocked MTX efflux transport in multidrug resistance 1 (MDR1)-MDCK (Madin Darby canine kidney cells) and multidrug resistance-associated protein 2 (MRP2)-MDCK cell monolayers, indicating that MDR1 and MRP2 in the gut were the targets of drug interaction during the absorption process. Damage to intestine due to MTX was not increased by RSV therapy.

RSV inhibited P-gp, MRP2, organic anion transporters (OAT)1, and OAT3, increasing MTX absorption in the gut and decreasing MTX renal elimination. RSV reduced the effects of MTX on the kidneys without raising the risk of intestinal toxicity [100].

Coadministration of PIP increased anti-inflammatory activity of RSV in arthritic rats. RSV and PIP coadministration significantly reduces paw swelling and improves histological changes. The combination treatment significantly lowered serum tumor necrosis factor-alpha, IL-1b, TBARS, and NO_x levels. Furthermore, coadministration of PIP with RSV resulted in a largely negative expression of NF-kB p65 in synovial tissue. Results of combined treatment were comparable to those of the diclofenac treatment [101].

Combination of RES and Naringin (NAR) provided significant protection against IRI-induced myocardial damage as compared to the RSV alone treatment. A pharmacokinetic interaction produced results that are similar to those of a pharmacodynamic interaction. The combination of RSV and NAR demonstrated significant restoration of biomarker, antioxidant, and heart rate compared to a treatment of RSV per se. There was a significant increase in bioavailability and half-life, as well as a considerable reduction in clearance, when RSV, in combination, was compared to RSV per se [102].

The effect of RSV, alone, or in combination with PIP on cerebral blood flow measurements and cognitive function in healthy human subjects shows that adding PIP to the polyphenol RSV improves its bioefficacy. While oral therapy with 250 mg *trans*-RSV had no effect on the overall cerebral blood flow (total Hb) during cognitively demanding activities, coadministration with 20 mg PIP led to a significantly greater CBF (cerebral blood flow) for the 40-min post-dose task duration.

Findings of the RSV supplementation study are broadly similar to a prior study's dosage–response pattern of CBF seen following resveratrol administration, in which a dose of 250 mg was mainly ineffective. Despite the PIP -mediated augmentation of resveratrol's CBF effects, neither active treatment showed significant differences in cognitive task performance, blood pressure/heart rate, or mood ratings [103].

12.5 Recent advances in bioenhancers

Ayurveda is a 5,000-year-old Hindu traditional medicine that originated in India. Yogvahi (synergism) is a technique for increasing bioavailability, tissue dispersion, and drug efficiency that is frequently referenced in Ayurveda [104].

Mean particle size of the drug-loaded solid lipid nanoparticles (SLNs) decreased for Ketoprofen-loaded SLNs made from beeswax and carnauba wax mixed with Tween 80 and egg lecithin, and with increased total surfactant concentration. The capacity of SLNs to integrate a low water-soluble medication like ketoprofen was indicated by their high drug entrapment efficiency of 97%. After 45 days of storage, differential scanning calorimetry thermograms and HPLC analysis revealed that nanoparticles were stable with minimal drug leakage. Nanoparticles, with more beeswax in their core, released drugs more quickly than those with having more carnauba wax [105].

Easy, consistent, specific, and precise reversed-phase HPLC technique for deciding camptothecin (CPT, current procedural terminology) in creature organs after administration in SLNs was created and validated. This technique can be utilized to decide CPT sum in rodent organ specimens after IV administration of CPT in suspension [106].

Investigation of preparation and characterization of SLNs stacked with curcuminoids showed that at ideal method environments, lyophilized curcuminoids stacked SLNs delivered round particles with a mean molecule size of 450 nm and a polydispersity file of 0.4, with up to 70% (w/w) curcuminoids stacked SLNs, displaying circular particles with a mean particle size of 250 nm [107].

Overall percentages of residual curcumin, bisdemethoxycurcumin, and demethoxycurcumin were found to be 91, 96, and 88, respectively, after 6 months of storage in the absence of daylight [107].

A result of investigation of Trikatu on the pharmacokinetic profile of indomethacin in rabbits showed that Trikatu-enhanced absorption of indomethacin, which was believed to be because of an increase in the gastrointestinal blood stream and a quicker pace of transport through the gastrointestinal mucosa [108].

PIP metabolism showed that the maximum level in the abdomen and small intestine was found at the 6th h. PIP traces were found in the spleen, kidney, and serum from 0.5 to 24 h [109].

Gallic acid has a synergistic impact, in combination with PIP. This resulted in a greater therapeutic potential for decreasing beryllium-induced hepato-renal dysfunction and oxidative stress. Individual dose of gallic acid and PIP moderately altered the increased metabolic variables, according to the researchers. The combination of these, on the other hand, was found to totally reverse the beryllium-induced metabolic changes and oxidative stress effects [110].

Use of a solvent evaporation approach to encapsulate the plant-derived antioxidant quercitrin on PLA (poly-D, L-lactide) nanoparticles increased its solubility, permeability, and stability. Quercitrin-PLA nanoparticles have a diameter of 68 nm. According to HPLC and antioxidant assay results, nano-encapsulated quercitrin has a 40% encapsulation effectiveness. Under physiological conditions, the in vitro release kinetics of quercitrin demonstrate an early burst followed by persistent release. These characteristics of quercitrin nanomedicine open up new possibilities for the development of improved therapeutics for intestinal anti-inflammatory effects and nutraceutical substances using a fewer beneficial but extremely active antioxidant molecule [111]. Some patents available for bioenhancers as veterinary phytomedicines are enlisted in Table 12.4.

Table 12.4: List of patents of bioenhancer as veterinary phytomedicines.

Name of bioenhancer	Patent name	Patent number	References
Stevia	Bioavailability/bioefficacy enhancing activity of *Stevia rebaudiana* and extracts and fractions and compounds thereof.	US2004/ 0073060A1	[24]
Piperine	Use of piperine as a bioavailability enhancer.	US5744161A	[27]
Cumin	Bioavailability/bioefficacy enhancing activity of *Cuminum cyminum* and extracts and fractions thereof.	US 7.514105 B2	[42]
Ginger	Bioavailability enhancing activity of *Zingiber officinale* Linn and its extracts/fractions thereof.	EP1465646a1	[66]
Lysergol	Antibiotic pharmaceutical composition with lysergol as bioenhancer and method of treatment.	US20070060604A1	[79]

12.6 Future perspective of bioenhancers

The use of bioenhancers allows for a reduction in dosage while also reducing the risk of drug resistance. Reduced dosage reduces medication toxicity, like taxol, which is especially important for anticancer treatments. Additionally, there are ecological benefits. Taxol is derived from the bark of the pacific yew tree, one of the globally slowest growing tree, and is used to treat ovarian and breast cancer. Currently, six trees between 25 and 100 years old must be cut down to cure a single patient. Fewer will be destroyed only due to bioenhancers.

List of abbreviations

%	Percent
°C	Degree Celsius
AKBA	Acetyl-11-keto-β-boswellic acid
anti-TB	Antitubercular
API	Active pharmaceutical ingredient
AUC	Area under the curve
BCRP	Breast cancer resistance protein
CBF	Cerebral blood flow
Cl_B	Total body clearance
C_{max}	Maximum plasma concentration
CP	Cyclophosphamide
CPT	Current procedural terminology
CYP3A4	Cytochrome P450 3A4
HPLC	High-performance liquid chromatography
I.V.	Intravenous
IL-1b	Interleukin 1 beta
IRI	Ischemia reperfusion injury
Kg	Kilogram
MDCK	Madin Darby canine kidney cells
MDR1	Multidrug resistance 1
mg	milligram
MRP2	Multidrug resistance-associated protein 2
MTX	Methotrexate
MW	Molecular weight
NAR	Naringin
nm	Nanometer
NO_x	Total nitrate/nitrite
OAT	Organic anion transporters
Pg/dL	Picogram/deciliter
P-gp	P-glycoprotein
PIP	Piperine
RRL	Regional Research Laboratory
RSV	Resveratrol
SD	Sprague Dawley
$t_{1/2}$	Half-life
TBARS	Thiobarbituric acid-reactive substances
T_{max}	Time taken to reach the maximum concentration
UDP	Uridine 5′-diphosphate
w/w	Weight by weight

References

[1] Atal N, Bedi KL, Bioenhancers: Revolutionary concept to market, Journal of Ayurveda and integrative medicine, 2010, 1(2), 96–99.

[2] Upadhyay RK, Drug delivery systems, CNS protection, and the blood brain barrier, Bio Medicine Research International, 2014, 2014, 869269.

[3] Currie GM, Pharmacology, part 2: Introduction to Pharmacokinetics, Journal of nuclear medicine technology, 2018, 46(3), 221–230.

[4] Alexander A, Qureshi A, Kumari L, Vaishnav P, Sharma M, Saraf S, Saraf S, Role of herbal bioactives as a potential bioavailability enhancer for active pharmaceutical ingredients, Fitoterapia, 2014, 97, 1–4.

[5] Randhawa GK, Jagdev Singh Kullar R, Bioenhancers from Mother Nature and their applicability in modern medicine, International Journal of Applied and Basic Medical Research :, 2011, 1(1), 5.

[6] Zutshi RK, Singh R, Zutshi U, Johri RK, Atal CK, Influence of piperine on rifampicin blood levels in patients of pulmonary tuberculosis, The Journal of the Association of Physicians of India, 1985, 33(3), 223–224.

[7] Singh A, Deep A, Piperine: A bioenhancer, International Journal of Pharmacy Research & Technology., 2011, 1(1), 01–5.

[8] Riviere JE, Absorption, distribution, metabolism, and elimination, Journal of Veterinary Pharmacology, 2009, 9, 11–46.

[9] Kesarwani K, Gupta R, Bioavailability enhancers of herbal origin: An overview, Asian Pacific journal of tropical biomedicine, 2013, 3(4), 253–266.

[10] Patra JK, Das G, Franceto LF, Campos EV, Del Pilar Rodriguez-torres M, Acosta-Torres LS, Diaz-Torres LA, Grillo R, Swamy MK, Sharma S, Habtemariam S, Nano based drug delivery systems: Recent developments and future prospects, Journal of Nanobiotechnology, 2018, 16 (1), 1–33.

[11] Ud Din F, Aman W, Ullah I, Qureshi OS, Mustapha O, Shafique S, Zeb A, Effective use of nanocarriers as drug delivery systems for the treatment of selected tumors, International Journal of Nanomedicine, 2017, 12, 7291.

[12] Jeevanandam J, Barhoum A, Chan YS, Dufresne A, Danquah MK, Review on nanoparticles and nanostructured materials: History, sources, toxicity and regulations, Beilstein journal of nanotechnology, 2018, 9(1), 1050–1074.

[13] Ekor M, The growing use of herbal medicines: Issues relating to adverse reactions and challenges in monitoring safety, Frontiers in Pharmacology, 2014, 4, 177.

[14] David AV, Arulmoli R, Parasuraman S, Overviews of biological importance of quercetin: A bioactive flavonoid, Pharmacognosy Reviews, 2016, 10(20), 84.

[15] Lakhanpal P, Rai DK, Quercetin: A versatile flavonoid, Internet Journal of Medical Update, 2007, 2(2), 22–37.

[16] Choi JS, Piao YJ, Kang KW, Effects of quercetin on the bioavailability of doxorubicin in rats: Role of CYP3A4 and P-gp inhibition by quercetin, Archives of Pharmacal Research, 2011, 34 (4), 607–613.

[17] Umathe SN, Dixit PV, Bansod KU, Wanjari MM, Quercetin pre-treatment increases the bioavailability of pioglitazone in rats: Involvement of CYP3A inhibition, Biochemical Pharmacology, 2008, 75(8), 1670–1676.

[18] Choi JS, Jo BW, Kim YC, Enhanced paclitaxel bioavailability after oral administration of paclitaxel or prodrug to rats pretreated with quercetin, European Journal of Pharmaceutics and Biopharmaceutics, 2004, 57(2), 313–318.

[19] Choi JS, Han HK, The effect of quercetin on the pharmacokinetics of verapamil and its major metabolite, norverapamil, in rabbits, Journal of Pharmacy and Pharmacology, 2004, 56(12), 1537–1542.

[20] Choi JS, Piao YJ, Kang KW, Effects of quercetin on the bioavailability of doxorubicin in rats: Role of CYP3A4 and P-gp inhibition by quercetin, Archives of Pharmacal Research, 2011, 34 (4), 607–613.

[21] Li X, Choi JS, Effects of quercetin on the pharmacokinetics of etoposide after oral or intravenous administration of etoposide in rats, Anticancer Research, 2009, 29(4), 1411–1415.

[22] Abdullateef RA, Osman M, Studies on effects of pruning on vegetative traits in stevia rebaudiana bertoni (Compositae), International Journal of Biological, 2012, 4(1), 146.

[23] Arumugam B, Subramaniam A, Alagaraj P, Stevia as a natural sweetener: A review, Cardiovascular & Hematological Agents in Medicinal Chemistry (Formerly Current Medicinal Chemistry-Cardiovascular & Hematological Agents), 2020, 18(2), 94–103.

[24] Gokaraju GR, Gokaraju RR, D'souza C, Frank E, inventors; Laila Impex, assignee. Bio-availability/bio-efficacy enhancing activity of Stevia rebaudiana and extracts and fractions and compounds thereof. United States patent application US 12/610,. 2010, 502.

[25] Tiwari A, Mahadik KR, Gabhe SY, Piperine: A comprehensive review of methods of isolation, purification, and biological properties, Medicine in Drug Discovery, 2020, 7, 100027.

[26] Dudhatra GB, Mody SK, Awale MM, Patel HB, Modi CM, Kumar A, Kamani DR, Chauhan BN, A comprehensive review on pharmacotherapeutics of herbal bioenhancers, World Journal, 2012, 2012, 637953.

[27] Majeed M, Badmaev V, Rajendran R, inventors; Sabinsa Corp, assignee. Use of piperine as a bioavailability enhancer. United States patent US 5,744, 1998, 161.

[28] Pattanaik S, Hota D, Prabhakar S, Kharbanda P, Pandhi P, Effect of piperine on the steady-state pharmacokinetics of phenytoin in patients with epilepsy, Phytotherapy Research, 2006, 20(8), 683–686.

[29] Vijayakumar RS, Surya D, Nalini N, Antioxidant efficacy of black pepper (Piper nigrum L.) and piperine in rats with high fat diet induced oxidative stress, Redox Report : Communications in Free Radical Research, 2004, 9(2), 05–10.

[30] Dama MS, Varshneya C, Dardi MS, Katoch VC, Effect of trikatu pretreatment on the pharmacokinetics of pefloxacin administered orally in mountain Gaddi goats, Journal of Veterinary Science, 2008, 9(1), 25–29.

[31] Badmaev V, Majeed M, Norkus EP, Piperine, an alkaloid derived from black pepper increases serum response of beta-carotene during 14-days of oral beta-carotene supplementation, Nutrition research reviews, 1999, 19(3), 381–388.

[32] Kapil RS, Zutshi U, Bedi KL, Singh G, Johri RK, Dhar SK, Kaul JL, Sharma SC, Pahwa GS, Kapoor N, Tickoo AK. Pharmaceutical compositions containing piperine and an antituberculosis or antileprosy drug. European Patent Number, EP0650728B1,2002.

[33] Hiwale AR, Dhuley JN, Naik SR, Effect of coadministration of piperine on pharmacokinetics of β-lactam antibiotics in rats, Indian Journal of Experimental Biology, 2002, 409(3), 277–281.

[34] El-Saber Batiha G, Magdy Beshbishy A, G Wasef L, Elewa YH, A Al-Sagan A, El-Hack A, Mohamed E, Taha AE, M Abd-Elhakim Y, Prasad Devkota H, Chemical constituents and pharmacological activities of garlic (Allium sativum L.): A review, Nutrients, 2020, 12(3), 872.

[35] Bayan L, Koulivand PH, Gorji A, Garlic: A review of potential therapeutic effects, Avicenna Journal of Phytomedicine, 2014, 4(1), 1.

[36] Yurdakok-Dikmen B, Turgut Y, Filazi A, Herbal bioenhancers in veterinary phytomedicine, Frontiers in Veterinary Science, 2018, 5, 249.

[37] Ogita A, Hirooka K, Yamamoto Y, Tsutsui N, Fujita KI, Taniguchi M, Tanaka T, Synergistic fungicidal activity of Cu2+ and allicin, an allyl sulfur compound from garlic, and its relation to the role of alkyl hydroperoxide reductase 1 as a cell surface defense in Saccharomyces cerevisiae, Toxicol, 2005, 215(3), 205–213.

[38] Ogita A, Fujita KI, Tanaka T, Enhancing effects on vacuole-targeting fungicidal activity of amphotericin B, Frontiers in Microbiology, 2012, 3, 100.

[39] Asdaq SM, Inamdar MN, Pharmacodynamic and pharmacokinetic interactions of propranolol with garlic (*Allium sativum*) in rats. Evid. –based complement, Alternative Medicine, 2011, 2011.

[40] Asdaq SM, Inamdar MN, Pharmacodynamic interaction of garlic with captopril Amalraj in ischemia-reperfusion induced myocardial damage in rats, Pharmacologyonline, 2008, 2, 875–888.

[41] Singh RP, Gangadharappa HV, Mruthunjaya K, Cuminum cyminum-A popular spice: An updated review, Pharmacognosy Journal, 2017, 9, 3.

[42] Qazi GN, Bedi KL, Johri RK, Tikoo MK, Tikoo AK, Sharma SC, Abdullah ST, Suri OP, Gupta BD, Suri KA, Satti NK, inventors. Bioavailability/bioefficacy enhancing activity of Cuminum cyminum and extracts and fractions thereof. United States patent US 7,514,. 2009, 105.

[43] Johri RK, *Cuminum cyminum* and *Carum carvi*: An update, Pharmacognosy Reviews, 2011, 5 (9), 63.

[44] Choudhary N, Khajuria V, Gillani ZH, Tandon VR, Arora E, Effect of *Carum carvi*, a herbal bioenhancer on pharmacokinetics of antitubercular drugs: A study in healthy human volunteers, Perspectives in Clinical Research, 2014, 5(2), 80.

[45] Qazi G, Bedi K, Johri R, Tikoo M, Tikoo A, Sharma S, Abdullah T, Suri O, Gupta B, Suri K, Satti N, inventors. Bioavailability enhancing activity of *Carum carvi* extracts and fractions thereof. United States patent application US 11/360, 2006 Nov 16, 839.

[46] Ahmad A, Husain A, Mujeeb M, Khan SA, Najmi AK, Siddique NA, Damanhouri ZA, Anwar F, A review on therapeutic potential of *Nigella sativa*: A miracle herb, Asian Pacific Journal of Tropical Biomedicine, 2013, 3(5), 337–352.

[47] Forouzanfar F, Bazzaz BS, Hosseinzadeh H, Black cumin (Nigella sativa) and its constituent (thymoquinone): A review on antimicrobial effects, Iranian Journal of Basic Medical Sciences, 2014, 12, 929.

[48] Ali B, Amin S, Ahmad J, Ali A, Ali M, Mir SR, Bioavailability enhancement studies of amoxicillin with Nigella, The Indian Journal of Medical Research, 2012, 135(4), 555.

[49] Amalraj A, Pius A, Gopi S, Gopi S, Biological activities of curcuminoids, other biomolecules from turmeric and their derivatives–A review, Traditional and Complementary Medicine, 2017, 7(2), 205–233.

[50] Hatcher H, Planalp R, Cho J, Torti FM, Torti SV, Curcumin: From ancient medicine to current clinical trials, Cellular and molecular life sciences : CMLS, 2008, 65(11), 1631–1652.

[51] Chakraborty M, Bhattacharjee A, Kamath JV, Cardioprotective effect of curcumin and piperine combination against cyclophosphamide-induced cardiotoxicity, Indian Journal of Pharmacology, 2017, 49(1), 65.

[52] Abo-El-Sooud K, Samar MM, Fahmy MA, Curcumin ameliorates the absolute and relative bioavailabilities of marbofloxacin after oral administrations in broiler chickens, Wulfenia, 2017, 24(3), 284–297.

[53] Tatiraju DV, Bagade VB, Karambelkar PJ, Jadhav VM, Kadam V, Natural bioenhancers: An overview, Journal of Pharmacognosy and Phytochemistry, 2013, 2(3), 55–60.

[54] Zhang W, Tan TM, Lim LY, Impact of curcumin-induced changes in P-glycoprotein and CYP3A expression on the pharmacokinetics of peroral celiprolol and midazolam in rats, Drug Metabolism and Disposition, 2007, 35(1), 110–115.

[55] Peterson B, Weyers M, Steenekamp JH, Steyn JD, Gouws C, Hamman JH, Drug bioavailability enhancing agents of natural origin (bioenhancers) that modulate drug membrane permeation and pre-systemic metabolism, Pharmaceutics, 2019, 11(1), 33.

[56] Sharma A, Magotra A, Nandi U, Singh G, Enhancement of paclitaxel oral bioavailability in swiss mice by four consecutive days of pretreatment with curcumin, Indian Journal of Pharmaceutical Education and Research, 2017, 51, 566–570.

[57] Pavithra BH, Prakash N, Jayakumar K, Modification of pharmacokinetics of norfloxacin following oral administration of curcumin in rabbits, Veterinary Science, 2009, 10(4), 293.

[58] Yan YD, Kim DH, Sung JH, Yong CS, Choi HG, Enhanced oral bioavailability of docetaxel in rats by four consecutive days of pre-treatment with curcumin, International Journal of Pharmaceutics, 2010, 399(1–2), 116–120.

[59] Martinelli A, Rodrigues LA, Cravo P, Plasmodium chabaudi: Efficacy of artemisinin+ curcumin combination treatment on a clone selected for artemisinin resistance in mice, Experimental Parasitology, 2008, 119(2), 304–307.

[60] Patial V, Mahesh S, Sharma S, Pratap K, Singh D, Padwad YS, Synergistic effect of curcumin and piperine in suppression of DENA-induced hepatocellular carcinoma in rats, Environmental Toxicology and Pharmacology, 2015, 40(2), 445–452.

[61] Rinwa P, Kumar A, Garg S, Suppression of neuro-inflammatory and apoptotic signaling cascade by curcumin alone and in combination with piperine in rat model of olfactory bulbectomy induced depression, PLoS One, 2013, 8(4), e61052.

[62] Jhanwar B, Gupta S, Biopotentiation using herbs: Novel technique for poor bioavailable drugs, International Journal of PharmTech Research, 2014, 6(2), 443–454.

[63] Mao QQ, Xu XY, Cao SY, Gan RY, Corke H, Li HB, Bioactive compounds and bioactivities of ginger (*Zingiber officinale* Roscoe), Foods, 2019, 8(6), 185.

[64] Prasad S, Tyagi AK, Ginger and its constituents: Role in prevention and treatment of gastrointestinal cancer, Gastroenterology Research and Practice, 2015, 2015.

[65] Zafar N, Herbal bioenhancers: A revolutionary concept in modern medicine, World Journal of Pharmaceutical Research, 2017, 6(16), 381–397.

[66] Qazi G, Bedi K, Johri R, Tikoo M, Tikoo A, Sharma S, Abdullah S, Suri O, Gupta B, Suri K, Satti N, inventors. Bioavailability enhancing activity of *Zingiber officinale* Linn and its extracts/ fractions thereof. United States patent application US 10/318, 2003, 314.

[67] Singh R, Devi S, Patel J, Patel U, Bhavsar S, Thaker A, Indian herbal bioenhancers: A review, Pharmacognosy Reviews, 2009, 3(5), 90.

[68] Santos Braga S, Ginger: Panacea or Consumer's Hype?, Applied Sciences, 2019, 9(8), 1570.

[69] Chavhan SA, Shinde SA, Gupta IN, Current trends on natural bio enhancers: A review, International Journal of Pharmacognosy & Chinese Medicine, 2018, 2, 2576–4772.

[70] Imai T, Sakai M, Ohtake H, Azuma H, Otagiri M, Absorption-enhancing effect of glycyrrhizin induced in the presence of capric acid, International Journal of Pharmaceutics, 2005, 294 (1–2), 11–21.

[71] Chen L, Yang J, Davey AK, Chen YX, Wang JP, Liu XQ, Effects of diammonium glycyrrhizinate on the pharmacokinetics of aconitine in rats and the potential mechanism, Xenobiotica, 2009, 39(12), 955–963.

[72] Saxena V, Singh A, An update on bio-potentiation of drugs using natural options, Asian Journal of Pharmaceutical and Clinical Research, 2020, 13(11), 25–32.

[73] Oladimeji FA, Adegbola AJ, Onyeji CO, Appraisal of bioenhancers in improving oral bioavailability: Applications to herbal medicinal products, Journal of pharmaceutical research international, 2018, 29, 1–23.

[74] Vinson JA, Al Kharrat H, Andreoli L, Effect of *Aloe vera* preparations on the human bioavailability of vitamins C and E, Phytomedicine, 2005, 12(10), 760–765.

[75] Panahi Y, Aslani J, Hajihashemi A, Kalkhorani M, Ghanei M, Sahebkar A, Effect of aloe vera and pantoprazole on gastroesophageal reflux symptoms in mustard gas victims: A randomized controlled trial, Pharmaceutical Sciences, 2016, 22, 190–194.

[76] Yang MS, Yu CP, Huang CY, Chao PD, Lin SP, Hou YC, Aloe activated P-glycoprotein and CYP 3A: A study on the serum kinetics of aloe and its interaction with cyclosporine in rats, Food & function, 2017, 8(1), 315–322.

[77] Saxena V, Singh A, An update on bio-potentiation of drugs using natural options, Asian Journal of Pharmaceutical and Clinical Research, 2020, 13(11), 25–32.

[78] Jain G, Patil UK, Strategies for enhancement of bioavailability of medicinal agents with natural products, International Journal of Pharmaceutical Sciences Research, 2015, 6(12), 5315–5324.

[79] Khanuja S, Arya J, Srivastava S, Shasany A, Kumar TS, Darokar M, Kumar S, inventors; Council of Scientific, Industrial Research (CSIR), assignee. Antibiotic pharmaceutical composition with lysergol as bio-enhancer and method of treatment. United States patent application US 11/395,527. 2007.

[80] Verma CP, Verma S, Ashawat MS, Pandit V, An overview: Natural bio-enhancer's in formulation development, ce and Technology, 2019, 9(6), 201–205.

[81] Patil S, Dash RP, Anandjiwala S, Nivsarkar M, Simultaneous quantification of berberine and lysergol by HPLC-UV: Evidence that lysergol enhances the oral bioavailability of berberine in rats, Biomedical chromatography : BMC, 2012, 26(10), 1170–1175.

[82] Shukla M, Malik MY, Jaiswal S, Sharma A, Tanpula DK, Goyani R, Lal J, A mechanistic investigation of the bioavailability enhancing potential of lysergol, a novel bioenhancer, using curcumin, RSC advances, 2016, 6(64), 58933–58942.

[83] Verma S, Rai S, Bioenhancers from mother nature: A paradigm for modern medicines, Scholars Academic Journal of Pharmacy, 2018, 7(10), 442–451.

[84] Bouraoui A, Brazier JL, Zouaghi H, Rousseau M, Theophylline pharmacokinetics and metabolism in rabbits following single and repeated administration of Capsicum fruit, European Journal of Drug Metabolism and Pharmacokinetics, 1995, 20(3), 173–178.

[85] Bouraoui A, Toumi A, Brazier JL, Effects of capsicum fruit on theophylline absorption and bioavailability in rabbits, Drug-nutrient interactions, 1988, 5(4), 345–350.

[86] Sumano-López H, Gutiérrez-Olvera L, Aguilera-Jiménez R, Gutiérrez-Olvera C, Jiménez-Gómez F, Administration of ciprofloxacin and capsaicin in rats to achieve higher maximal serum concentrations, Arzneimittel forschung, 2007, 57(05), 286–290.

[87] Gutierrez OL, Sumano LH, Zamora QM, Administration of enrofloxacin and capsaicin to chickens to achieve higher maximal serum concentrations, The Veterinary record, 2002, 150 (11), 350–353.

[88] Gerber W, Steyn D, Kotzé A, Svitina H, Weldon C, Hamman J, Capsaicin and piperine as functional excipients for improved drug delivery across nasal epithelial models, Planta medica, 2019, 85(13), 114–123.

[89] Chan K, Liu ZQ, Jiang ZH, Zhou H, Wong YF, Xu HX, Liu L, The effects of sinomenine on intestinal absorption of paeoniflorin by the everted rat gut sac model, Journal of ethnopharmacology, 2006, 103(3), 425–432.

[90] Balakrishnan A, Therapeutic uses of peppermint-a review, International journal of pharmaceutical sciences research, 2015, 7(7), 474.

[91] Chivte VK, Tiwari SV, Nikalge AP, Bioenhancers: A brief review, Advanced journal of pharmacie and life science research, 2017, 2, 1–8.

[92] Sindhoora D, Bhattacharjee A, Shabaraya AR, Bioenhancers: A comprehensive review, International Journal of Pharmaceutical Sciences Review and Research, 2020, 60(1), 126–131.

[93] Majumdar SH, Kulkarni AS, Kumbhar SM, Yogvahi (Bioenhancer): An ayurvedic concept used in modern medicines, International Research Journal of Pharmacy, 2018, 1(1), 20–25.

[94] Wacher VJ, Wong S, Wong HT, Peppermint oil enhances cyclosporine oral bioavailability in rats: Comparison with d-α-tocopheryl poly (ethylene glycol 1000) succinate (TPGS) and ketoconazole, Journal of Pharmaceutical Sciences, 2002, 91(1), 77–90.

[95] Mahboubi M, Caraway as important medicinal plants in management of diseases, Natural Products and Bioprospecting, 2019, 9(1), 1–1.

[96] Stavri M, Piddock LJ, Gibbons S, Bacterial efflux pump inhibitors from natural sources, The Journal of antimicrobial chemotherapy, 2007, 59(6), 247–260.

[97] Stermitz FR, Tawara-Matsuda J, Lorenz P, Mueller P, Zenewicz L, Lewis K, 5 '-Methoxyhydnocarpin-D and pheophorbide A: Berberis species components that potentiate berberine growth inhibition of resistant staphylococcus a ureus, Journal of natural products, 2000, 63(8), 1146–1149.

[98] Stermitz FR, Lorenz P, Tawara JN, Zenewicz LA, Lewis K, Synergy in a medicinal plant: Antimicrobial action of berberine potentiated by 5′-methoxyhydnocarpin, a multidrug pump inhibitor, Proceedings of the National Academy of Sciences of the United States of America, 2000, 97(4), 1433–1437.

[99] Bedada SK, Yellu NR, Neerati P, Effect of resveratrol treatment on the pharmacokinetics of diclofenac in healthy human volunteers, Phytotherapy research, 2016, 30(3), 397–401.

[100] Jia Y, Liu Z, Wang C, Meng Q, Huo X, Liu Q, Sun H, Sun P, Yang X, Ma X, Liu K, P-gp, MRP2 and OAT1/OAT3 mediate the drug-drug interaction between resveratrol and methotrexate, Toxicology and Applied Pharmacology, 2016, 306, 27–35.

[101] El-Ghazaly MA, Fadel NA, Abdel-Naby DH, Abd El-Rehim HA, Zaki HF, Kenawy SA, Potential anti-inflammatory action of resveratrol and piperine in adjuvant-induced arthritis: Effect on pro-inflammatory cytokines and oxidative stress biomarkers, Egypt Rheumatol, 2020, 42(1), 71–77.

[102] Chakraborty M, Bhattacharjee A, Ahmed MG, Es SP, Shahin H, Taj T, Effect of Naringin on myocardial potency of Resveratrol against ischemia reperfusion induced myocardial toxicity in rat, Synergy, 2020, 10, 100062.

[103] Wightman EL, Reay JL, Haskell CF, Williamson G, Dew TP, Kennedy DO, Effects of resveratrol alone or in combination with piperine on cerebral blood flow parameters and cognitive performance in human subjects: A randomised, double-blind, placebo-controlled, cross-over investigation, The British journal of nutrition, 2014, 112(2), 203–213.

[104] Pandey MM, Rastogi S, Rawat AK, Indian traditional ayurvedic system of medicine and nutritional supplementation, Evidence-Based Complementary and Alternative medicine, 2013, 2013.

[105] Kheradmandnia S, Vasheghani-Farahani E, Nosrati M, Atyabi F, Preparation and characterization of ketoprofen-loaded solid lipid nanoparticles made from beeswax and carnauba wax, Nanomedicine: Nanotechnology, Biology and Medicine, 2010, 6(6), 753–759.

[106] Martins SM, Wendling T, Gonçalves VM, Sarmento B, Ferreira DC, Development and validation of a simple reversed-phase HPLC method for the determination of camptothecin in animal organs following administration in solid lipid nanoparticles, Journal of Chromatography. B, Biomedical Applications, 2012, 880, 100–107.

[107] Tiyaboonchai W, Tungpradit W, Plianbangchang P, Formulation and characterization of curcuminoids loaded solid lipid nanoparticles, International Journal of Pharmaceutics, 2007, 337(1–2), 299–306.

[108] Kang MJ, Cho JY, Shim BH, Kim DK, Lee J, Bioavailability enhancing activities of natural compounds from medicinal plants, Journal of Medicinal Plant Research, 2009, 13, 1204–1211.

[109] Bhat BG, Chandrasekhara N, Studies on the metabolism of piperine: Absorption, tissue distribution and excretion of urinary conjugates in rats, Toxicology, 1986, 40(1), 83–92.

[110] Zhao JQ, Du GZ, Xiong YC, Wen YF, Bhadauria M, Nirala SK, Attenuation of beryllium induced hepatorenal dysfunction and oxidative stress in rodents by combined effect of gallic acid and piperine, Archives of Pharmacal Research, 2007, 30(12), 1575–1583.

[111] Kumari A, Yadav SK, Pakade YB, Kumar V, Singh B, Chaudhary A, Yadav SC, Nanoencapsulation and characterization of Albizia chinensis isolated antioxidant quercitrin on PLA nanoparticles, Colloids and Surfaces B: Biointerfaces, 2011, 82(1), 224–232.

Keyword index

https://doi.org/10.1515/9783110746808-013

www.ingramcontent.com/pod-product-compliance
Lightning Source LLC
Chambersburg PA
CBHW080708220326
41598CB00033B/5343